FLORA ZAMBESIACA

Flora terrarum Zambesii aquis conjunctarum

VOLUME EIGHT: PART TWO

FLORA ZAMBESIACA

MOZAMBIQUE

MALAWI, ZAMBIA, ZIMBABWE

BOTSWANA

VOLUME EIGHT: PART TWO

Edited by
E. LAUNERT & G. V. POPE

on behalf of the Editorial Board:

G. Ll. LUCAS
Royal Botanic Gardens, Kew

E. LAUNERT
Natural History Museum

M. L. GONÇALVES
*Centro de Botânica, Instituto de Investigação
Científica Tropical, Lisboa*

G. V. POPE
Royal Botanic Gardens, Kew

Published by the Managing Committee on behalf of
the contributors to Flora Zambesiaca
1990

Typeset at the Royal Botanic Gardens, Kew, by
Pam Arnold, Christine Beard, Brenda Carey,
Margaret Newman, Helen O'Brien
and Pam Rosen

Printed in Great Britain by
Whitstable Litho Printers Ltd.,
Whitstable, Kent

ISBN 0 950 7682 8 6

CONTENTS

FAMILY INCLUDED IN VOLUME VIII, PART 2   *page* vi

NEW NAME PUBLISHED IN THIS WORK   vi

GENERAL SYSTEMATIC TEXT   1

INDEX TO VOLUME VIII PART 2   175

FAMILY INCLUDED IN VOLUME VIII, PART 2

120. Scrophulariaceae*

NEW NAME PUBLISHED IN THIS WORK

Stemodiopsis glandulosa Philcox, sp. nov.   *page* 42

*The Orobanchaceae is no longer maintained as a separate family and is now incorporated in Scrophulariaceae. As a result the Flora Zambesiaca family number 121, originally allocated to the Orobanchaceae, will now not appear.

120. SCROPHULARIACEAE

By D. Philcox*

Annual or perennial, terrestrial, amphibious or aquatic herbs, or leafless parasitic herbs lacking chlorophyll, shrubs or undershrubs, rarely trees, glabrous, variously pubescent or glandular-viscid. Leaves opposite, alternate or verticillate, entire, toothed or variously lobed or dissected, or reduced to scales, estipulate. Flowers solitary axillary or in terminal spikes, racemes, heads or panicles, arranged in racemes or cymes. Pedicels ebracteolate or bibracteolate. Flowers hermaphrodite, usually irregular, at times subregular. Calyx inferior, persistent, usually 5-lobed, occasionally 4-, or rarely 3-lobed, usually more or less united, campanulate, tubular, or shortly so almost lacking distinct tube, equal, unequal, valvate, variously overlapping or open in bud. Corolla tubular, campanulate, cylindric or ventricose or enlarged above; tube straight or variously curved to geniculate, at times basally produced into 1 or 2 spurs or sacs; limb usually 5- or 4-lobed, rarely 3–8-lobed, lobes subequal, more or less spreading, or clearly bilabiate with upper lip entire, emarginate or bilobed, erect, concave or galeate or at times flat or spreading, lower lip 3- (rarely 4-) lobed, spreading. Stamens 4, didynamous or equal, or 2 with occasionally 2 reduced to staminodes, or 5 with fifth subrudimentary, rarely perfect. Filaments inserted in corolla tube or at throat, filiform or dilated, occasionally with lower variously appendaged at base. Anthers 1- or bilocular, free or frequently coherent, cells similar or with one smaller and sterile, connective often produced into 2 branches, each bearing cell. Ovary superior, sessile, entire, 2-, or rarely 3-celled, placentation central; style simple, entire or shortly cleft at apex; stigma small, capitate or clavate. Ovules many to numerous in each cell. Fruit usually capsular, septicidal or loculicidal, sometimes both, occasionally dehiscing by pores, rarely indehiscent. Seeds small, numerous, variously shaped, pitted, ridged or ribbed, rarely smooth.

1. Plants totally parasitic, lacking chlorophyll - - - - - - - - - 2
 – Plants not totally parasitic, chlorophyll present, but some hemiparasitic, blackening on drying - - - - - - - - - - 3
2. Calyx tubular with rounded or obtuse lobes; plant fleshy - - - **42. Cistanche**
 – Calyx campanulate with acute teeth; plant not fleshy - - - **43. Orobanche**
3. Plants climbing - - - - - - - - - - **35. Buttonia**
 – Plants not climbing - - - - - - - - - - 4
4. Flowers spurred, saccate or gibbous - - - - - - - - - 5
 – Flowers not spurred - - - - - - - - - - 7
5. Flowers gibbous at base - - - - - - - - - **5. Misopates**
 – Flowers spurred at base - - - - - - - - - 6
6. Erect herbs or undershrubs - - - - - - - - **3. Nemesia**
 – Creeping or prostrate herbs - - - - - - - - **4. Diclis**
7. Leaves alternate, never all radical - - - - - - - - - 8
 – Leaves opposite or at least subopposite with at times upper alternate, or verticillate or radical - - - - - - - - - 12
8. Stamens two, staminodes absent - - - - - - - **2. Anticharis**
 – Stamens four, rarely five - - - - - - - - - 9
9. Leaves circular, small creeping herb - - - - - - **24. Sibthorpia**
 – Leaves not circular, plant erect, prostrate or tufted - - - - - 10
10. Plant erect - - - - - - - - - - 11
 – Plant prostrate, tufted or less commonly erect; corolla up to 1.5(2) cm. long - - - - - - - **1. Aptosimum**
11. Plant sturdy; corolla 4–5 cm. long; leaves 7–15 × 2–6.5 cm. - - **26. Digitalis**
 – Plant slender; corolla 1.5–2 cm. long; leaves 0.2–0.3 × c. 0.05 cm., scale-like - - - - - - - - - **39. Micrargeriella**

* *Craterostigma* and *Striga* by F.N. Hepper

1

12. Stamens two, with or without two staminodes　-　-　-　-　-　-　-　-　- 13
 - Stamens four　-　-　-　-　-　-　-　-　-　-　-　-　-　- 16
13. Leaves much divided; plant aquatic, staminodes 0; cleistogamous flowers with two stamens only; chasmogamous flowers with four stamens　-　-　-　-　*Limnophila ceratophylloides*
 - Leaves not divided; plant terrestrial, creeping or erect　-　-　-　-　-　- 14
14. Corolla tube short, limb subrotate; stamens exserted　-　-　-　- **27. Veronica**
 - Corolla with distinct tube, usually bilabiate; stamens included　-　-　-　- 15
15. Staminodes present, with distinct appendanges with long slender sterile filaments　-　-　-　-　-　-　-　-　-　-　-　-　- **22. Lindernia**
 - Staminodes reduced to two small protruberances or absent　-　- **19. Dopatrium**
16. Trees shrubs or robust undershrubs, not small herbaceous undershrubs　-　-　- 17
 - Herbs or small herbaceous undershrubs　-　-　-　-　-　-　-　-　- 20
17. Fruit baccate, indehiscent　-　-　-　-　-　-　-　-　-　-　-　- 18
 - Fruit capsular, dehiscent　-　-　-　-　-　-　-　-　-　-　-　- 19
18. Calyx 3–5-cleft; corolla tube rather long, 18–33 mm. long　-　-　-　- **6. Halleria**
 - Calyx 5-partite; corolla tube cylindrical, shorter, up to 5 mm. long　-　-　- **7. Teedia**
19. Leaves 3–5(10) × 1–1.5(3.5) mm.; corolla tube 2.5–4 mm. long　**9. Antherothamnus**
 - Leaves 10–35 × 5–7 mm.; corolla tubes c. 10 mm. long　-　-　-　- **8. Freylinia**
20. Leaves all radical; small tufted herbs with or without stolons　-　-　- **23. Limosella**
 - Leaves not all radical; more or less erect herbs or undershrubs, not small,tufted or aquatic 21
21. Anthers monothecal, strictly or by confluence　-　-　-　-　-　-　-　- 22
 - Anthers clearly bithecal　-　-　-　-　-　-　-　-　-　-　-　- 32
22. Anthers monothecal by confluence　-　-　-　-　-　-　-　-　-　- 23
 - Anthers strictly monothecal　-　-　-　-　-　-　-　-　-　-　- 26
23. Calyx regularly 5-lobed　-　-　-　-　-　-　-　-　-　-　-　- 24
 - Calyx 5-lobed, bilabiate or appearing so　-　-　-　-　-　-　-　-　- 25
24. Leaves often in basal rosette; flowers often cymose; corolla tube slender, straight, not dilated or gibbous at apex　-　-　-　-　-　-　-　-　-　-　- **13. Manulea**
 - Leaves not in basal rosette; flowers usually axillary or simple-racemose; corolla tube often curved, usually wider and dilated or gibbous at apex　-　-　-　- **12. Sutera**
25. Plant turning black on drying; corolla tube greatly exceeding calyx　**10. Zaluzianskya**
 - Plant not turning black on drying; corolla tube small, up to 2 mm. long, equalling or only slightly longer than calyx　-　-　-　-　-　-　-　-　-　-　- **11. Polycarena**
26. Fruit indehiscent, enclosed in calyx, often separating into two cocci　-　-　- 27
 - Fruit dehiscent, not enclosed in calyx　-　-　-　-　-　-　-　-　-　- 29
27. Calyx spathaceous　-　-　-　-　-　-　-　-　-　- **44. Hebenstretia**
 - Calyx 3- or 5-lobed　-　-　-　-　-　-　-　-　-　-　-　-　- 28
28. Calyx 3-lobed, frequently with dorsal lobe smaller, calyx then appearing bilobed　-　-　-　-　-　-　-　-　-　-　-　-　- **45. Walafrida**
 - Calyx 5-lobed　-　-　-　-　-　-　-　-　-　-　-　-　- **46. Selago**
29. Corolla tube cylindric, straight or sometimes curved　-　-　-　-　-　- 30
 - Corolla tube sharply bent or curved above middle　-　-　-　-　- **32. Striga**
30. Capsule oblong or ellipsoid, straight; calyx tubular　-　-　-　-　-　- 31
 - Capsule mostly obliquely ovoid, long-beaked; calyx campanulate or at times tubular-campanulate　-　-　-　-　-　-　-　-　-　- **33. Rhamphicarpa**
31. Corolla often small, lobes equal or nearly so; fruit capsular, regular, dry, not rostrate　-　-　-　-　-　-　-　-　-　-　-　-　- **31. Buchnera**
 - Corolla often large, upper lobes united, smaller than lower; fruit capsular or baccate, if capsular, then capsule rostrate, winged, dehiscing along upper suture only　-　- **34. Cycnium**
32. Calyx markedly unequally lobed　-　-　-　-　-　-　-　-　- **18. Bacopa**
 - Calyx not markedly unequally lobed, though sometimes bilabiate or apparently so　 33
33. Stamens, at least some, inserted in throat of corolla　-　-　-　-　-　- 34
 - Stamens all inserted in corolla tube　-　-　-　-　-　-　-　-　-　- 36
34. Lower filaments arched with small boss-like appendage at base; calyx slightly bilabiate, narrowly winged along nerves　-　-　-　-　-　-　-　-　-　- **21. Torenia**
 - Lower filaments sharply angled with large boss-like appendage at base; calyx not markedly winged　-　-　-　-　-　-　-　-　-　-　-　-　- 35
35. Calyx deeply to shallowly 5-lobed or 5-partite; flowers solitary or in lax spike　**22. Lindernia**
 - Calyx 5-toothed; flowers in rather dense or very dense heads or spikes　**20. Craterostigma**
36. Anthers with two equal or subequal fertile thecae　-　-　-　-　-　-　- 37
 - Anthers with one fertile theca and one modified theca or appendage　-　-　- 50
37. Leaves pinnatisect or pinnately or variously divided　-　-　-　-　-　- 38
 - Leaves all entire or toothed, not deeply divided　-　-　-　-　-　-　- 39
38. Terrestrial perennial with woody rhizome　-　-　-　-　-　-　- **37. Gradaria**
 - Aquatic or amphibious annual or perennial lacking woody rhizome　**17. Limnophila**
39. Calyx large, inflated in fruit　-　-　-　-　-　-　-　-　-　- **28. Melasma**
 - Calyx not becoming inflated in fruit　-　-　-　-　-　-　-　-　-　- 40
40. Corolla 4-lobed, regular, rotate; calyx deeply 4–5-lobed　-　-　-　- **25. Scoparia**
 - Corolla 5-lobed; tube funnel-shaped or campanulate; calyx 5-toothed or 5-lobed　 41

41. Anther thecae separate, stipitate - - - - - - - - - - - 42
 – Anther thecae distinct, contiguous - - - - - - - - - - - 43
42. Terrestrial; plant strongly glandular-pubescent, viscid, foetid - - - **15. Stemodia**
 – Aquatic or amphibious; plant glabrous to subglabrous - - - **17. Limnophila**
43. Flowers distinctly pedicellate - - - - - - - - - - - 44
 – Flowers sessile or apparently so - - - - - - - - - - - 48
44. Fruit reflexed at maturity - - - - - - - - - - **16. Stemodiopsis**
 – Fruit not reflexed - - - - - - - - - - - - - 45
45. Bracts and bracteoles absent - - - - - - - - - - - 46
 – Bracts and bracteoles present - - - - - - - - - - - 47
46. Pedicels 10–20 mm. long; corolla clearly bilabiate; plant drying green **14. Mimulus**
 – Pedicels up to 5 mm. long; corolla subrotate but not truly bilabiate; plant drying
    blackish - - - - - - - - - - - - - - **41. Hedbergia**
47. Only bracts present; pedicels 4–14 mm. long - - - - - - **40. Gerardiina**
 – Both bracts and bracteoles present; pedicels 2–4 mm. long; plants drying
    blackish - - - - - - - - - - - **38. Micrargeria**
48. Bracteoles absent; corolla neither thin nor conspicuously veined, not marcescent; seeds
    winged - - - - - - - - - - - - - - **41. Hedbergia**
 – Bracteoles present - - - - - - - - - - - - - 49
49. Hemiparasitic; stems rather stout; leaves usually broad, toothed or at times reduced to scales;
    flowers crowded; corollas thin, conspicuously veined, marcescent - - **29. Alectra**
 – Not parasitic; stems very slender, almost filiform; leaves narrowly linear; flowers not
    crowded - - - - - - - - - - - **38. Micrargeria**
50. Plant climbing - - - - - - - - - - - - **35. Buttonia**
 – Plant not climbing - - - - - - - - - - - - - 51
51. Calyx teeth white-woolly within, frequently seen as white fringe on margin of teeth; corolla
    subrotate; leaves not reduced to scales; plant not parasitic - - - - - **36. Sopubia**
 – Calyx teeth not woolly within or on margin; corolla tubular; leaves – at least above – reduced to
    scales; plant hemiparasitic, drying black - - - - - - **30. Harveya**

1. APTOSIMUM Burch. ex Benth.

Aptosimum Burch. ex Benth. in Lindl., Bot. Reg. sub. t. 1882 (1836) *nom. cons.*

Prostrate, erect or suberect, or densely tufted undershrubs or woody herbs with long tap-roots. Leaves alternate, often crowded, entire, 1-nerved. Flowers solitary, axillary or short axillary cymose, sessile to subsessile, bibracteolate. Calyx 5-lobed; lobes linear to ovate; tube campanulate. Corolla 5-lobed; tube much longer than calyx, narrow at base widening into long throat; lobes subequal, flat, rounded. Stamens 4, didynamous, included; filaments filiform, inserted near base of corolla; anthers transverse, ciliate to subglabrous. Style long, filiform. Ovary bilocular with several to many ovules. Capsule ovoid-conical, acute or sub-acute to broadly ovoid-cylindrical, obtuse or emarginate, subcompressed. Seeds small, numerous, obovoid or compressed globose.

A genus of about 40 species native to tropical and southern Africa.

 1. Plant erect or suberect - - - - - - - - - - - - - 2
 – Plant prostrate - - - - - - - - - - - - - - 5
 2. Leaves spinose or rigid-apiculate; capsule broadly oblong to obovate-oblong truncate to
    emarginate - - - - - - - - - - - - - - - 3
 – Leaves not spinose or rigid-apiculate; capsule ovoid-cylindrical to conical, subacute to
    subobtuse - - - - - - - - - - - - - - - 4
 3. Leaves with white thickened margins; calyx lobes c. 4.5 mm. long; corolla
    14–20 mm. long - - - - - - - - - - - 1. *albomarginatum*
 – Leaves not thickened at margins; calyx lobes 2–2.5 mm. long; corolla tube
    10–12 mm. long - - - - - - - - - - - 2. *marlothii*
 4. Calyx 3–7.5 mm. long, subequal or longer than narrow base of corolla; plant profusely
    leafy - - - - - - - - - - - - - - 3. *lugardiae*
 – Calyx 2–3.5 mm. long, much shorter than narrow base of corolla; plant virgate, almost
    leafless - - - - - - - - - - - - - 4. *junceum*
 5. Leaves linear to narrowly oblanceolate, often 7.5–10 cm. long; margin thickened, persistent at
    least towards base, spinescent - - - - - - - - - 5. *lineare*
 – Leaves not linear, rarely exceeding 5 cm. long; midrib not specially thickened or
    spinescent - - - - - - - - - - - - - - - 6

6. Stems long subflexuous- hirsute; leaves 1–2 cm. long, covered, at least beneath, with forward directed, curved, simple or gland-tipped hairs - - - - - - - 6. *elongatum*
- Stems usually densely short, patent, glandular-pubescent; leaves (1.2)3–5 cm. long, subglabrous to minutely glandular-pubescent - - - - - - - - - 7. *decumbens*

1. **Aptosimum albomarginatum** Marl. & Engl. in Engl., Bot. Jahrb. **10**: 249 (1889). —Hiern in F.C. **4**, 2: 126 (1904). —Merxm. & Roessler in Merxm., Prodr. Fl. SW. Afr. **126**: 15 (1967). Type from S. Africa.

Erect or decumbent undershrub up to 30 cm. tall, densely branched or compact from base; stem woody, white to pale brown glossy, glabrous to sparsely setose-pilose; branches densely leafy. Leaves 4–25 × (1)2(3.5) mm., linear-oblanceolate, acute or obtuse, spinous pointed, glabrous above, glabrous beneath or with few long, white hairs in area of midrib, margin thickened, white cartilaginous. Flowers axillary, sessile to subsessile; bracts or attendant leaves c. 5 mm. long, 1.5 mm. wide, similar to leaves. Calyx 7 mm. long, lobes c. 1 mm. wide, joined at base for about 2.5 mm., linear-lanceolate, acute, externally sparsely long-setose mainly on median nerve, margin long ciliate-setose. Corolla pale blue to purple; tube 14–20 mm. long, narrowed towards base, sparsely to subdensely short pilose without; lobes broadly ovate to subrotund. Lower stamens about half as long as corolla tube, anthers sterile; all anthers glabrous, subreniform. Style minutely pubescent below, exceeding corolla tube. Capsule 5.5–6 mm. long, c. 4 mm. broad, obovoid-cylindrical, subcompressed, prominently parallel-nerved. Seeds c. 1.5 × 1 mm.; testa reticulately wrinkled.

Botswana. N: Groot Laagte (East), fl. 15.iii.1980, *Smith* 3189A (K). SW: c. 35 km. Takatswane on road to Lehututu, fl. & fr. 21.ii.1960, *de Winter* 7437 (K; PRE; SRGH). SE: Letlhakeng Distr., fr. v.1958, *Patterson* 10 (PRE; SRGH).
Also known from Namibia and S. Africa. Damp areas surrounding pans; 1150–1200 m.

2. **Aptosimum marlothii** (Engl.) Hiern in F.C. **4**, 2: 127 (1904). —Merxm. & Roessler in Merxm., Prodr. Fl. SW. Afr. **126**: 17 (1967). Type from S. Africa.
Peliostomum marlothii Engl. in Bot. Jahrb. **10**: 251 (1889). Type as above.

Erect or decumbent undershrub up to c. 40 cm. tall, much branched; branches densely spiny to sparsely so, woody, subglabrous, minutely glandular-pilose when young, leafy. Leaves (2)5–11 × 0.75–1.25 mm., alternate but often clustered on small branches, linear-spathulate with small apiculus, narrowing towards sessile base, fleshy, young leaves minutely glandular-pilose, denser so towards base, glabrescent with age. Flowers sessile to subsessile, axillary or on much reduced lateral shoots. Calyx 5–9 mm. long, with lobes 2.0–2.5(4) mm. long, lanceolate, acute, subdensely glandular-pilose, strongly nerved especially in fruit. Corolla bluish violet, darker at centre; tube 10–12 mm. long, narrow base within calyx, broader above, shortly glandular-pilose without; lobes subcircular. Anthers minutely pubescent, oblong-reniform. Capsule 6–7 × 4.5–7 mm., broadly oblong to square, laterally compressed, subtruncate-emarginate, deep median depression on both surfaces, pubescent at apex and in depression. Seeds closely minutely tuberculate, grey.

Botswana. SW: Takatswane, c. 155 km. N. Kang on road to Ghanzi, fr. 20.ii.1960, *de Winter* 7401 (K; PRE; SRGH).
Also known from Namibia and S. Africa; about 1200 m.

3. **Aptosimum lugardiae** (N.E.Br.) Phillips in Journ. S. Afr. Bot. **16**: 22 (1950) as "lugardae". TAB. 1. Type: Botswana, Kgwebe Hills, c. 1000 m., fl. & fr. 24.i.1898, *Lugard* 124 (K).
Peliostomum linearifolium Schinz ex Kuntze, Rev. Gen. Pl. 3, **3**: 238 (1898) nom. nud.
Peliostomum leucorrhizum var. *linearifolium* Weber in Bull. Herb. Boiss., Sér. 2, **3**: 904 (1903). —Hemsl. & Skan in F.T.A. **4**, 2: 274 (1906). Type from S. Africa.
Peliostomum lugardiae N.E. Br. ex Hemsl. & Skan in F.T.A. **4**, 2: 275 (1906) as "lugardae". Type as above.

Erect, rigid, much-branched undershrub or perennial herb, 15–38 cm. tall, usually with thick rootstock; stem above longitudinally ridged or somewhat quadrangular, minutely glandular-pubescent; stem below, base of branches and rootstock often covered with white corky layer. Leaves (5)15–33(38) × 1–1.5 mm., narrowly linear-lanceolate or linear, acute or subobtuse, subglabrous to minutely puberulent, entire. Flowers solitary, axillary-pedicellate; bracteoles 1–2 mm. long, 0.1–0.25 mm. wide, arising at or towards base of

Tab. 1. APTOSIMUM LUGARDIAE. 1, habit (× ⅔); 2, flower (× 3); 3, corolla opened showing androecium (× 3); 4, flower, longitudinal section showing gynoecium (× 3), 1–4 from *Standish-White* 19; 5, fruits (× 6), from *De Hoogh* 106.

pedicel, linear, glabrous. Pedicels 1–1.75 mm. long. Calyx 3–5(7.5) mm. long, lobes joined only at base, 0.25–0.5 mm. wide, linear, acute, minutely puberulent with short glandular hairs. Corolla blue to purple or mauve, tube 5–8(10.5) mm. long, funnel-shaped, narrowed at base, much wider above, mouth slightly oblique, sparsely minutely stipitate-glandular to glabrous without; lobes subcircular. Anthers sparsely pubescent. Capsule (5)7(8.5) × 2.5–3 mm., ovoid-oblong, acute to subobtuse, minutely papillose, not strongly and markedly reticulate-veined.

Botswana. N: between Odiakwe and Kanyu, on Francistown to Maun Rd., fr. 9.iii.1965, *Wild & Drummond* 6837 (BR; K; LISC; SRGH). SW: Kang pan, 300 km. W. of Gaborone, c. 1060 m., fl. & fr. 4.iii.1977, *Mott* 1125 (K; SRGH). SE: Boteti Delta area, NE. Mopipi, 900 m., fl. 16.iv.1973, *MacDonald* 2 (K; SRGH).
Also in S. Africa. Calcareous soils, sandveldt; 550–1050 m.

4. **Aptosimum junceum** (Hiern) Philcox in Kew Bull. **40**: 606 (1985). Type from S. Africa.
    Peliostomum leucorrhizum var. *junceum* Hiern in F.C. **4**, 2: 135 (1904). Type from S. Africa.

Erect, virgate undershrub up to 55 cm. tall, much-branched from base. Stem and branches glabrous, ribbed, often white-corky at base. Leaves up to 8(12) × 0.3–0.5 mm., narrowly obovate-oblong, glabrous where present, more usually absent or reduced to short leaf-like processes up to c. 2 mm. long. Flowers solitary, pedicellate. Pedicels 4.5–6 mm. long, glabrous. Calyx lobes 2–3 mm. long, joined only at base for c. 0.5 mm., linear to narrowly linear-ovate, glabrous without but occasionally short-ciliate, minutely glandular-pubescent within. Corolla purplish-blue, tube (10)15–18 mm. long, funnel-shaped, narrow base much longer than calyx, wider above, sparsely, minutely stipulate-glandular to glabrous without; lobes broadly ovate-circular, to 5.5 mm. diam. Capsule 5.5–7 × 2.75–3 mm., obvoid-oblong to ovoid-conical, acute to subobtuse.

Botswana. SW: c. 21 km. SE. of Groot-Kalk, Nossob R., c. 975 m., fl. & fr. 21.iv.1960, *Leistner* 1865 (K; SRGH).
Also in S. Africa and Namibia. Riversides and sandy pan areas; 450–1200 m.

5. **Aptosimum lineare** Marl. & Engl. in Engl., Bot. Jahrb. **10**: 25 (1889). —Hiern, Cat. Afr. Pl. Welw. **1**: 755 (1898). —Hiern in F.C. **4**, 2: 129 (1904). —Hemsl. & Skan in F.T.A. **4**, 2: 269 (1906). —Merxm. & Roessler in Merxm., Prodr. Fl. SW. Afr. **126**: 17 (1967). Type from Namibia.
    Aptosimum randii S. Moore in Journ. Bot. **37**: 171 (1899). Type: Zimbabwe, Bulawayo, fl. Dec. 1897, *Rand* 180 (BM, holotype).

Plant usually prostrate, branched or rarely densely tufted under-shrub; branches 2–15(40) cm. long, densely leafy. Leaves 1.5–10 × 0.1–0.65(0.9) cm., linear to linear-oblong or narrowly oblanceolate, apex obtuse, short apiculate, attenuate to subsessile or sessile base, sparsely long glandular-hirsute above, very sparsely hispid on major nerve beneath, ciliate especially below; midrib thickened, prominent beneath, at least lower part persistent, spinescent. Flowers sessile to subsessile; bracts to 7 mm. long, 1 mm. wide, narrowly linear. Calyx c. 8 mm. long, lobes 5–7 mm. long, narrowly linear-lanceolate, acute, prominently 1-nerved, long pilose-ciliate. Corolla up to 2.5 cm. long, bright deep blue to violet, darker at throat, sparsely and usually shortly glandular-pilose without, lobes subequal, broadly obovate. Lower stamens more than half as long as corolla tube; anthers pilose; style hairy below, slender, exceeding corolla tube. Ovary compressed-ovoid, glabrous. Capsule c. 5 × 4.5 mm. laterally compressed-obovoid, pubescent; seeds black.

Botswana. N: Boteti R., Toromoja, fl. 10.xii.1978, *Smith* 2587 (K; PRE; SRGH). SW: Deception Pan, Central Kalahari Game Reserve, fl. & fr. 5.iv.1975, *Owens* 81 (K; PRE; SRGH). SE: Lobatse, 64 km. SE. of Gaborone, fl. 5.v.1974, *Mott* 248 (K; SRGH). **Zimbabwe**. W: Plumtree Distr., Tegwani, fl. & fr. 8.i.1946, *McCosh* 11 (K; SRGH). C: Marondera (Marandellas), fl. 10.i.1950, *West* 3138 (K; SRGH). E: Lower Sabi Distr., Devuli R., c. 1200 m., fl. 1.ii.1948, *Wild* 2469 (BR; K; SRGH). S: Gwanda Distr., Tzibizini Dam, Doddieburn Ranch, c. 725 m., fl. & fr. 5.v.1972, *Pope* 647 (BR; K; LISC; PRE; SRGH). **Mozambique**. Z: Limpopo, Guija, 20 km. on road to Massingire, fl. & fr. 22.v.1948, *Torre* 7913 (LISC). GI: Caniçado, fl. & fv. 11.vi.1975, *Marques* 2801 (MO).
Also known from Angola, Namibia and S. Africa. Grasslands bordering riverine woodlands; 900–1200 m.

There is a specimen in the Kew Herbarium collected by Alfred Owen purportedly from "N. Rhodesia and country beyond". By the distribution as seen above, Zambia seems unlikely as the locality; more probably it was collected from Botswana.

6. **Aptosimum elongatum** Engl., Bot. Jahrb. **10**: 249 (1888). —Hemsl. & Skan in F.T.A. **4**, 2: 273 (1906). Type from S. Africa.

Aptosimum pubescens Weber in Bull. Herb. Boiss., Sér. 2, **3**: 903 (1903). Type from Namibia.

Aptosimum depressum var. *elongatum* Hiern in F.C. **4**, 2: 131 (1904). Type from S. Africa.

Aptosimum procumbens var. *elongatum* (Hiern) Codd in Bothalia **14**, 1: 80 (1982).

Procumbent undershrub, widely spreading; branches slender, up to 60 cm. long, densely covered with long, patent, simple, or slender, gland-tipped white hairs. Leaves (7)10–18 × 3.5–5(7.5) mm., sessile, oblanceolate to broadly obovate, obtuse, mucronate, tapering below into petiole-like base, subglabrous to subdensely pilose above covered with regularly forward-directed, curved, simple or glandular hairs, lower surface with similar indumentum, midrib prominent, margin frequently somewhat thickened, subciliate. Flowers shortly pedicellate, bracteate, solitary. Pedicel 1.5–2.5 mm. long, 0.75–1.25 mm. wide, linear or linear-lanceolate, very minutely papillose within, subdensely simple or glandular-hirsute with forward-directed hairs up to 0.5 mm. long, subciliate, prominently 1-nerved. Corolla deep blue to purple, tube 11–15 mm. long, shortly glandular-pubescent, wide at mouth; lobes c. 3 mm. in diam., subcircular, subequal, reticulately veined at least when dry. Anthers c. 2 mm. broad, 1.5 mm. long, with ridge of long, white, simple hairs at apex. Seeds 1 × 0.6 mm., cylindrical ellipsoid, colliculate, black.

Botswana. N: between Odiakwe and Kanuy, Francistown to Maun road, fl. 9.iii.1965, *Wild & Drummond* 6828 (K; PRE; SRGH). SW: Kang, 320 km. W. of Gaborone, c. 1060 m., fl. & fr. 20.x.1975, *Mott* 762 (GAB; K; PRE; SRGH). SE: c. 5 km. N. of Lobatsi, fl. 16..1960, *Leach & Noel* 132 (K; LISC; PRE; SRGH).

Also known from Namibia and S. Africa. Scrub and open wooded grassland, mainly on Kalahari Sand, 900–1100 m.

It should be noted here that when Hiern used the epithet *elongatum* for a variety of *A. depressum*; he gave no indication that he based it on *A. elongatum* Engl.

Despite frequent misidentifications in the past, the plant hitherto known as *A. depressum* Burch. ex Benth. nom. illegit. (1836) is now correctly known as *A. procumbens* (Lehm.) Steud. (1841) and is a native of S. Africa and does not occur in the Flora Zambesiaca area.

7. **Aptosimum decumbens** Schinz in Verh. Bot. Ver. Brand. **31**: 184 (1890). —Hemsl. & Skan in F.T.A. **4**, 2: 272 (1906). —Merxm. & Roessler in Merxm., Prodr. Fl. SW. Afr. **126**: 16 (1967). Type from Namibia.

Decumbent undershrub, widely spreading, much branched; branches up to 1 m. long, densely short-pubescent. Leaves (12)30–50 × 2.5–12 mm., usually rather crowded, oblanceolate-oblong, obtuse or rounded or shortly apiculate, tapering at acute base, sessile or very shortly petiolate, glabrous to sparsely, short glandular-pubescent above, denser so beneath. Flowers sessile or subsessile, axillary, solitary or few, clustered, bracteate. Pedicel where present c. 1 mm. long. Bracts c. 5.5 mm. long, narrowly linear, acute, pubescent above and below, shortly ciliate towards base. Calyx 6.25–7.5 mm. long, shortly tubular with lobes 5.0–6.5 mm. long, linear-lanceolate, acute, sparsely short glandular-pubescent, very long ciliate, obscurely 1-nerved. Corolla purple to deep blue, tube 12–13.5 mm. long, densely short glandular-pubescent without, prominently nerved; lobes c. 4–4.5 mm. diam., broadly obovate, subequal. Anthers with small tuft of hairs at apex. Capsule c. 4.25 × 5.5 mm., laterally somewhat compressed, very broadly obovoid, retuse, densely short-pubescent, reticulate veined, margin of valves densely papillose.

Botswana. N: Thamalakane R., Maun, fl. & fr. 21.ii.1974, *Smith* 849 (K; MO; PRE; SRGH). SE: Gaborone, fl. & fr. 29.xi.1972, *Kelaole* A62 p.p. (MO). **Zambia**. B: Sesheke, fl. & fr. i.1924, *Borle* s.n. (PRE). S: near Katonta Pool, 1 km. S. of Eastern Cateracts, Victoria Falls Trust Area, c. 875 m., fr. 20.v.1963, *Bainbridge* 774 (K; LISC; NY; SRGH). **Zimbabwe**. W: Hwange Distr., Kazuma Range, 1000 m., fl. & fr. 10.v.1972, *Russell* 1932 (K; LISC; SRGH).

Also known from Angola, Namibia and S. Africa. Grassy open woodland on various soils from sand to basalt clay; 850–1100 m.

2. ANTICHARIS Endl.

Anticharis Endl., Nov. Stirp. Dec.: 22 (1839).

Erect or suberect, glandular-pubescent herbs or small undershrubs. Leaves alternate, entire. Flowers solitary, axillary, pedicellate, often bi-bracteolate. Calyx 5-lobed, lobes

8

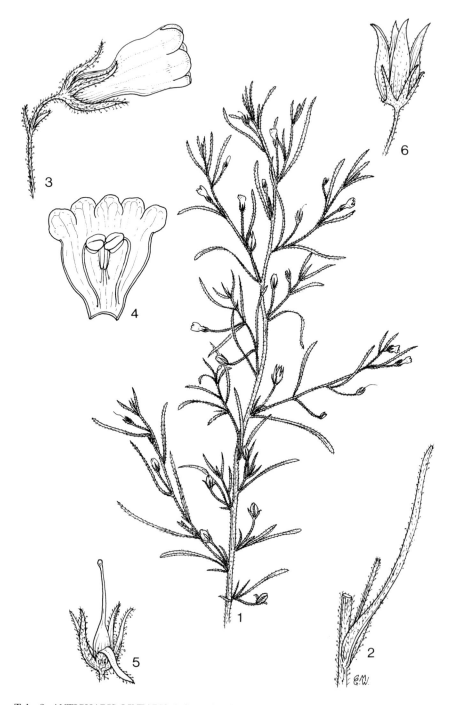

Tab. 2. ANTICHARIS LINEARIS. 1, flowering branch (× ⅓); 2, leaf (× 4), 1–2 from *Bally* B 5217; 3, flower (× 4); 4, corolla, opened showing androecium (× 4); 5, gynoecium and calyx (× 4); 6, dehisced fruit (× 4), 3–6 from *Glover & Samuel* 2984. (Illustrations made from Kenyan plants).

narrowly lanceolate. Corolla 5-lobed; tube much longer than calyx, dilated above, narrow at base; lobes flat, spreading, subequal. Stamens 2, anticous, included; filaments filiform, inserted above base of corolla tube; anthers transverse, glabrous or hairy, thecae confluent; staminodes absent. Style filiform, subclavate. Ovary bilocular; ovules many. Capsule ovoid or cylindrical, subacuminate, loculicidal and septicidal; valves on dehiscence bend marginally inwards exposing placentiferous column. Seeds small, numerous, ovoid or cylindrical, striate, rugose or not.

A genus of some 10 species from southwest Africa to Arabia and India.

Anticharis linearis (Benth.) Hochst. ex Ascherson in Monatsber. Akad. Wiss. Berl. **1866**: 882 (1866). —Hiern, Cat. Afr. Pl. Welw. **1**: 756 (1898). —Hemsl. & Skan in F.T.A. **4**, 2: 276 (1906). —Merxm. & Roessler in Merxm., Prodr. Fl. SW. Afr. **126**: 11 (1967). TAB. **2**. Type from northeast Africa.
    Doranthera linearis Benth. in DC., Prodr. **10**: 347 (1846). Type as above.
    Anticharis arabica Hochst. ex Benth. in DC., Prodr. **10**: 347 (1846) non Endl.
    Gerardiopsis fischeri Engl., Pflanzenw. Ost-Afr., **C**: 359 (1895); in Engl., Bot. Jahrb. **23**: 507 (1897). Type from East Africa.

Erect, much-branched herb up to 50 cm. tall, glandular-pubescent throughout. Leaves up to 40 × 0.5–1.5(3.5) mm., sessile, linear to narrowly lanceolate, acute apex, base attenuate, occasionally glabrescent with age. Pedicels 6–18 mm. long, finely filamentous; bracteoles 0.6–1.5 mm. long, opposite or subopposite, fine, subulate, usually in upper⅓ of pedicel. Calyx 3–4 mm. long, lobes joined only at base, narrowly lanceolate. Corolla pale mauve, about 10 mm. long, tube funnel-shaped, narrowed at base, wider above; lobes subequal. Anthers ciliate. Style shorter than corolla tube. Stigma bifid. Capsule 6–8.5 mm. long, 4 mm. broad, pale brown, strongly reticulate.

Botswana. N: near Toromoja-Mopipi road, fl. & fr. 17.ii.1980, *Smith* 3080 (K; PRE; SRGH). **Zimbabwe**. S: Beitbridge Distr., Shashi Drift, Thuli (Tuli), fl. & fr. 21.iii.1959, *Drummond* 5890 (K; LISC; PRE; SRGH).
Also in West, Northeast and East Africa, India and Asia Minor and Namibia. Open grasslands or bushland, on dry clayey or sandy soils liable to seasonal flooding or waterlogging.

3. NEMESIA Vent.

Nemesia Vent. in Jard. Malmaison **1**: 41, t. 41 (1803).

Annual or perennial herbs or undershrubs. Leaves opposite, entire or toothed. Flowers in terminal racemes at end of branches or solitary-axillary, bracteate, pedicellate, bracteoles absent. Calyx 5-lobed. Corolla bilabiate; tube short, produced at front into sac or spur; posterior lip 4-lobed, anterior lip entire or emarginate with palate at base. Stamens 4, didynamous, inserted in mouth of corolla tube, filaments of anterior pair usually bent round posterior pair at base; anthers usually cohering in parts, thecae confluent. Style filiform. Ovary bilocular, ovules numerous. Capsule laterally compressed, septicidal. Seeds numerous in one or two series, cylindrical-ellipsoid with entire or interrupted membranous wing.

A genus of about 60 species mostly from southern Africa.

1. Leaves ovate, 0.6–1.8(3.4) cm. wide - - - - - - - 1. *zimbabwensis*
 – Leaves narrowly lanceolate to linear-lanceolate, up to 0.6 cm. wide - - - - 2
2. Leaves entire to laxly serrate-dentate; calyx 2.5–3 mm. long; upper and lower lips of corolla c. 6 mm. long; capsule usually 9–10.5 mm. long - - - - - 2. *fruticans*
 – Leaves shallowly dentate; calyx 2–2.5 mm. long; upper lip of corolla 3–5 mm. long, lower 3–4 mm. long; capsule 5–7 mm. long - - - - - - - - - 3. *lilacina*

1. **Nemesia zimbabwensis** Rendle in Journ. Bot. **70**: 95 (1932). —Norlindh in Bot. Not. **1951**, 2: 99, fig. 1a (1951). TAB. **3**. Type: Zimbabwe, Great Zimbabwe, 11.viii.1929, *Rendle* 329 (BM, holotype).
    Nemesia montana Norlindh in Bot. Not. **1951**, 2: 100, fig. 1b & 2 (1951). Type: Zimbabwe, Nyanga (Inyanga) Distr., Mt. Inyangani, c. 2400 m., 7.xii.1930, *Fries, Norlindh & Weimark* 3586 (LD, holotype; K; PRE, isotypes).

Erect annual or biennial herb to 65 cm. tall, branched from base. Stem and branches quadrangular, sparsely pubescent or glandular-pubescent to glabrous. Leaves sub-sessile

Tab. 3. NEMESIA ZIMBABWENSIS. 1, habit (× ⅔); 2, flower (× 4); 3, flower, longitudinal section (× 4); 4, corolla opened (× 4); 5, dehiscing fruit (× 4), all from *Robinson* 1979.

to shortly petiolate, 1.2–3.5(6.0) × 0.6–1.8(3.5) cm., ovate, acute apex, base rounded to broadly, unequally cuneate, serrate, glavrous or sparsely furnished with short, flexuous white hairs; petiole 0.5–3(6) mm. long. Racemes up to c 12 cm. long in fruit. Bracts small, leaf-like. Pedicels 4–9 mm. long, to 15(25) mm. in fruit, minutely pubescent or glandular-pubescent to subglabrous. Calyx 2.2–4(4.5) mm. long, lobes 0.5–1.2 mm. wide linear or linear-oblong, subobtuse, sparsely glandular-pubescent, ciliate. Corolla white to pale mauve with purple striations; upper lip 4–5 mm. long, lower lip 5 mm. long, emarginate; spur 2–3(4.5) mm. long. Capsule cylindrical, 5–12.5 × 3–7.3 mm., shallowly emarginate, base truncate.

Zimbabwe. E: Nyanga (Inyanga) Distr., summit of Mt. Inyangani, c. 2400 m., fl. & fr. 25.x.1946, *Sturgeon* in GHS 16956 (K; SRGH). S: Masvingo Distr., Great Zimbabwe, fl. & fr. 4.x.1949, *Wild* 3036 (K; SRGH). **Mozambique**. MS: Gogogo Mt., Gorongosa (Gorongoza), c. 1760 m., fl. & fv. 4.vii.1955, *Schelpe* 444 (BM).

Known only from Zimbabwe and Mozambique; 900–2600 m.

Nemesia montana and *N. zimbabwensis* seem to be conspecific. They appear to differ, according to the original descriptions, in that *N. montana* has sessile to shortly petiolate leaves with the petiole, where present, up to 3 mm. long, whereas *N. zimbabwensis* has petioles from 3–7 mm. long. *N. montana* is perennial; *N. zimbabwensis* is an annual.

From the specimens at Kew the leaves not only appear to fit the above, but both sessile and petiolate leaves occur and in some cases on the same plant e.g. *Greatrex* in GHS 14772, *Robinson* 1979. Even *Norlindh* 4709, named by Norlindh the authority of *N. montana* has sessile leaves. I prefer to consider them of the same taxon sensu lato and under the earlier name *N. zimbabwensis* Rendle.

2. **Nemesia fruticans** (Thunb.) Benth. in Hook., Comp. Bot. Mag. **2**: 22 (1836); in DC., Prodr. **10**: 263 (1846). Type from S. Africa.
   Antirrhinum fruticans Thunb., Prodr. Pl. Cap.: 105 (1800); in Fl. Cap., ed. Schultes: 483 (1823). —Benth. in DC., Prodr. **10**: 263 (1846). Type as above.
   Antirrhinum capense Thunb., Prodr. Pl. Cap.: 105 (1800) non Burm.f., 1768 nom. illegit.
   Nemesia foetens Vent., Jard. Malm. **1**: sub t. 41 (1803). Type from S. Africa.
   Linaria fruticans (Thunb.) Spreng., Syst. Veg. **2**: 789 (1825). Type as for *Nemesia fruticans*.
   Linaria capensis Spreng., Syst. Veg. **2**: 796 (1825). Type as for *Antirrhinum capense* Thunb.
   Nemesia divergens Benth. in Hook., Comp. Bot. Mag. **2**: 22 (1836). Type from S. Africa.
   Nemesia capensis (Spreng.,) Kuntze, Rev. Gen. Pl. **3**: 237 (1898). Type as for *Linaria capensis*.
   Nemesia fruticans var. *divergens* (Benth.) Norlindh in Bot. Not. **1951**, 2: 97 (1951). Type as for *Nemesia divergens*.

Erect annual herb to 40 cm. tall, branched, sometimes profusely, from base. Stem and branches quadrangular, glabrous or rarely sparsely glandular, especially immediately within inflorescence. Leaves 1.5–3 × 0.15–0.5 cm., sessile to shortly petiolate, narrowly lanceolate, apex obtuse, rounded to cuneate at base, entire to laxly serrate-dentate, glabrous. Racemes 2–5 cm. long, up to 17 cm. long in fruit. Bracts 2–3 mm. long, broadly lanceolate. Pedicels (2.5–4)9–13 mm. long, laxly glandular-pubescent. Calyx 2.5–3 mm. long, lobes 0.6–1.5 mm. wide, lanceolate-oblong, obtuse, shortly glandular-pubescent. Corolla pale pink or mauve or white, throat orange-yellow; upper lip c. 6 mm. long; spur 2–3.4 mm. long, cylindrical, straight. Capsule (5)9–10.5 × (3.6)4.8–7 mm., very broadly oblong, ovate, subtruncate at apex, very slightly emarginate, slightly bicornute, more markedly so when young.

Zimbabwe. W: Bulawayo Distr., Hillside, c. 1400 m., fl. & fr. viii.1903, *Eyles* 1229 (PRE; SRGH). C: Harare Distr., c. 1400 m., fl. & fr. 8.v.1927, *Eyles* 4943 (K; SRGH). E: Nyanga (Inyanga) Distr., Rhodes Nyanga Estate, c. 1800–2100 m., fl. & fr. 16–20.xi.1931, *Brain* 6948 (K; SRGH). S: Gwanda Distr., banks of Umzingwane R., fl. & fr. 18.xii.1956, *Davies* 2371 (PRE; SRGH).

One specimen is tentatively recorded here from Botswana, (*Mogg* 8031 (PRE)), but as this is only in flower, and fruits are needed for a more certain determination, it has not been included in the above distribution.

Also in Namibia and S. Africa. Stream or riverbank vegetation and roadsides.

3. **Nemesia lilacina** N.E. Br. in Bull. Misc. Inf., Kew **1909**: 376 (1909). Type from Namibia.

Annual herb, erect, 15–35 cm. tall, much branched especially from base. Stem and branches subterete to angular, glandular-pubescent to subglabrous. Leaves 1.5–3.5 × 0.3–0.6 cm., narrowly lanceolate or linear-lanceolate, acute or obtuse, shallowly dentate, glandular-pubescent. Racemes (4)9–28 cm. long, lax, glandular-pubescent. Bracts 2–5 mm. long, sessile, cordate-ovate, acute. Pedicels 7–12 mm. long, slender to filiform. Calyx 2–2.5 mm. long, lobes 0.5–1 mm. wide, oblong or linear-oblong, subacute. Corolla lilac to violet; upper lip 3.5 mm. long, 5–7 mm. broad; lower lip 3–4 mm. long, 5–6 mm. broad; spur 3–3.5 mm. long, subcompressed-cylindric, straight. Capsule 5–7 × 3.5–4 mm., oblong, emarginate apex.

Botswana. SW: Okwa Valley, fl. & fr. 21.ix.1976, *Skarpe* S-77 (PRE; SRGH).
Also recorded from Namibia and S. Africa. Grassy sandy areas, and in or near dried-up watercourses.

4. DICLIS Benth.

Diclis Benth. in Hook., Comp. Bot. Mag. **2**: 23 (1836).

Annual or perennial herbs, creeping, prostrate or decumbent, rarely erect. Leaves opposite or upper alternate, toothed to subentire. Flowers solitary, axillary, pedicellate, ebracteate. Calyx 5-lobed. Corolla bilabiate; tube short, produced at front into spur; posterior lip bilobed, anterior lip 3-lobed. Stamens 4, didynamous, filaments of anterior pair circumflex at base; anthers cohering in pairs, thecae confluent. Style slender. Ovary bilocular, ovules numerous. Capsule subglobose clearly or obscurely emarginate at apex, loculicidal. Seeds ovoid, reticulate-foveolate.

A genus of about 9 species from tropical east and southern Africa and Madagascar.

1. Plant softly villous; leaves coarsely serrate-dentate; corolla spur distinctly curved, 2–8 mm. long - - - - - - - - - - - - 1. *tenella*
  - Plant not villous, pubescent to short-glandular-stipitate - - - - - 2
2. Corolla up to 4 mm. long, spur 0.4–1.8 mm. long, slender - - - - 2. *ovata*
  - Corolla 4.5–9 mm. long, spur 1–2.5 mm. long, subdeltoid - - - - 3. *petiolaris*

1. **Diclis tenella** Hemsl. in Bull. Misc. Inf., Kew **1896**: 163 (1896). —Hemsl. & Skan in F.T.A. **4**, 2: 287 (1906). TAB. **4**, fig. A. Type: Malawi, Mt. Chiradzulu, *Whyte* s.n. (K, holotype).

Prostrate or creeping annual herb, much branched; branches 5–25 cm. long, slender, softly villous occasionally rooted at nodes. Leaves petiolate; lamina (5)15–25(42) × (4)9–22(34) mm., very broadly ovate to subcircular, shortly subcuneate at base, coarsely serrate-denate, sparsely to subdensely villous above and beneath; petiole (5)10–18(28) mm. long, slender, variously villous. Pedicels (10)25–65 mm. long, capillary to filiform, shortly villous to subglabrous especially above. Calyx 1.5–2.8 mm. long, lobes c. 0.5 mm. wide, unequal, ovate-oblong, shortly pubescent to subglabrous. Corolla white to occasionally pale pinkish-mauve 6–9(10.5) mm. long including spur; spur 2–6(8) mm. long, distinctly curved; upper lip bilobed, lobes rounded; lower lip 3-lobed, lobes unequal. Capsule 1.6–2.4 mm. long, 2.2–2.5 mm. wide, subglobose, laterally compressed, emarginate, minutely glandular-pubescent.

Zimbabwe. C: Makoni Distr., fl. & fr. 18.ii.1946, *Wild* 798 (K; SRGH). E: Mutare Distr., Vumba Mts., W. of Castle Beacon, c. 1650 m., fl. & fr. 13.v.1956, *Chase* 6119 (BM; K; SRGH). S: Mwenezi (Nuanetsi), Bubi (Bubye) R., c. 840 m., fl. & fr. 8.v.1958, *Drummond* 5692 (BR; COI; K; LD; LISC; PRE; SRGH). **Malawi**. N: Mzimba Distr., Mzuzu, Marymount, c. 1350 m., fl. & fr. 14.ii.1974, *Pawek* 8110 (P). C: Dedza Distr., Chencherere Hill, Chongoni Forest Reserve, 1675–1800 m., fl. & fr. 23.iv.1970, *Brummitt* 10048 (K; SRGH). S: Mulanje Distr., Mulange Mt., foot of Great Ruo Gorge, 870–1060 m., fl. & fr. 18.iii.1970, *Brummitt & Banda* 9213 (K). **Mozambique**. MS: Chimanimani Mts., c. 1530 m., fl. & fr. 17.iv.1960, *Goodier* 993 (K; LISC; SRGH).
Known also from Tanzania. Usually in the shelter of overhanging rocks, in crevices and mouths of caves both in open grassland and forest areas; 850–1900 m.

2. **Diclis ovata** Benth. in Hook., Comp. Bot. Mag. **2**: 23 (1836); in DC., Prodr. **10**: 265 (1846). —Hemsl. & Skan in F.T.A. **4**, 2: 287 (1906). TAB. **4**, fig. B. Type from Madagascar.
  Linaria veronicoides A. Rich., Tent. Fl. Abyss. **2**: 114 (1851). Type from Ethiopia.
  Anarrhinum veronicoides (A. Rich.) Kuntze in Jahrb. Bot. Gart. Berl. **4**: 269 (1886). Type as above.
  Simbuleta veronicoides (A. Rich.) Kuntze, Rev. Gen. Pl. **2**: 465 (1891). Type as above.

Prostrate or decumbent annual herb, sparingly branched; branches 4–30 cm. long, slender, minutely pubescent. Leaves petiolate; lamina 10–25(35) × 3–13(24) mm., broadly ovate, obtuse, cuneate at base, shallowly dentate to denticulate, sparsely, minutely

13

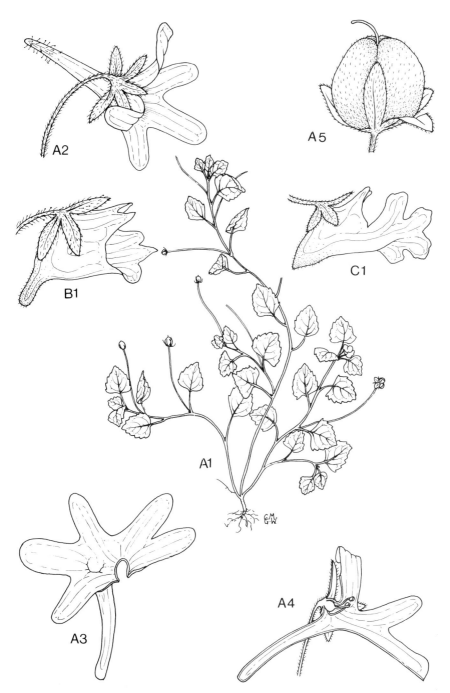

Tab. 4. A.—DICLIS TENELLA. A1, habit (× ⅔); A2, flower (× 4); A3, corolla (× 4); A4, corolla, longitudinal section showing spur (× 4); A5, capsule (× 8), A1–5 from *Brummitt* 9673. B.— DICLIS OVATA. B1, flower (× 8), from *Fanshawe* 5773. C.—DICLIS PETIOLARIS. C1, flower (× 3), from *Clark* 304.

glandular-pubescent to subglabrous above, slightly denser beneath especially on major nerves; petiole (2)6–10(16) mm. long, shortly stipitate glandular. Pedicels 10–24(55) mm. long, capillary to rarely filiform, stipitate glandular. Calyx 1.4–2.4 mm. long, lobes 0.6–1 mm. wide, unequal, ovate-oblong, sparsely glandular stipitate. Corolla 2–3(4) mm. long including spur, white to cream, occasionally upper lip pink-tinged; spur 0.4–1(1.8) mm. long, straight, glandular; upper lip bifid, lobes short, acute; lower lip trilobed, lobes rounded. Capsule 2–2.5 × 2–3 mm., compressed globose, shallowly emarginate, sparsely and finely stipitate glandular.

Zambia. B: edge of Mashi R., Shangombo, c. 1050 m., fl. & fr. 8.viii.1952, Codd 7691 (PRE). N: Mbala Distr., Itimbwe Gorge, 1500 m., fl. & fr. 3.i.1960, Richards 12053 (K). W: Kitwe, fl. & fr. 3.iii.1968, Mutimushi 2513 (K; SRGH). C: Chilanga, c. 1200 m., fl. & fr. 8.vi.1958, Stewart 124 (K). S: Livingstone Distr., Victoria Falls, fl. & fr. 3.viii.1947, Brenan & Greenway 7633 (K; P). Zimbabwe. N: Makonde Distr., Mashunganyendi Pools, c. 425 m., fr. 9.ix.1950, Whellan 469 (SRGH). W: Hwange Distr., Victoria Falls, 880 m., fr. 6.viii.1979, Mshasha 216 (SRGH). C: Marondera (Marandellas), fl. & fr. 13.iv.1948, Corby 93 (K; SRGH). E: Nyanga (Inyanga) Distr., Pungwe Hills, c. 1860 m., fl. & fr. 23.x.1946, Wild 1489 (BR; K; SRGH). S: Chibi Distr., c. 6 km. N. of Rundi (Lundi) R. Bridge, fl. & fr. 3.v.1962, Drummond 7913 (K; PRE; SRGH). Malawi. N: Rumphi Distr., Nyika Plateau, Kafwimba Forest, 1875 m., fl. & fr. 28.iv.1973, Pawek 6694 (K). C: Dedza Distr., Chongoni Forestry School, fl. & fr. 27.i.1967, Salubeni 522 (K; SRGH). S: Blantyre Distr., Limbe, 1140 m., fl. & fr. 24.ii.1970, Brummitt 8737 (K; SRGH). Mozambique. Z: Milanje Distr., Serra Turbina, c. 900 m., fl. & fr. 17.i.1971, Hilliard & Burtt 6296 (E). GI: Gaza Distr., Bilene between Chissano and Licilo, fl. & fr. 5.viii.1958, Barbosa & Lemos 8297 (K). M: Maputo (Lourenco Marques), Jardim Vasco da Gama, fl. & fr. 3.iv.1971, Balsinhas 1939 (K; LISC).

Known also from West Africa, Ethiopia and Sudan, East Africa, Madagascar, the Mascarene Islands and Angola. Weed of cultivation, damp woodland tracks, riversides and occasionally open savannas; 750–2400 m.

3. **Diclis petiolaris** Benth. in DC., Prodr. **10**: 265 (1846). —Hemsl. & Skan in F.T.A. **4**, 2: 287 (1906). TAB. **4**, fig. C. Type from S. Africa.

  Anarrhinum pechuelii Kuntze in Jahrb. Bot. Gart. Berl. **4**: 269 (1886). Type from S. Africa.
  Diclis viridis Marloth ex Engl., Bot. Jahrb. **10**: 253 (1888) nomen nudum.
  Simbuleta pechulii (Kuntze) Kuntze, Rev. Gen. Pl. **2**: 465 (1891). Type as above.

Prostrate or decumbent, much branched herb; branches 4–25 cm. long, short stipitate-glandular to subglabrous. Leaves petiolate; lamina (7)15–27 × (5)8–18 mm., circular, elliptic to obovate, apex rounded, cuneate at base, entire or more rarely obscurely dentate but not coarsely so, glandular above and beneath to varying degrees; petiole 9–17 mm. long, stipitate-glandular. Pedicels (10–30)50–65 mm. long, capillary to filiform, glandular. Calyx 1.2–2.5 mm. long, lobes 0.6–0.8 mm. wide, unequal, oblong to ovate-lanceolate, minutely glandular-stipitate. Corolla white with purple markings at throat, 4.5–9(13) mm. long including spur; spur 1–2.5 mm. long, broad, subdeltoid; upper lip bilobed, lobes subdeltoid; lower lip 3-lobed, lobes broad, rounded. Capsule c. 2 mm. long, 2.5 mm. in diam., depressed globose, emarginate.

Botswana. N: Okavango Delta, Xudum R., fl. & fr. 24.vii.1975, Hiemstra 244 (BR; K; MO; PRE; SRGH). SE: Gaborone Dam, c. 960 m., fl. & fr. 2.vi.1974, Mott 277 (K; SRGH). Zambia. S: Livingstone Distr., Victoria Falls, 770 m., fl. & fr. 5.viii.1972, Kornaś 1980 (K). Zimbabwe. W: Victoria Falls, fl. & fr. 19.vii.1947, Keay in GHS 21393 (K; SRGH). S: Gwanda Distr., Umzingwane R., 710 m., fl. & fr. 14.v.1972, Pope 796 (K; LISC; P; PRE; SRGH).

Also in Namibia and S. Africa. Damp lowland woods, streambanks, riversides and marshy areas in grassland or rain-forest; 700–1900 m.

5. MISOPATES Rafin.

Misopates Rafin., Autikon Bot.: 158 (1840).

Herbs, annual, simple or branched, glabrous or glandular-pilose. Leaves simple, entire, usually opposite below, becoming alternate above. Flowers zygomorphic in terminal racemes, or solitary-axillary. Calyx 5-lobed; lobes longer than corolla tube, markedly unequal. Corolla tubular; tube broadly cylindric, wide mouthed, gibbous at base, villous within. Stamens 4, didynamous, included. Ovary bilocular; ovules many. Capsule ovoid, gibbous, opening with 2 pores. Seeds somewhat flattened, one face finely tuberculate with a wide, raised, papillose border, the other smooth, keeled and produced into a narrow wing.

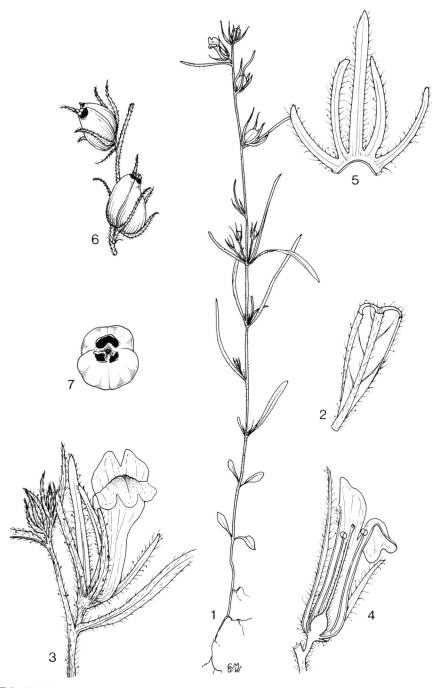

Tab. 5. MISOPATES ORONTIUM. 1, habit (× ⅔); 2, leaf base (× 2); 3, flowering branch (× 4);
4, flower, longitudinal section (× 4); 5, calyx, opened out (× 4), 1–5 from *Polhill & Paulo* 1206;
6, fruiting branch (× 2); 7, fruit, apical view (× 4), 6–7 from *Jackson* 2074.

A genus of 3 species occurring naturally from the Mediterranean through Asia Minor to Northwest India.

Misopates orontium (L.) Rafin., Autikon Bot.: 158 (1840). —Pennell in Bull. Torr. Bot. Club **48**: 95 (1921). TAB. **5**. Type from Europe.

    Antirrhinum orontium L., Sp. Pl.: 617 (1753). —Hiern in F.C. **4**, 2: 205 (1904). —Hemsl. & Skan in F.T.A. **4**, 2: 294 (1906). Type as above.

Erect or ascending herb to 60 cm. tall, simple or branched, usually glabrous to sparsely long-hirsute at base, becoming glabrous above, then further above glandular-pubescent. Leaves 12–30 × 0.3–3 mm. linear to linear-lanceolate, sub-acute, entire, subsessile, margins frequently revolute (at least when dry), glabrous to rarely and remotely ciliate. Inflorescence a lax-flowered raceme, or lower flowers appearing solitary and axillary. Pedicel 0.5–1.5 mm. long in flower becoming slightly longer in fruit, glandular-pubescent. Calyx 4–8 mm. long, lobes to 0.3 mm. wide, linear, the longest half again as long as shortest, glandular-pubescent. Corolla 7–9 mm. long, whitish-pink to pink-purple with darker purple longitudinal venation, throat long-yellow-villous. Capsule 7–8 mm. long, brown, glandular-pubescent. Seeds greyish-brown, c. 1 × 0.5 mm.

    Malawi. C: Dedza Distr., Chongoni Forest Reserve, fl. 30.iii.1968, *Salubeni* 1033 (K; LISC; SRGH).

    In addition to its occurrence naturally in southern Europe and North Africa, this plant is widely spread to the Atlantic islands and further west where it has been recorded from Haiti. It also occurs eastwards to India and south through eastern Africa to the Cape; collections have also been made from Queensland and New South Wales in Australia.

6. HALLERIA L.

Halleria L., Sp. Pl.: 625 (1753).

Shrubs or small trees, glabrous. Leaves opposite, petiolate, ovate or elliptic, dentate or subentire. Flowers subfascicled in axillary clusters. Pedicels bibracteolate. Calyx 3–5-lobed, subrotate or cup-shaped, persistent. Corolla 4–5-lobed, bilabiate, tube curved or straight, inflated above to funnel-shaped, upper lip bilobed, lower lip shortly 3-lobed, lobes subequal, all much shorter than tube. Stamens 4, staminode absent, didynamous, subequal, usually exserted, filaments filiform, inserted subcentrally on corolla tube; anther cells diverging, at length divaricate. Style filiform, persistent, usually exserted. Ovary bilocular, ovoid; ovules numerous. Fruit a berry. Seeds compressed, narrowly winged.

A genus of about 10 species from tropical Southern Africa and the Malagasy Republic.

Corolla tube curved, slightly gibbous at base, mouth oblique; leaf lamina
    4.5–10 cm. long - - - - - - - - - - - - - - - 1. *lucida*
Corolla tube straight, not gibbous at base, mouth regular; leaf lamina usually
    2–3.5 cm. long - - - - - - - - - - - - - - 2. *elliptica*

1. **Halleria lucida** L., Sp. Pl.: 625 (1753) excl. var. B. —Sims in Bot. Mag. t. 1744 (1815). —Benth. in Hook., Comp. Bot. Mag. **2**: 54 (1836); in DC., Prodr. **10**: 301 (1846). —A. Rich., Tent. Fl. Abyss. **2**: 116 (1850). —Engl., Bot. Jahrb. **30**: 401 (1902). —Hiern in F.C. **4**, 2: 207 (1904). —Hemsl. & Skan in F.T.A. **4**, 2: 295 (1906). TAB. **6**, fig. B. Type from S. Africa.

    Halleria abyssinica Jaub. & Spach, Illustr. Pl. Orient **5**: 65, 66, tab. 459 & 460 (1855). Type from Ethiopia.

Small tree or erect to straggling shrub up to 12 m. tall; branches sub-quadrangular. Leaves petiolate; lamina 4.5–10 × 2.0–6.2 cm., broadly ovate, acuminate to caudate-acuminate, rounded to broadly cuneate at base, coriaceous, shortly serrate to serrate-crenate, glabrous, minutely glandular-punctate beneath; petiole 4–12 mm. long. Pedicels 10–14 mm. long with pair of bracteoles below the middle; very sparse, irregularly hirsute. Bracteoles 1–1.8 × 0.4–0.5 mm., lanceolate. Calyx 2.4–3.5 mm. long, lobes 3.8–4.2 mm. wide, joined to beyond halfway, rounded, subglabrous. Corolla orange-yellow to brownish-red, 25–33 mm. long, tube curved, widening above to 6–10 mm. diam., slightly gibbous at base minutely glandular-pubescent without, mouth oblique, unequally 4-lobed; lobes not marginally pubescent. Stamens long exserted. Style 20–38 mm. long. Berry 12–18 mm. long, 8.5–14 mm. in diam., ovoid to subglobose, blackish-purple when ripe.

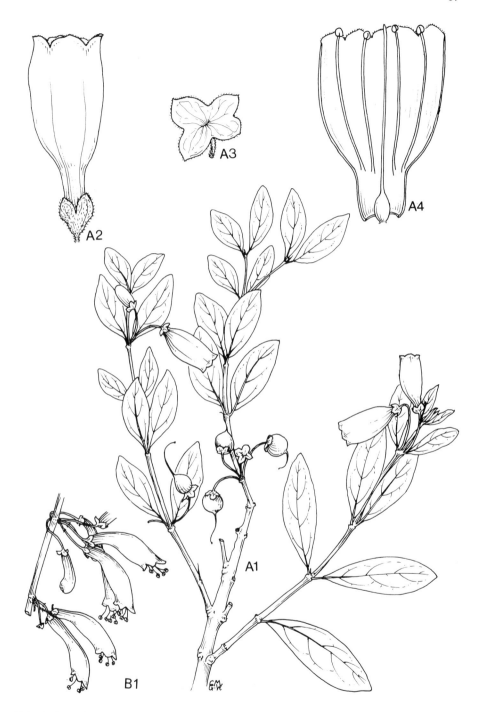

Tab. 6. A.—HALLERIA ELLIPTICA. A1, flowering and fruiting branch (×⅔); A2, flower (× 2); A3, calyx (× 4); A4, corolla opened to show gynoecium and androecium (× 2), A1–4 from *Pawek* 3779. B.—HALLERIA LUCIDA. B1, flowering portion of branch (×⅔), **from** *Phillips* 3987.

Botswana. SE: Pharing Gorge, Kanye, st. iii.1941, *Miller* B275 (PRE). **Zambia**. W: near Solwezi, fl. 12.ix.1952, *White* 3226 (BM; BR; COI; K; PRE). **Zimbabwe**. W: Matobo Distr., c. 1430 m., fl. & fr., *Miller* 1883 (SRGH). C: Makoni Distr., 1500 m., fl. 20.ix.1946, *Hopkins* in GHS 15388 (K; SRGH). E: Nyanga (Inyanga) Distr., Pungwe Source, 1920 m., fl. 20.x.1946, *Wild* 1434 (K; LD; SRGH). S: Mberengwa (Belingwe) Distr., Buhwa Hill, east slopes, c. 1200 m., st. 5.vii.1968, *Müller* 785 (SRGH). **Malawi**. N: Chitipa Distr., Nyika Plateau, 4 km. WNW. Muzengapakweru, 2210 m., fl. 26.viii.1972, *Synge* WC 318 (K; SRGH). C: Viphya, Luwawa, c. 1500 m., fl. 16.xii.1961, *Chapman* 1506 (SRGH). S: Zomba Distr., Mulunguzi Bridge, Zomba Plateau, fl. 11.x.1979, *Banda & Salubeni* 1561 (MO; SRGH). **Mozambique**. Z: Gúruè, Mt. Namuli, fl. & fr. 12.viii.1949, *Andrada* 1844 (COL; LISC). MS: Gorongoza, fl. & fr. 1884–1885, *Carvalho* s.n. (COI).

Also in Angola and Yemen, and from Ethiopia southwards to S. Africa. Montane forests and slopes to river margins and gallery forests.

2. **Halleria elliptica** Thunb. in Nova Acta R. Soc. Scient. Upsal. **6**: 39 (1799). —Benth. in Hook., Comp. Bot. Mag. **2**: 54 (1836); in DC., Prodr. **10**: 302 (1846). —Hiern in F.C. **4**, 2: 209 (1904). —Hemsl. & Skan in F.T.A. **4**, 2: 296 (1906). TAB. **6**, fig. A. Type from S. Africa.

Erect, compact, shrub or small, slender tree to 6 m. tall; branches sub-quadrangular. Leaves petiolate: lamina 20–35(50) × 8.5–22(26) mm., elliptic to broadly elliptic-ovate, obtuse, shortly apiculate, coriaceous, minutely, remotely denticulate to occasionally serrate, clearly to obscurely glandular-punctate beneath, margin slightly revolute; petiole 3–6 mm. long. Pedicels (6)10–12 mm. long, slender with pair of bracteoles below middle; bracteoles 0.8–1.3 × 0.2–l0.3 mm., lanceolate. Calyx 2.3–4.5 mm. long, lobes 2.5–3.6 mm. wide, joined to more than halfway, broadly rounded, very shortly mucronate. Corolla orange to deep or brownish-red, (12)18–25 mm. long; tube straight at base, widening above to (5)7–9(12) mm. in diam., minutely glandular-pubescent below, 4-lobed; lobes c. 1–1.5 mm. long, subequal, broadly rounded, marginally densely, short glandular-pubescent. Stamens barely exserted. Style up to c. 18 mm. long, exserted. Berry 8–9 mm. in diam., globose or subglobose.

Malawi. S: Mt. Mulanje, between Thuchila (Tuchila) and Ruo Valley, c. 2000 m., fl. & fr. 13.vii.1956, *Jackson* 1887 (BR; K; FHO). **Mozambique**. Z: Namuli, Makua, fl., *Last* s.n. (K). MS: Gorongoza, Gogogo Mt., c. 1780 m., fl. 5.vii.1955, *Schelpe* 465 (BM; LISC; SRGH).

Also in S. Africa. Forest margins and among rocks; 1600–2300 m.

7. TEEDIA Rudolphi

Teedia Rudolphi in Schrad. Journ. Bot. **2**: 289 (1799).

Shrubs or coarse herbs. Leaves opposite, subovate, denticulate. Inflorescence cymose, axillary or terminal. Calyx 5-toothed, lobes linear-lanceolate, slightly overlapping in bud. Corolla deciduous; tube cylindrical, equalling to exceeding calyx; limb 5-lobed, spreading, subregular; lobes imbricate in bud, flat when open. Stamens 4, perhaps with a rudimentary fifth; filaments short, filiform, inserted near base of corolla tube; anthers short, bithecal, included. Ovary bilocular, subglobose; style short, fleshy; stigma ovoid, capitate, cleft at apex; ovules numerous. Fruit bilocular, subglobose, indehiscent. Seeds numerous, ovoid, rugose.

A genus of 4 species from S. Africa with one species extending into the Flora Zambesiaca area.

Teedia lucida (Solander) Rudolphi in Schrad. Journ. Bot. **2**: 288 (1799). —Edwards in Bot. Reg. **2**, tab. 209 (1817). —Benth. in Hook. Comp. Bot. Mag. **2**: 54 (1836); in DC., Prodr. **10**: 334 (1836). TAB. **7**. Type from S. Africa.
    Capraria lucida Solander in Aiton, Hort. Kew **2**: 353 (1789). Type as above.

Woody herb to 1 m. tall, glabrous, foetid; stems tetragonous, smooth. Leaves 3–13 × (0.8)1.8–6 cm., oval, ovate to broadly elliptic, obtuse or acute, abruptly or gradually narrowed in the short petiolate, subauriculate base, serrulate or denticulate; petioles up to 1.75 cm. long, winged, decurrent at base. Inflorescence cymose, terminal or axillary, dense, leafy; peduncles axillary, 1–3.5 cm. long, 3–7-flowered; pedicels 2–6 mm. long; bracts c. 2–4 mm. long, opposite, lanceolate. Flowers up to 6 mm. long, blue through lilac to mauve. Calyx lobes up to 4.5 mm. long, lanceolate-oblong, obtuse. Corolla tube to 5 mm. long, almost straight, shortly exceeding calyx, cylindric; limb patent, to 7 mm. in diam., lobes rounded. Ovary subglobose, longitudinally somewhat compressed, c. 1.8–2

Tab. 7. TEEDIA LUCIDA. 1, branch (× ⅔); 2, flowering stem (×3); 3, flower, longitudinal section showing gynoecium (×4); 4, corolla opened out showing androecium (×4), 1–4 from *Wild* 4579; 5, fruiting branch (× ⅔), from *Biegel* 4129.

mm. in diam. Fruits 0.5–1 cm. in diam., globose, brown to purplish-black, occasionally style-base semi-persistent. Seeds black.

Zimbabwe. E: Chimanimani Mts., fl. 20.viii.1954, *Wild* 4579 (K; SRGH).
Also known more commonly in S. Africa. In Zimbabwe it appears to favour the rocky outcrops and crags of the Eastern Districts; occurring at high altitudes up to 2500 m. in Nyanga (Inyanga) Distr.

8. FREYLINIA Colla

Freylinia Colla, Hort. Rupul.: 56 (1823).

Shrubs or subshrubs. Leaves opposite, verticillate or scattered, sessile to shortly petiolate, entire or toothed in upper half. Inflorescence of terminal and axillary cymes. Calyx 5-partite, campanulate, lobes ovate, elliptic or lanceolate, imbricate in bud. Corolla 5-lobed; tube funnel-shaped or subcylindric, straight, longer than calyx; lobes spreading, almost flat, imbricate in bud. Stamens four, didynamous, occasionally with a fifth rudimentary and sterile stamen present, included; filaments terete or linear; anthers shortly oblong, with two parallel thecae. Ovary bilocular, ellipsoid; ovules many; stigma capitate, globose or ovoid; style thick, linear. Fruit bilocular, ovoid, capsular. Seeds discoid, with membranous testa and winged margin.

A genus of five species from tropical and South African montane habitats.

Freylinia tropica S. Moore in Journ. Linn. Soc., Bot. **40**: 152 (1911). TAB. **8**. Type: Zimbabwe, Chimanimani (Melsetter), c. 1828 m. fl. 23.ix.1906, *Swynnerton* 608 (BM, holotype; K, isotype).

Shrub 0.6–2(4) m. tall; stems tetragonous, glabrous, viscid when young. Leaves 1–3.5 × 0.5–0.75 cm., obovate to oblanceolate, obtuse, attenuate at base into short petiole, margins thickened, frequently with few teeth in upper half; petiole 0.5–2 mm. long. Inflorescence of axillary cymes, each with 2–5 flowers, somewhat viscid; bracts minute, 1–1.5 mm. long. Flowers purple through lilac to pale blue and white. Calyx 3–3.5 mm. long, lobes 2–2.5 × 0.5–1 mm., oblong, obtuse, shortly to more or less densely minutely stipitate-glandular. Corolla tube 10–11 mm. long, subcylindric, more or less densely stipitate-glandular, up to 2.5 mm. wide at the middle, broadening slightly to the 3.5 mm. wide mouth; limb spreading, lobes to about 3 × 3.5 mm., rounded. Ovary shortly ovoid, basally compressed. Fruit 3.5–5 × 3 mm., broadly ellipsoid, dull yellow.

Zimbabwe. E: Nyanga Village, 2000 m., fl. 15.ix.1960, *Rutherford-Smith* 107 (K; SRGH).
Mozambique. MS: Nyamkwarara Valley, Gorongo Mt., fl. 2.xi.1967, *Mavi* 453 (K; LISC; SRGH).
Known also from Tanzania and S. Africa (Transvaal), submontane, inhabiting wet places such as riverbanks and streamsides; 1800 m.

9. ANTHEROTHAMNUS N.E. Br.

Antherothamnus N.E. Br. in Hook. Ic. Pl. **31**: 3007 (1915).

Bushy, glabrous shrub; branches slender. Leaves alternate-fasciculate. Calyx 5–7-lobed. Corolla tubular, infundibuliform; lobes 5, flatly spreading. Stamens 4, didynamous, included, staminode 1; anthers unithecal. Stigma small; style filiform, slightly thickened above; ovary bilocular with many ovules. Capsule ovoid.

A monotypic genus recorded only from S. Africa and Namibia outside the Flora Zambesiaca area.

Antherothamnus pearsonii N.E. Br. in Hook. Ic. Pl. **31**: 3007 (1915). —Merxm. in Merxm., Prodr. Fl. SW. Afr. **126**: 8 (1967). TAB. **9**. Types from S. Africa.
Sutera rigida L. Bolus in Ann. S. Afr. Mus. **9**: 267 (1915). Type from S. Africa.
Selaginastrum rigidum (L. Bolus) Schinz & Thell. in Viert. Naturf. Ges. Zürich **74**: 121 (1929). Type as above.
Antherothamnus rigida (L. Bolus) Phillips in Bothalia **3**: 271 (1937). Type as above.

Bushy shrub up to 3 m. tall; branches slender, pale orange-brown, shortly glandular-pubescent when young becoming dark brownish-grey, glabrescent with age. Leaves

Tab. 8. FREYLINIA TROPICA. 1, flowering branch (× ⅔); 2, flower (× 3); 3, corolla opened out showing androecium (× 3); 4, flower, longitudinal section showing gynoecium (× 3); 5, dehisced capsule (× 4), all from *Crook* M46.

Tab. 9. ANTHEROTHAMNUS PEARSONII. 1, flowering branch (× ⅔), from *Giess* 10333; 2, flowering shoot (× 4); 3, corolla opened showing androecium and filiform style (× 8); 4, gynoecium (× 8), 2–4 from *Miller* B/501.

3–5(10) × 1–1.5(3.5) mm., alternate-fasciculate, obovate cuneate, obtuse, somewhat succulent, shortly glandular-pubescent when young, glabrescent, minutely dark punctate. Inflorescence with racemes 1–1.5 cm. long, arising from upper fascicles. Flowers alternate, pedicellate. Pedicels 1.5–2.75 mm. long, slender, bibracteolate. Bracteoles opposite, 0.6–1.4 mm. long, linear. Calyx 5–7-lobed; lobes 1–2.5 × c. 0.5 mm., obtuse, linear, glabrous. Corolla white with yellow throat, fragrant; tube 2.5–4 mm. long, 1.5–3.25 mm. wide at mouth, narrowly infundibuliform; lobes 1.5–2.5 mm. in diam., subcircular, lower pair slightly smaller. Filaments filiform; anthers subglobose-ellipsoid. Style curved below apex. Capsule 3–3.5 × 2 mm., ovoid, glandular-pubescent when young, glabrescent.

Botswana. SE: by Livingstone's Cave, Molepolole, 56 km. NW. of Gaborone, 1160 m., fl. 21.iv.1974, *Mott* 239 (K; SRGH). **Zimbabwe**. W: Matobo Distr., Matopos, fr. 6.xi.1925, *Eyles* 7024 (SRGH).

Also known from S. Africa and Namibia; rocky hillsides; 960–1220 m.

10. ZALUZIANSKYA F.W. Schmidt

Zaluzianskya F.W. Schmidt in Neue Selt. Pflanzen: 11 (1793) *nom. conserv.*
non *Zaluzianskia* Necker.

Annual or perennial herbs, subviscid, turning black on drying. Stems erect or ascending, simple or branched, sometimes cushion- or mat-forming, becoming woody at base and into tap-root with crown often bearing many vegetative buds or not. Leaves opposite, or alternate above, simple, dentate or entire. Flowers in terminal spikes or occasionally axillary, bracteate. Calyx 5-toothed, bilabiate, oblong to tubular-ovoid. Corolla 5-lobed, tubular with tube greatly exceeding calyx, often splitting at base, regular or bilabiate; lobes shorter than tube, subequal or not, entire, emarginate or bifid. Stamens 4, didynamous or rarely 2 by abortion of anterior pair, posterior pair usually included. Filaments short, inserted near apex of tube; anthers unithecal by confluence, anterior pair smaller or sterile. Ovary bilocular; style filiform, entire, somewhat clavate at apex; ovules numerous. Capsule ovoid-oblong, septicidally bivalved. Seeds numerous.

A complex genus of about 40 species from southern Africa. TAB. **10**.

Hitherto, *Zaluzianskya* in the Flora Zambesiaca area was accommodated under two names, namely *Z. maritima* and *Z. capensis* both of which were included by Hiern in Flora Capensis (1904). Hilliard & Burtt (Notes R.B.G. Edinb. **41**: 1–43 (1983)) have made a special study of this complex as related to south-eastern Africa and conclude that these two names, over the years, have become a depository for a number of taxa which are not related one with the other. They also state clearly that neither species occur as far north as Natal and by that observation, not in Zimbabwe from where the bulk of our material comes. They consider that before concise specialised studies can be made on the complex in the Flora Zambesiaca area, much more material needs to be collected especially of root systems, along with further studies of the plants in the wild.

It is not my intention here to try and unravel the speciation of our material although Hilliard & Burtt suggest that up to three taxa are probably present. According to these authors, none of these appear to fit any species already described, and from the material available for this work vital organs essential in species delimitation, such as root-systems and subterranean basal buds are sadly lacking.

For the purpose of this work, the following gives only a representation of the genus within the Flora Zambesiaca area.

Zimbabwe. N: Mazowe (Mazoe) Distr., Iron Mask Hill, 1550 m., fl. & fr. v.1907, *Eyles* 548 (BM; K; SRGH). W: Matobo Distr., Quaringa Farm, c. 1450 m., fl. & fr. v.1954, *Miller* 2412 (K; SRGH). C: Makoni Distr., Rusape Rd., fl. 13.ii.1961, *Rutherford-Smith* 529 (K; SRGH). E: Nyanga (Inyanga) Distr., Mt. Inyangani summit, 2500 m., fl. 7.iii.1981, *Philcox, Leppard et al.* 8909 (K). S: Mberengwa (Belingwe) Distr., Mt. Buhwa, c. 1200 m., fl. 27.iv.1973, *Pope* 959 (K; SRGH). **Mozambique**. MS: Tsetserra, 1980 m., fl. 2.iii.1954, *Wild* 4476 (K; SRGH).

Genus also known from Southern Africa with one species in Uganda; hillsides and montane grasslands; up to 2500 m.

11. POLYCARENA Benth.

Polycarena Benth. in Hook., Comp. Bot. Mag. **1**: 371 (1836).

Annual, or more rarely perennial, herbs, sub-viscid-pilose. Leaves opposite, or

24

Tab. 10. ZALUZIANSKYA 1, flowering stem (× ⅔), from *Boughey* 549; 2, rootstock (× ⅔); 3, flower opened out showing androecium and gynoecium (×2); 4, young fruit (× 2), 2–4 from *Philcox et al.* 8909; 5, dehisced fruit (× 3), from *Chase* 5253.

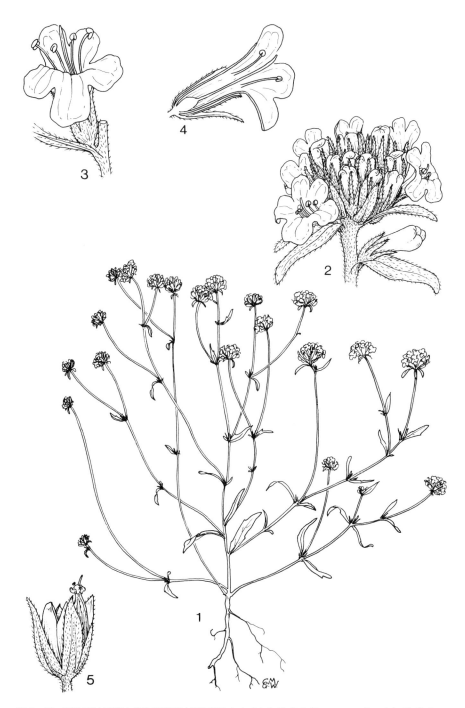

Tab. 11. POLYCARENA TRANSVAALENSIS. 1, habit (× ⅔); 2, inflorescence head (× 4); 3, flower (× 6); 4, flower, longitudinal section showing gynoecium (× 6); 5, dehiscing fruit (× 6), all from *Hansen* 3439.

alternate especially above, entire to serrate or dentate. Flowers usually spicate, small, unibracteate. Bracts adnate below to pedicel and or base of calyx. Calyx 5-lobed, shallowly or deeply-cleft, or appearing bilabiate. Corolla 5-lobed, tubular; tube slender, strict, mouth occasionally oblique; lobes more or less regular, entire or slightly retuse, shorter than tube. Stamens 4, didynamous, exserted; filaments filiform inserted on corolla tube, glabrous. Style filiform, exserted. Ovary bilocular, ovules numerous. Capsule oblong, ovoid or subglobose, subcompressed, obtuse.

A genus of 40 species from southern Africa, one of which occurs in the Flora Zambesiaca area.

Polycarena transvaalensis Hiern in F.C. **4**, 2: 330 (1904). TAB. **11**. Types from S. Africa.
   Phyllopodium calvum Hiern in F.C. **4**, 2: 315 (1904). Type from S. Africa.
   Polycarena calva (Hiern) Levyns in Journ. S. Afr. Bot. **5**, 1: 37 (1939). Type as above.

Erect herb, 5–8(16) cm. tall, simple to much-branched, minutely glandular-pubescent occasionally sparsely mixed with simple white hairs. Leaves opposite, or alternate above; lower leaves 7–17 mm. long, lamina 3.5–11 × 0.8–4 mm., obovate to broadly elliptic-lanceolate, obtuse at apex, tapering to narrow, petiole-like base up to 7 mm. long, entire to more rarely laxly broad-toothed; upper leaves 4.5–17(24) × 0.5–2.3 mm., linear to lanceolate-linear, acute, entire to laxly dentate with up to 2 teeth on each margin. Flowers sessile to subsessile in short, terminal, spherical or slightly elongated spicate heads. Bracts 1.6–2.4 × 0.5–0.6 mm., oblong, obtuse, apex rounded, attached at base to pedicel where present and to base of calyx. Pedicel 0–0.25 mm. long. Calyx 1.9–2.2 mm. long, deeply 5-lobed, regular or apparently bilabiate possibly due to different depth of free portion of lobes; lobes 0.25–0.5(0.8) mm. wide. Corolla white to pale pinkish-mauve with yellow-orange throat; tube up to 2 mm. long, 0.7 mm. in diam., strict, widening at mouth; lobes c. 1.5 mm. long, broadly obovate, rounded at apex. Ovary c. 1 mm. long, 0.5 mm. in diam., ellipsoid. Style 2.8–3 mm. long, exserted. Anthers globular, exserted. Capsule c. 2.5 mm. long.

Botswana. SE: Gaborone, 1020 m., fl. 22.viii.1978, *Hansen* 3439 (K; SRGH). **Zimbabwe.** W: 8 km. SE. of Bulawayo, c. 1460 m., fl. vi.1961, *Miller* 7973 (K, as 7974; SRGH).
   Also in S. Africa (Transvaal). Mainly grasslands on Kalahari Sand in the Flora Zambesiaca area; 1000–1500 m.
   Smith 3511 (SRGH) from Xaraga, Quangwa in Botswana is an immature specimen but appears to belong to *Polycarena*. It is included here to indicate the probable generic distribution.

12. SUTERA Roth

Sutera Roth, Bot. Bremerk.: 172 (1807) non Roth, Nov. Pl. Sp.: 291 (1821).

Annual or perennial herbs or small shrubs, erect, decumbent or ascending, pubescent, hispid, glandular-viscid or more rarely glabrous, frequently becoming black on drying. Leaves usually opposite or nearly so, dentate, crenate, lobed or incised, rarely entire, petiolate or sessile. Bracts usually similar to leaves but smaller, free from pedicels. Flowers solitary-axillary or in terminally racemose or simple, cymose or spicate inflorescences, pedicellate, bracteolate or not. Calyx 5-partite or 5-lobed; lobes linear, oblong, lanceolate or rarely ovate. Corolla tubular, deciduous; tube longer or shorter than calyx, cylindrical or infundibuliform, straight or curved towards apex, at times dilated at throat; limb 5-lobed, spreading, regular or bilabiate; lobes entire, emarginate or shortly bifid, subequal. Stamens 4, didynamous, exserted or wholly or partly included. Filaments filiform, inserted in corolla tube. Anthers perfect, reniform, unithecal by confluence of cells. Ovary bilocular. Style filiform, included or exserted, somewhat clavate above. Stigma obtuse. Capsule septicidal, valves bifid. Seeds numerous, small, rugose.

A genus of about 130 species from tropical and southern Africa and the Canary Islands.

1. Corolla tube up to 8.5 mm. long - - - - - - - - - - -	2
– Corolla tube 9–24 mm. long - - - - - - - - - - -	7
2. Perennial herbs or undershrubs - - - - - - - - - -	3
– Annual herbs - - - - - - - - - - -	1. *hereroensis*
3. Calyx 4–8.5 mm. long - - - - - - - - - - -	4
– Calyx 2–3.5 mm. long - - - - - - - - - - -	5

4. Leaves narrowly lanceolate, irregularly pinnatifid, never undivided, segments dentate,
pinnatifid or lobed   -   -   -   -   -   -   -   -   -   - 2. *elegantissima*
– Leaves broadly ovate to subcircular, coarsely dentate   -   -   -   -   - 11. *floribunda*
5. Leaves rarely exceeding 2 mm. in width, usually linear or rarely
obovate-lanceolate   -   -   -   -   -   -   -   -   -   -   -   - 3. *palustris*
– Leaves up to 20 mm. wide, never linear   -   -   -   -   -   -   -   -   -   - 6
6. Corolla orange-red to red   -   -   -   -   -   -   -   -   -   -   - 4. *aurantiaca*
– Corolla yellow or white, pink or blue, never red   -   -   -   -   -   - 5. *micrantha*
7. Leaves fasciculate or subfasciculate   -   -   -   -   -   -   -   -   -   -   - 8
– Leaves not as above   -   -   -   -   -   -   -   -   -   -   -   -   -   - 10
8. Leaves very small, 1–4 mm. long, linear to oblanceolate, usually entire; calyx
2–3 mm. long   -   -   -   -   -   -   -   -   -   -   -   - 6. *atropurpurea*
– Leaves larger, spathulate, cuneate-oblong or obovate, entire to incised dentate, or pinnatifid;
calyx 3–8 mm. long   -   -   -   -   -   -   -   -   -   -   -   -   -   - 9
9. Leaves spathulate, entire or with few teeth towards apex; calyx 3–4.5 mm. long   7. *brunnea*
– Leaves oblong or obovate, incised-dentate or pinnatifid; calyx 6–8 mm. long   8. *burkeana*
10. Pedicels minute, up to 0.5 mm. long; leaves narrowly oblong-elliptic, crenate, bullate, densely to
somewhat densely tomentose; margins strongly revolute   -   -   -   -   - 9. *fodina*
– Pedicels 2–6 mm. or more long; leaves not as above   -   -   -   -   -   -   -   - 11
11. Corolla tube 15–24 mm. long; leaves ovate to ovate-lanceolate, crenate-serrate   10. *carvalhoi*
– Corolla tube 5–9 mm. long; leaves broadly ovate-elliptic to subcircular, coarsely
dentate   -   -   -   -   -   -   -   -   -   -   -   -   - 11. *floribunda*

1. **Sutera hereroensis** (Engl.) Skan in F.T.A. **4**, 2: 301 (1906). TAB. **12**, fig. B. Type from Namibia.
Chaenostoma hereroense Engl., Bot. Jahrb. **19**: 150 (1894). Type as above.

Erect to spreading annual herb, 12–15 cm. tall, slender, branched or occasionally simple, sparsely to subdensely pilose; branches 8–20(28) cm. long. Leaves 9–30 × 4.5–22 mm., broadly ovate, usually acute, truncate or slightly connate at base, coarsely dentate to bidentate, pilose, sparsely so above. Petiole 2.5–7(11) mm. long, pilose. Inflorescence racemose, terminal, few-flowered, slender glandular-pilose. Bracts similar to leaves but smaller. Pedicels 3.5–10(16) mm. long, slender, pilose. Calyx (3)5–6 mm. long; lobes 0.15–0.25 mm. wide, linear, acute, glandular-pilose. Corolla pale blue, pink mauve or white with yellow throat, minutely pilose; tube 6–8 mm. long, infundibuliform, pilose within; limb spreading; lobes c. 2 mm. in diam., circular to obovate. Stamens included. Capsule c. 3 × 1.5 mm., narrowly ovoid.

Zimbabwe. W: Matobo Distr., Besna Kobila Farm, c. 1460 m., fl. & fr. vii.1956, *Miller* 3601 (K; SRGH). C: Ruzawi Distr., near Marondera (Marandellas), fl. & fr. 2.iv.1944, *Greatrex* in GHS 13245 (K; SRGH). E: Chipinge Distr., Vermont Farm, fl. 17.v.1974, *Cannell* 605 (K; SRGH). S: 16 km. N. of Masvingo (Fort Victoria) on Chirumanzu (Chilimanzi) road, fl. & fr. 4.v.1962, *Drummond* 7940 (K; SRGH).
Known also from Namibia and S. Africa; in shade of rocks on granite outcrops (kopjes) and rocky woodlands; up to 1460 m.

2. **Sutera elegantissima** (Schinz) Skan in F.T.A. **4**, 2: 302 (1906). TAB. **12**, fig. A. Type from Namibia.
Lyperia elegantissima Schinz in Verh. Bot. Ver. Brandenb. **31**: 192 (1890). Type as above.

Erect perennial herb or undershrub up to 1 m. tall, glandular-pilose, much branched; branches erect or spreading, varying in length up to 45 cm. long. Leaves 12–30 × 3.5–9 mm., narrowly lanceolate, irregularly pinnatifid, short petiolate, densely to somewhat densely pilose, aromatic, segments irregularly dentate, pinnatifid or lobed, obtuse. Petiole up to 5 mm. long. Flowers solitary in axils of upper leaves. Pedicels (6)10–15 mm. long, spreading to almost erect. Calyx 5–8.5 mm. long, deeply 5-lobed to base; lobes linear, acute or subobtuse. Corolla yellow, minutely glandular-pubescent on underside of lobes; tube 7.5–9 mm. long, infundibuliform, minutely glandular-pubescent below; limb spreading, up to c. 4 mm. in diam.; lobes c. 1 mm. in diam. or somewhat more, rounded, wavy margined (when dry), subequal. Capsule c. 5.5 × 2–2.5 mm., narrowly ovoid.

Caprivi Strip. c. 37 km. NW. of Ngoma Ferry, c. 1100 m., fl. & fr. 16.vii.1952, *Codd* 7081 (BM; K). **Botswana**. N: Shorobe, 38 km. NE. of Maun, fl. & fr. 19.iii.1965, *Wild & Drummond* 7180 (BR; K; LISC; SRGH). **Zambia**. B: 72 km. W. Nangweshi, 1035 m., fl. & fr. 6.viii.1952, *Codd* 7417 (BM; K; SRGH). C: near Mumbwa, fl. & fr. 1911, *Macaulay* 484 (K). S: Livingstone Distr., banks of Zambezi near Boat Club, fl. & fr. 31.viii.1947, *Greenway & Brenan* 8033 (K). **Zimbabwe**. N: Binga Distr., near Binga, edge of Zambezi, c. 450 m., fl. & fr. 6.xi.1958, *Phipps* 1364 (BR; K; P; SRGH).
Also known from Namibia; grasslands and open woodland, bordering lakes and rivers, and other wet areas; 425–1150 m.

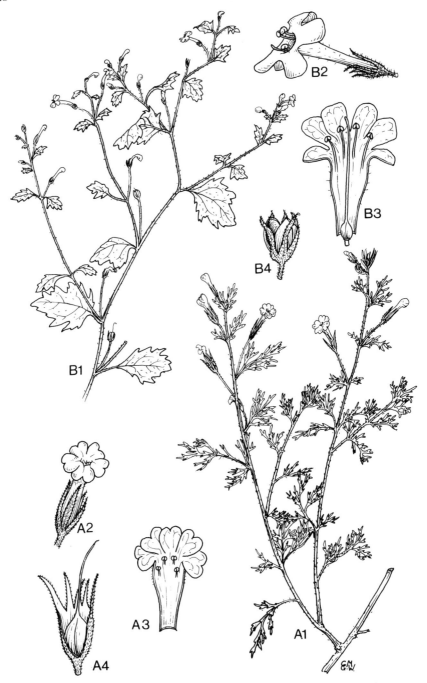

Tab. 12. A.—SUTERA ELEGANTISSIMA. A1, flowering branch (× ⅔), from *Biegel et al.* 5102; A2, flower (× 3); A3, corolla opened showing androecium (× 3); A4, fruit (× 3). A2–4 from *Lambrecht* 401. B.—SUTERA HEREROENSIS. B1, flowering branch (× ⅔); B2, flower (× 3); B3, flower opened showing androecium and gynoecium (× 3), B1–3 from *Cannell* 605; B4, dehisced fruit (× 3), from *Drummond* 7940.

3. **Sutera palustris** Hiern in F.C. **4**, 2: 256 (1904). Type from S. Africa.

Erect, much-branched herbaceous shrub up to 60 cm. tall; branches almost opposite, spreading or ascending, slender, glandular-pubescent to subglabrous. Leaves 5–20 × 0.5–1(7) mm., opposite or almost opposite, often appearing fasciculate, linear or rarely obovate-lanceolate, obtuse, sessile, glandular-puberulent, entire or slightly, remotely dentate, margins revolute. Inflorescence terminally racemose. Bracts similar to leaves, becoming smaller above, linear to subulate. Pedicels 5–12 mm. long at maturity, slender to filamentous, glandular-puberulent. Calyx 2–2.75 mm. long, campanulate, 5-lobed to about or below midway, glandular-pubescent; lobes 0.25–0.4 mm. wide, triangular-subulate or lanceolate. Corolla pink or blue, throat yellow, infundibuliform, sparsely glandular-puberulent without; tube 2–3 mm. long; limb c. 3 mm. in diam.; lobes 0.9–1.1 mm. long, c. 0.5 mm. wide, obovate. Capsule c. 3 × 1 mm., ovoid, minutely glandular-papillose above.

Botswana. SE: Ootse, 1350–1450 m., fl. 9.x.1977, *Hansen* 3218 (K).
Also known from S. Africa. Rocky ground in dry bushland.

4. **Sutera aurantiaca** (Burch.) Hiern, Cat. Afr. Pl. Welw. **1**: 757 (1898). —Hiern in F.C. **4**, 2: 292 (1904). —Hemsl. in F.T.A. **4**, 2: 302 (1906). Type from S. Africa.
    Buchnera aurantiaca Burch., Trav. S. Afr. **1**: 388 (1822). Type as above.
    Lyperia multifida Benth. in Hook., Comp. Bot. Mag. **1**: 380 (1936); in DC., Prodr. **10**: 361 (1846). —Schinz in Bull. Trav. Soc. Bot. Genéve **6**: 70 (1891). Type from S. Africa.

Erect to ascending or decumbent perennial herb or undershrub up to c. 30 cm. tall, glandular-pubescent; branches spreading, up to 40 cm. long. Leaves up to 25 × 10 mm., ovate to subcircular, shortly petiolate, bipinnatisect, segments entire, linear or deeply pinnatifid. Flowers solitary in axils of upper leaves. Pedicels (3)6–12(20) mm. long, spreading to almost erect. Calyx 2.5–3.5 mm. long, deeply 5-lobed almost to base; lobes c. 0.5 mm. wide, linear spathulate, obtuse or subacute, entire, 1-nerved, densely glandular-pubescent. Corolla orange-red to red, throat yellow, tube 5–5.5 mm. long, densely short-glandular without, arcuate below limb; limb 7.5–10 mm. in diam.; lobes c. 4.5 mm. in diam., rounded, subequal. Capsule 3.5–4.5 × c. 2 mm., ovoid, shortly glandular-pubescent.

Botswana. SE: Kgatleng Distr., 8 km. NE. of Modipane, fl. & fr. 13.viii.1978, *Hansen* 3429 (K).
Also known from S. Africa, Lesotho and Namibia; rocky hillsides, grasslands, riverbanks; 1220–1750 m.

5. **Sutera micrantha** (Klotzsch) Hiern in F.C. **4**, 2: 263 (1904). —Skan in F.T.A. **4**, 2: 303 (1906). Type: Mozambique, Rios de Sena, *Peters* s.n. (B†), near Lupata, Oct. 1858, *Kirk* s.n. (K, neotype chosen here).
    Lyperia micrantha Klotzsch in Peters, Reise Mossamb. Bot.: 222 (1861). Type as above.
    Chaenostoma micrantha (Klotzsch) Engl. ex Diels in Engl., Bot. Jahrb. **23**: 489 (1897). Type as above.
    Sutera fissifolia S. Moore in Journ. Bot. **38**: 467 (1900). Type: Zimbabwe, Bulawayo, early January 1898, *Rand* 155 (BM, holotype; BR, isotype).
    Sutera blantyrensis Skan in F.T.A. **4**, 2: 304 (1906). Syntypes: Malawi, *Buchanan* in Herb. *Wood* 6630 (K), *Sharpe* 96 (K).

Erect or ascending, prostrate or scrambling perennial herb or undershrub, branched usually towards base, up to 30(60) cm. tall, glandular-pubescent; branches slender, 14–25(50) cm. long, spreading. Leaves 18–30(42) × 7–12(20) mm., ovate or ovate-oblong, obtuse, pinnatifid-lobed, opposite or alternate, petiolate; lobes entire or crenate-serrate. Petiole up to c. 8–10 mm. long, slender, narrowly winged. Flowers either solitary-axillary or in terminal racemes up to 15 cm. or more long. Bracts: lower leaf-like but smaller, upper 3-lobed, with uppermost linear. Pedicels (7)10–20(42) mm. long. Calyx 2.5–3.5(4.5) mm. long, deeply 5-lobed; lobes 0.6–0.8(1) mm. wide, linear to oblong-spathulate, obtuse or subacute. Corolla pale pink or blue or more usually yellow or white, all with yellow throat; tube c. 4 mm. long, narrowly cylindric at base becoming wider at throat, barely curved; limb c. 6 mm. in diam., lobes ovate, obtuse. Capsule 3.5–4 × c. 1.8 mm., elliptic-ovoid, minutely glandular to subglabrous.

Botswana. N: Quangwa (Kangwa), 27 km. NE. of Aha Hills, fl. & fr. 12.iii.1965, *Wild & Drummond* 6943 (K; LISC; SRGH). SW: Ghanzi, 1065 m., fl. & fr. 20.x.1969, *Brown* 6737 (K; SRGH). SE: Gaborone, 975 m., fl. & fr. 2l.ix.1974, *Mott* 361 (K; SRGH). **Zambia**. C: Chilanga Distr., Quien Sabe Farm, 1100 m., fl. & fr. 28.ix.1929, *Sandwith* 161 (K). S: Lusitu, fl. 26.ix.1959, *Fanshawe*

5227 (SRGH). **Zimbabwe**. N: Binga Distr., near Binga, c. 450 m., fl. & fr. 6.xi.1958, *Phipps* 1360 (BR; K; P; SRGH). W: Hwange Distr., Kazuma Range, 1000 m., fl. & fr. 9.v.1972, *Russell* 1909 (P; SRGH). C: Makoni Distr., Maidstone, about 9 km. from Rusapi, 1450 m., fl. & fr. 29.xi.1930, *Fries, Norlindh & Weimarck* 3287 (K). E: Chipinge Distr., Rupembe Gorge, fl. & fr. 21.ii.1957, *Goodier & Phipps* 97 (LISC). S: Beitbridge Distr., Shashi R., fl. & fr. 1.ii.1973, *Ngoni* 183 (K; SRGH). **Malawi**. C: Kasungu, 1000 m., fl. & fr. 28.viii.1946., *Brass* 17449 (K). **Mozambique**. MS: near Lupata, sand-banks, Zambezi R., October 1858, *Kirk* s.n. (K). GI: Chibuto, between Maniquenique and Cicacate, margin of Limpopo, fl. & fr. 6.viii.1958, *Barbosa & Lemos* 8307 (K). M: Namaacha, fl. & fr. 5.vii.1974, *Marques* 2495 (SRGH).

Also known from Swaziland and S. Africa; widespread weed of cultivation, riverbanks, seasonally dry stream- and riverbeds; up to 1500 m.

The type of *Lyperia micrantha* Klotzsch, Peters s.n. from Rios de Sena, Mozambique, was destroyed in Berlin by war action and I have not been able to trace any isotype specimens. I have chosen the un-numbered collection by *Kirk* made in 1858 from Lupata in the same district as Peters' material, as the neotype of that name.

Having studied most material available in herbaria and also the complex in the field, I consider that plants described as, and hitherto accepted under, the name *S. blantyrensis* Skan, illustrate nothing more than a coarser, larger state of *S. micrantha*. Other than by size, I can see no reason for giving *S. blantyrensis* distinct taxonomic rank at any level and so reduce it here into synonymy under the earlier name, *S. micrantha* (Klotzsch) Hiern.

6. **Sutera atropurpurea** (Benth.) Hiern in F.C. **4**, 2: 306 (1904). —Skan in F.T.A. **4**, 2: 308 (1906). Type from S. Africa.

   Lyperia atropurpurea Benth. in Hook., Comp. Bot. Mag. **1**: 380 (1836). Type as above.
   Lyperia aspalathoides Benth. in tom. cit. 381 (1836); in DC., Prodr. **10**: 362 (1846). Type from S. Africa.
   Lyperia crocea Benth. in DC., Prodr. **10**: 361 (1846). Type from S. Africa.
   Sutera aspalathoides (Benth.) Hiern in F.C. **4**, 2: 308 (1904). Type as above.

Erect much branched ericoid shrub up to 1 m. tall; branches many, pale brown or pale grey, shortly glandular-pubescent to subglabrous. Leaves almost opposite to alternate-fasciculate, 1–4 × 0.4–1 mm., linear or narrowly oblanceolate, obtuse, cuneate or scarcely so into narrower base, fleshy, glabrous or minutely glandular-pubescent, usually entire but occasionally with one or two minute blunt teeth towards apex. Inflorescence terminal or axillary, laxly racemose. Pedicels 3–8 mm. long (up to 20 mm. or more in fruit) erect to spreading, slender, glandular-pubescent to subglabrous. Calyx 2–3 mm. long; lobes 0.75–1 mm. wide, lanceolate, acute, densely to obscurely glandular, rarely subglabrous. Corolla pink to purple, brown and deep orange-yellow; tube 8–14 mm. long, 1–1.5 mm. wide for greater part of length, curved and widened above to c. 2 mm. at throat, sparsely finely glandular without; limb to c. 9 mm. in diam.; lobes 2–5.5 × 1.75–2 mm., unequal, oblong to circular. Capsule 4.5–6.5 × 2.5–3 mm., ovoid, narrowed above, obscurely to finely glandular-pubescent, drying pale brown.

Botswana. N: 27 km. NNW. of Rakops, fr. 22.iii.1965, *Wild & Drummond* 7220 (K; LISC; SRGH). SW: Kang, 300 km. W. of Gaborone, fr. 20.x.1975, *Mott* 774 (K; SRGH). SE: Letlhakeng (Lettaking) Valley, fl. & fr. 16.ii.1960, *Wild* 4966 (BM; K; SRGH). **Zambia**. C: 8 km. E. of Lusaka, fl. & fr. 27.xii.1957, *Noak* 295 (K; P; SRGH). **Zimbabwe**. W: Shangani Distr., 1400 m., fl. & fr. iii.1918, *Eyles* 949 (SRGH). W: Bulawayo, fl. & fr. v.1898, *Rand* 370 (BM).

Known also from S. Africa and Namibia, grasslands bordering lakes and rivers; 525–1750 m.

After intensive study of material hitherto regarded as *S. aspalathoides* and *S. atropurpurea* including the type specimens, I have failed to find any constant character by which to separate them clearly. The characters used by Bentham, are by no means constant when applied to material more recently collected than that available to him when writing his account for De Candolle's Prodromus (1846). For example the leaf-size can vary greatly within one collection and even on one individual specimen. If one uses leaf-size and shape as he does, then his range of sizes of the corolla tube hardly ever apply to his distinctions with any constancy. It may happen that a monographer of the genus will be able to separate them, but for the purposes of this work I prefer to keep them together.

7. **Sutera brunnea** Hiern in F.C. **4**, 2: 305 (1904). Type from S. Africa.

   Sutera brunnea var. *macrophylla* Hiern in F.C. **4**, 2: 306 (1904). Type: Mozambique, between Lebombo Mts. and Komati R., Delagoa Province, c. 150 m., fl. & fr. viii.1886, *Bolus* 7609 (K, holotype).

Erect or procumbent undershrub up to 60 cm. tall; branches minutely pubescent to glandular-pubescent above, glabrous to subglabrous below. Leaves 2–9(20) × 1–1.5(6) mm., opposite or alternate, or subfasciculate, spathulate or narrowly oblanceolate, obtuse, markedly narrowing into sessile base, entire or with few teeth towards apex, glabrous or minutely glandular-papillose. Flowers axillary or terminally racemose at ends of branches, pedicellate. Pedicels 4.5–9(18) mm. long, often apparently accrescent in fruit, minutely papillose to subglabrous. Calyx 3–4.5 mm. long, deeply 5-lobed, lobes 0.5–1.25 mm. wide, broadly linear, finely papillose to subglabrous. Corolla deep brown; tube 9.5–14(20) mm. long, slender, straight, somewhat dilated below limb, finely glandular-papillose; limb c. 7 mm. in diam.; lobes 2.5–3.5 × 1.5–2 mm., broadly oblong, entire. Capsule 5–8 mm. long, 2–4 mm. broad, broadly cylindric-ovoid, minutely papillose.

Zambia. C: c. 10 km. SE. of Lusaka, 1280 m., fl. & fr. 22.v.1955, *King* 12 (BR; K). S: Kafue Distr., King Edward's Copper Mine, near Native Reserve, fl. & fr. 11.ix.1929, *Sandwith* 172 (K). **Zimbabwe**. N: Makonde Distr., near Rod Camp Mine, fl. & fr. 22.ii.1961, *Rutherford-Smith* 579 (SRGH). W: Bulawayo, Commonage, 1370 m., fl. & fr. vi.1904, *Eyles* 148 (SRGH). C: Kwekwe Distr., Mashaba (Mhlaba) Hills, 16 km. S. of Ngesi, Great Dyke, fl. & fr. 16.i.1962, *Wild* 5595 (K; SRGH). E: Mutare Distr., Odzi, Ordwell Mine, fl. & fr. 28.iii.1966, *Wild* 756l (K; LISC; SRGH). S: Mberengwe (Belingwe), near Otto Mine, fl. & fr. 17.iii.1964, *Wild* 6390 (BR; K; LISC; SRGH). **Mozambique**. M: Arredores de Moamba, fl. & fr. 26.ii.1948, *Torre* 7433 (LISC).
Known also from S. Africa. In the Flora Zambesiaca area it appears to have an affinity for serpentine and 'heavy metal' soils, where it occurs in grassland; up to c. 1500 m.
Since 1904, when Hiern described this species along with var. *macrophylla*, more has been added to collections especially from S. Africa. Amongst this material there is plenty showing evidence that the larger leaves and denticulate leaves that Hiern used to separate his variety var. *macrophylla*, are found quite commonly on most plants where the lower stems and root systems are included. In all other respects var. *macrophylla* compares with the typical variety and because of these factors, I am combining them under the one name.

8. **Sutera burkeana** (Benth.) Hiern in F.C. **4**, 2: 299 (1904). —Skan in F.T.A. **4**, 2: 308 (1906). Type from S. Africa.
   Lyperia burkeana Benth. in DC., Prodr. **10**: 310 (1846). Type as above.

Erect shrub up to 1.5 m. tall, divergently branched, densely glandular-pubescent, viscid. Leaves 3–15 × 1.5–5.5 mm., fasciculate, cuneate-oblong or obovate, incised dentate or pinnatifid, rigid, often recurved, shortly petiolate. Petiole up to 3 mm. long. Inflorescence terminally racemose, comparatively few-flowered. Bracts 2–3 mm. long, linear-subulate, acute. Pedicels 4.5–11 mm. long, stout. Calyx 6–8 mm. long; lobes c. 1 mm. long, linear, acute, subequal. Corolla white with dark brown throat or pink or mauve, densely glandular-pubescent without; tube (8)12–20 mm. long, cylindric below, slightly incurved, dilated at throat; limb 8–11 mm. in diam.; lobes 4–6 mm. in diam., unequal, circular to broadly oblong. Capsule 5–7 × c. 3 mm., narrowly ovoid, apically somewhat compressed, minutely pubescent.

Zambia. C: Mulungushi, fl. & fr. 9.ii.1964, *Fanshawe* 8259 (K). **Zimbabwe**. N: c. 1 km. N. of Gokwe, fl. & fr. 29.vi.1963, *Bingham* 761 (K; LISC; SRGH). W: Ndumba Hill, Inyathi (Inyati), fl. & fr. 18.iv.1947, *Keay* in GHS 21202 (K; SRGH). C: Shurugwi, Unkis Farm, fl. & fr. 19.iii.1964, *Wild* 6430 (BR; K; LISC; SRGH). E: Nyanga (Inyanga) Distr., Mika Hill, St. Swithin's Tribal Trust Land, fl. & fr. 20.iv.1972, *Wild* 7953 (K; LISC; SRGH). S: 32 km. N. of West Nicholson on Mberengwa road, fl. & fr. 17.iii.1964, *Wild* 6412 (K; LISC). **Malawi**. C: Dedza Distr., Chongoni Forest Reserve, 1675–1800 m., fl. & fr. 23.iv.1970, *Brummitt* 10777 (K; SRGH).
Also known from S. Africa; grasslands and open woodlands and scrub; 920–1800 m.

9. **Sutera fodina** Wild in Kirkia **5**: 79 (1965). Type: Zimbabwe, Guruve (Sipolilo) Distr., Impinge (Mpingi) Pass, Great Dyke, 1370 m., 17.v.1962, *Wild* 5775 (K, isotype; SRGH, holotype).

Erect, much-branched perennial herb or subshrub up to 1 m. tall; branches erect to spreading, whole plant glandular-pilose to tomentose. Leaves 7–16(22) × 2–7.5 mm., narrowly oblong-elliptic or elliptic, acute to subobtuse, cuneate at base, crenate, bullate above with impressed nervation, nerves prominent beneath, margins revolute, shortly petiolate. Petioles c. 1 mm. long. Inflorescence terminally short-racemose. Bracts c. 5 mm. long, narrow ovate. Pedicels to c. 0.5 mm. long, obscure. Calyx 5–6.5 mm. long, lobes 0.7–1 mm. wide, narrowly oblong, acute, margins slightly revolute. Corolla white or cream with yellow or orange throat; tube 12–15 mm. long, minutely glandular-pubescent to subglabrous below, curved above midway, shortly hirsute above; limb 11–14.5 mm. in diam.; lobes 5.5–7 × c. 6 mm., broadly obovate to oblong, truncate to wavy at apex, subequal. Capsule 3–3.5 × c. 1.5 mm., broadly ellipsoid-ovoid, minutely glandular-papillose.

Zimbabwe. N: Makonde Distr., near Rod Camp Mine, fl. 22.ii.1961, *Rutherford-Smith* 578 (BR; K; LISC; SRGH). C: Kwekwe (Queque), Sebakwe Nat. Park, fl. 15.xi.1970, *Wild* 7797 (E; K; SRGH). S: Chibi Distr., Mhindamukova Pass, 1200 m., fl. 10.v.1970, *Biegel* 3304 (K; LISC; P; SRGH).
Not known from elsewhere. It is restricted to the serpentine soils of the Great Dyke; 1200–1680 m.

10. **Sutera carvalhoi** (Engl.) Skan in F.T.A. **4**, 2: 307 (1906). TAB. **13**. Type: Mozambique, Gorongosa (Gorongoza), *Carvalho* s.n. (COI, holotype).
　　Cycnium carvalhoi Engl., Pflanzenw. Ost-Afr. **C**: 360 (1895) as '*carvalhi*'; Bot. Jahrb. **23**: 513 (1897). Type as above.

Erect, branched undershrub up to 2 m. tall, densely glandular-pubescent; main branches up to 30 cm. or more long, secondary branches much shorter, erect-spreading. Leaves 8–30(45) × 4–12(22) mm., opposite, ovate to ovate-lanceolate, obtuse or subacute, cuneate at base into short petiole, shortly glandular-pubescent, especially on nerves beneath, minutely hispid-pubescent above, serrate-crenate, teeth many, small, obtuse. Flowers loosely terminal-racemose; bracts similar to leaves but smaller. Pedicels 4–6 mm. long. Calyx 6–8 mm. long, densely glandular-pubescent; lobes 1–1.5 mm. wide, linear to linear-lanceolate. Corolla white to lilac-blue or mauve, with yellow to brown or brownish-purple throat; tube 15–24 mm. long, cylindric, enlarged towards apex, slightly curved below limb, shortly densely glandular-pubescent; limb 13–20 mm. in diam., somewhat bilabiate; lobes 6–8.25 mm. long, 4–7.25 mm. wide, broadly obovate-truncate or slightly emarginate. Capsule 6–7 × 2.75–3 mm., narrowly ovoid, minutely glandular-pubescent.

Zimbabwe. E: Nyanga (Inyanga) Distr., near Pungwe View, c. 2000 m., fl. 14.ix.1960, *Rutherford-Smith* 90 (BR; K; LISC; SRGH). **Mozambique**. MS: Gorongosa Mt., Gogogo Summit area, fl. vii.1970, *Tinley* 1966 (BR; K; LISC; SRGH).
From eastern Zimbabwe and adjacent Mozambique. Montane grasslands; up to about 2500 m.

11. **Sutera floribunda** (Benth.) Hiern in F.C. **4**, 2: 277 (1904). Type from S. Africa.
　　Chaenostoma floribundum Benth. in Hook., Comp. Bot. Mag. **1**: 376 (1836); in DC., Prodr. **10**: 356 (1846). Type as above.
　　Manulea floribunda (Benth.) Kuntze, Rev. Gen. **3**, 2: 235 (1898). Type as above.
　　Sutera pulchra Norl. & Weim. in Bot. Not. **1951**: 107, figs. 3 & 4 (1951). Type: Zimbabwe, Nyanga (Inyanga) Distr., Mt. Inyangani, c. 2450 m., fl. 14.ii.1931, *Norlindh & Weimarck* 4994 (LD, holotype; K, isotype).

Erect or ascending subherbaceous shrub, 20–60 cm. to 1 m. tall, much branched; branches opposite to subopposite, erect to spreading, minutely pubescent below, viscid glandular-pubescent above, leafy particularly below inflorescence. Leaves 5–11(38) × 2.5–16(20) mm., broadly ovate, ovate-elliptic to subcircular, apex obtuse, cuneate at base, coarsely dentate with obtuse or rounded teeth, subglandular-pubescent or shortly pubescent on both surfaces, petiolate. Petiole short or up to half length of leaf. Inflorescence racemose or cymose with simple or compound racemes in terminal leafy panicles. Pedicels 2–4(10) mm. long, slender, finely pubescent to glandular-pubescent. Bracts opposite, lanceolate, subulate, or lower similar to but smaller than leaves, acute, entire or sparingly dentate. Flowers blue, pinkish-mauve to purple or white, with yellow or orange centre. Calyx 4–8 mm. long, deeply 5-lobed, softly hispid to glandular-pubescent; lobes linear-lanceolate to subulate, acute. Corolla tube 5–9 mm. long, narrowly infundibuliform, minutely finely glandular-pubescent without; limb 4.5–5.5 mm. in diam.; lobes c. 2 mm. in diam., glandular-pubescent beneath, broadly oblong-ovate to subcircular. Capsule 3–4 mm. long, ovoid, glabrous.

Zimbabwe. N: Mazowe (Mazoe), Iron Mask Range, 1550–1580 m., fl. & fr. v.1907, *Eyles* 549 (BM; K; SRGH). C: Makoni Distr., Timarm, Makoni Reserve, Rusape, c. 1520 m., fl. v.1961, *Plowes* 2181 (K; SRGH). E: Mutare Distr., Engwa, 1980 m., fl. 10.ii.1955, *Exell & Mendonça* 354 (BM; LISC; SRGH). S: Mberengwa (Belingwe), Buhwa Mt. fl. 29.iv.1973, *Pope* 988 (K; SRGH).
Not known from elsewhere. In hillside grasslands; 1500–2450 m.

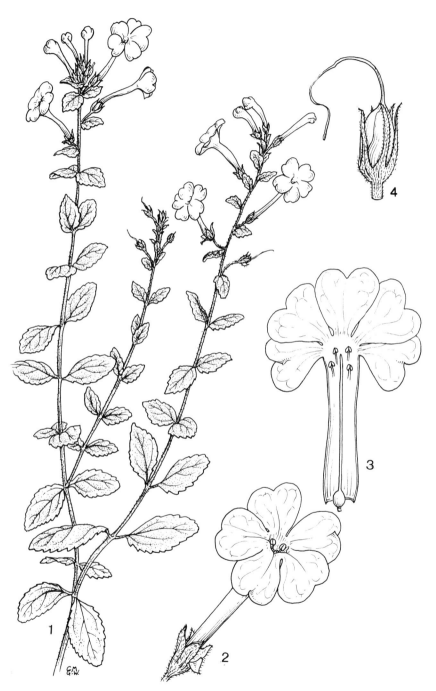

Tab. 13. SUTERA CARVALHOI. 1, flowering branch (×⅔), from *Wild* 3628; 2, flower (× 2); 3, corolla, longitudinal section showing androecium and gynoecium (× 2), 2–3 from *Methuen* 21; 4, fruit (× 3), from *Wild* 3628.

13. MANULEA L.

Manulea L., Mant. Pl.: 12 (1767).

Erect, annual or perennial herbs or rarely undershrubs, glabrous to variously pubescent. Leaves all radical and rosetted, or sometimes also cauline and opposite or uppermost alternate, entire or dentate. Flowers several to numerous, pedicellate or subsessile, bracteate or ebracteate. Inflorescence terminal, racemose, spicate or compound, thyrsoid or paniculate. Calyx 5-lobed. Corolla 5-lobed, lobes subequal; tube slender, rarely dilated at throat. Stamens 4, didynamous; filaments filiform, short; anthers unithecal, thecae confluent, anterior pair reniform, perfect or sterile, included or shortly exserted; posterior pair reniform or oblong, perfect, included. Style filiform, included or slightly exserted, stigma subclavate. Ovary bilocular, ovules numerous. Seeds rugose.

A genus of some 50 subtropical, southern and S. African species.

1. Inflorescence initially subcapitate - - - - - - - - - - 2
 – Inflorescence not subcapitate, laxly racemose to densely, many-flowered thyrsoid; calyx lobed to halfway or more, lobes linear-spathulate; leaves linear to linear-lanceolate, almost entire to sparsely dentate - - - - - - - - - - - 3. *rhodesiana*
2. Inflorescence becoming paniculate to laxly-flowered thyrsoid; calyx lobed almost to base, lobes linear-lanceolate; leaves serrate-dentate to subentire - - - - - 1. *crassifolia*
 – Inflorescence becoming simple spicate or compound spicate; calyx lobed to little more than halfway, lobes narrowly lanceolate; leaves remotely dentate to lobate-dentate 2. *conferta*

1. **Manulea crassifolia** Benth. in Hook., Comp. Bot. Mag. **1**: 382 (1836) excl. spec. Drege ex Witbergen. —Hiern in F.C. **4**, 2: 230 (1904). Type from S. Africa, Mooi Plaats, *Drege* 7919b (K, lectotype chosen here).

Annual or biennial herb up to 65 cm. tall, erect, simple or branched usually from or towards base, minutely, somewhat densely white papillose; branches terete, rarely subtetragonous. Leaves: cauline where present alternate, 12–50 × 1.5–9 mm., lanceolate-elliptic to narrowly oblong, obtuse, narrowed into petiole-like base, serrate-dentate to subentire, finely papillose; basal 40–90 mm. long, 8–12(20) mm. wide, similar to cauline. Inflorescence usually subcapitate, at first becoming thyrsoid to paniculate. Bracts 1.5–2.5(3.8) mm. long, 0.3–0.5 mm. wide, linear to linear-lanceolate, obtuse. Pedicel 0–0.8(2) mm. long, stout. Calyx 2–4 mm. long, deeply 5-lobed mostly almost to base; lobes 2.1–3.8 × 0.4–0.5 mm., glandular-pubescent to subglabrous, linear to linear-lanceolate. Corolla yellow or buff to orange; tube 5.5–7(9.5) mm. long, slender, gibbous above, glandular-pubescent; lobes 1.8–2.5 × 1–1.5 mm., spreading, margins not recurving on drying, subequal, oblong. Anterior stamens inserted in mouth of corolla, anthers small, sterile 0.23–3 mm. long, partly extruded, anthers of posterior fertile pair 1–1.25 mm. long. Capsule 3–3.6 × 1.5–2.5 mm., oblong in outline, laterally somewhat compressed.

Mozambique. GI: Chibuto, Posto de Alto Changane, fl. 15.vii.1944, *Torre* 6761 (LISC). M: Marracuene, between Rio Incomati and Lake Pati, fl. & fr. 24.iii.1954, *Barbosa & Balsinhas* 5455 (BM; LISC).
Also known from S. Africa. Damp grasslands; up to 2000 m. (S. Africa).

2. **Manulea conferta** Pilger in Engl., Bot. Jahrb. **48**: 437 (1912). —Merxm. in Merxm., Prodr. Fl. SW. Afr. **126**: 31 (1969). Type from Namibia.

Annual or perennial herb (15)30–55 cm. tall, erect, simple to moderately branched, densely short white glandular-pubescent-papillose; branches terete or subtetragonous. Leaves: upper cauline leaves 12–18 mm. long, 1–3 mm. wide, linear to narrowly linear-lanceolate, obtuse, sessile, remotely dentate or lobate-dentate; lower cauline leaves (22)35–50 mm. long, (3)4.5–8 mm. wide, sessile, similar to upper; basal leaves where present up to 40 × 10 mm., elliptic to elliptic-lanceolate, serrate-dentate, apex obtuse, gradually narrowed at base into slender petiole 12–15 mm. long. Inflorescence simple- or compound-spicate, 3–8 cm. long (extending to c. 17 cm. in fruit). Bracts (2.5)4.5–6 × 0.2–0.6 mm., linear to linear-lanceolate, acute or obtuse, glandular-pubescent. Calyx 4–4.5 mm. long, 5-lobed to little more than halfway in flower, lobes narrowly lanceolate, glandular-pubescent. Corolla yellow; tube (6)8.5–10.5 mm. long, slender, gibbous above, glandular-pubescent; lobes 2.5–4.5 × 1–1.3(2) mm., spreading, margins recurved when dry, linear-oblong, obtuse, subequal. Anterior stamens inserted in mouth of corolla, anthers small,

Tab. 14. MANULEA RHODESIANA. 1a, habit (×⅔); 1b, habit (×1/10); 2, flowers (×4); 3, flower (×4); 4, corolla opened showing gynoecium and androecium (×4); 5, fruits (×8), all from *Norlindh & Weimarck* 4623.

sterile, c. 0.5 mm. long, partly extruded, anthers of posterior fertile pair c. 1.5 mm. long. Capsule 2.5–3 × 2 mm., broadly ovoid-cylindric, laterally subcompressed.

Botswana. SW: Namibia border on Ghanzi-Gobabis road, within Botswana, fl. 2.viii.1955, *Story* 5082 (PRE).
Known also from Namibia. In sandy, grass scrubland; up to 1800 m.

3. **Manulea rhodesiana** S. Moore in Journ. Bot. **49**: 156 (1911). TAB. **14**. Type: Zimbabwe, Masvingo (Victoria), *Monro* 924 (B, holotype).

Annual or biennial herb up to 1 m. or more tall, simple or branched, minutely more or less densely white papillose-pubescent; branches tetragonous. Leaves: upper cauline alternate, 14–22 × 1.5–2.5 mm., sessile, linear to linear-lanceolate, subentire to sparsely denticulate; lower cauline opposite or alternate, 24–75 mm. long, 3.5–13 mm. wide, linear-lanceolate to narrowly oblanceolate, apex obtuse, narrowing into slender petiole-like base, sparsely to remotely dentate; basal (22)70–115 mm. long (4)10–12 mm. wide, clustered to rosette-like, oblong-lanceolate, obtuse at apex, narrowed into slender petiole, sparsely dentate. Inflorescence a few-flowered lax raceme to a many-flowered thyrse. Bracts 1–1.8 mm. long, linear, obtuse. Pedicel up to 1 mm. long, stout. Calyx (1.2)1.5–2(3) mm. long, lobed to more than half-way; lobes 0.2–0.3 mm. wide, linear to linear-spathulate, with spathulate appearance being due to cucculate apices. Corolla dull yellow to orange, or purple [?], tube (5.5)6–7(8) mm. long, slender, densely, minutely papillose, gibbous towards mouth; lobes 1.2–2 × 0.6–0.8(1) mm., spreading, margins recurved when dry, subequal, broadly oblong to obovate. Anterior stamens very shortly exserted. Style c. 7 mm. long, exserted. Capsule 1.8–2 × 1–1.25 mm., oblong, laterally somewhat compressed.

Zimbabwe. C: Marondera Distr., Macheke, c. 1525 m., fl. xii.1919, *Eyles* 2013 (K; PRE; SRGH). S: Masvingo Distr., Makoholi, fl. (purple) 17.iii.1978, *Senderayi* 285 (K; PRE; SRGH). E: Nyanga (Inyanga) Distr., Juliasdale, 1950 m., fl. 23.i.1973, *Biegel* 4165 (K; LISC; SRGH).
Not known from elsewhere. Wet grassland bordering rivers and streams, also weed of roadsides and cultivation; from c. 1000–1850 m.
After examination of a wide range of material collected from the Flora Zambesiaca area and housed in various herbaria under the names *Manulea parviflora* Benth. and *M. rhodesiana* S. Moore, I have concluded that only one of these taxa is represented here. *M. parviflora* is, as the epithet infers, a small flowered plant with a corolla tube barely exceeding 5 mm. in length, with the lobes averaging 1 mm. long and never exceeding 1.2 mm. The calyx at flowering time is very small and not larger than 1 mm. long with lobes more or less equal in length to the tube. It is widely distributed throughout S. Africa. On the other hand *M. rhodesiana* was described as having a longer corolla and this is evident as the tube length ranges between 6–8 mm. while the lobes vary from 1.2–2 mm. long. The calyx too is clearly different insofar as it measures from 1.5–2 mm. long and in the extreme 3 mm. with the lobes being considerably longer than the tube. It appears to occur only in Zimbabwe.

14. MIMULUS L.

Mimulus L. in Act. Soc. Sc. Uppsal.: 82 (1741).

Herbs, erect or decumbent, glabrous, pilose or viscid. Leaves opposite, simple, entire or dentate. Inflorescence of terminal, leafy racemes or flowers solitary, axillary. Calyx 5-toothed, generally 5-ribbed or 5-angled, lobes shorter than tube, somewhat unequal, ovate-acuminate, tube campanulate. Corolla bilabiate, tubular; tube cylindric to slightly dilated at mouth, villous within, longer than lips; posterior lip bilobed, exterior in bud; anterior lip trilobed usually with two proturberances in throat; lobes subequal, rounded. Stamens 4, didynamous, included or exserted; filaments filiform, variously inserted near base of corolla-tube; anthers bithecal, cells distinct or confluent at apex. Ovary bilocular, oblong with many ovules; style filiform with usually bilobed stigma. Capsule cylindric to ovoid, loculicidally dehiscent. Seeds numerous, small, ellipsoid.

A genus with over 100 species widely distributed throughout the world but mainly in temperate America. Only 1 species occurs in the Flora Zambesiaca area.

Mimulus gracilis R.Br., Prodr. Fl. Nov. Holl.: 439 (1810). —Hiern in F.C. **4**, 2: 354 (1904). —Hemsl. & Skan in F.T.A. **4**, 2: 310 (1906). TAB. **15**. Type from Australia.

Perennial to 0.45(1) m. tall, glabrous, erect, ascending or decumbent. Stem branching from or near base or simple, subquadrangular, narrowly winged. Leaves 2.0–4.7(6.0)

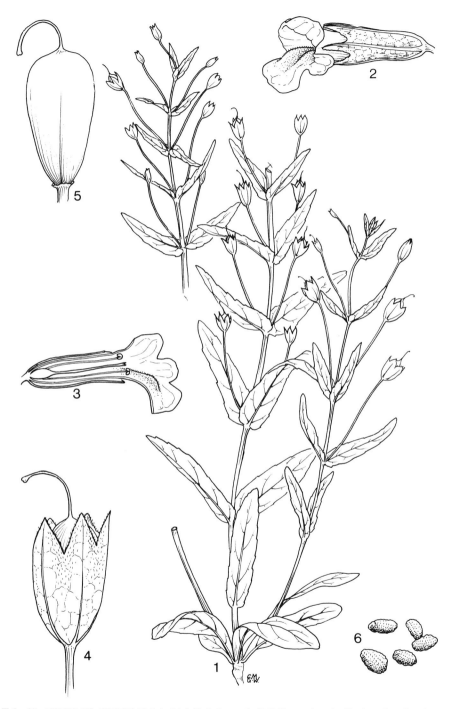

Tab. 15. MIMULUS GRACILIS. 1, habit (×⅔); 2, flower (×4); 3, flower, longitudinal section showing gynoecium (×4); 4, fruiting calyx (×4); 5, capsule with calyx removed (×4); 6, seeds (×44), all from *Robinson* 737.

× 0.3–1.6 cm., lanceolate to narrowly oblong, apex obtuse to subacute, subcordate or obtuse at base, almost entire or more usually laxly denticulate, glandular-punctate beneath, sessile. Flowers solitary in leaf axils on long pedicels forming loose, leafy racemes; pedicels 1.0–3.5 cm. long (up to 5 cm. long in fruit), slender to filiform. Calyx 4.5–8.5 × 4–6 mm., campanulate, lobes 1.3–2 × 0.5–0.7 mm., broadly lanceolate to deltoid, acute, minutely ciliate, ribs prominent especially in fruit. Corolla 7.5–12 mm. long, tubular, white or whitish, throat yellow with brown protuberances. Capsule ovoid to obovoid, 5–6 mm. long.

Botswana. N: Ngamiland, Okavango Swamp, NE. of Gwetshaa Island, fl. 23.ii.1973, *Smith* 420 (BR; K; MO; PRE; SRGH). SE: Gaborone Dam, fl. ix.1967, *Lambrecht* 335 (K; SRGH). **Zambia**. B: Machili, fl. 20.vii.1960, *Fanshawe* 5794 (K; PRE; SRGH). C: Ndola, Kabwe (Broken Hill) to Bwana Mkubwa, fl. x.1906, *Allen* 365 (K; SRGH). E: Lundazi, fr. 18.x.1967, *Mutimushi* 2247 (K; SRGH). S: Mapanza, fl. 9.v.1954, *Robinson* 737 (K; SRGH). **Zimbabwe**. C: Harare Distr., between Avondale West and Mabelreign, 1480 m., fl. 16.x.1955, *Drummond* 4907 (BR; K; PRE; SRGH). W: Maleme Dam, Matopos Nat. Park, fl. 25.ii.1981, *Philcox & Leppard* 8839 (K). E: Mutare Distr., Mutare Golf-course, c. 1100 m., fl. 20.xii.1956, *Chase* 6282 (K; PRE; SRGH). Malawi. C: Lilongwe Distr., Chilenje Mt., S. of Mkhoma, E. of Chilenje, c. 400 m., fl. 7.vii.1978, *Iwarsson & Ryding* 909 (K). S: Livulezi R., near Ntcheu, 1075 m., fl. 29.i.1959, *Robson* 1334 (BM; K; LISC; PRE; SRGH). **Mozambique**. MS: near Expedition Is., fl. viii.1858, *Kirk* s.n. (K). M: Incanhini, near Delagoa Bay, fl. 13.i.1898, *Schlechter* 12016 (BM; PRE).

Known also from West Africa, Yemen and Ethiopia; Lesotho, Namibia and S. Africa; India, China and Australia. Usually in or near water or in damp situations; 250–1500 m.

15. STEMODIA L.

Stemodia L., Syst. Nat., ed. 10: 1118 (1759).

Annual, glandular-pubescent, often aromatic herbs, much-branched, or subshrubs. Leaves opposite or verticillate, occasionally alternate below, simple, subentire to variously toothed. Flowers zygomorphic in lax terminal racemes or solitary-axillary, pedicellate. Pedicels uni- or bibracteolate. Calyx 5-lobed; lobes narrow, equal or subequal. Corolla tubular; tube cylindric; upper lip broad, emarginate to entire; lower lip trilobed. Stamens 4, didynamous, included; filaments slender; anther thecae stipitate, all fertile. Style usually bilobed. Capsule globose, ovoid, sometimes acuminate, valves 2, bifid or 4; usually loculicidally dehiscent. Seeds numerous, small, striate.

A tropical genus of some 30 species of both the Old and New Worlds.

Stemodia serrata Benth. in DC., Prodr. **10**: 381 (1846). —Hemsl. & Skan in F.T.A. **4**, 2: 314 (1906). —Hepper in F.W.T.A. ed. 2, **2**: 357 (1968). TAB. **16**. Syntypes from Ethiopia and West Africa.

Erect, much-branched herb, 15–35 cm. tall, strongly viscid, foetid, glandular-pubescent throughout to varying degrees. Leaves 15–50 × 6–20 mm., oblong to lanceolate, acute or obtuse, narrowed at base or broader, semi-amplexicaul, coarsely serrate. Inflorescence with flowers crowded into leafy spike-like raceme; pedicels about 1 mm. long; bracts 2, to about 3.25 × 0.2 mm., somewhat ciliate, narrowly linear. Calyx 4.5–6.5 mm. long, lobes 3.5–5.5 × 0.5–1.2 mm., narrowly linear-lanceolate, acuminate. Corolla about 5.5–6.5 mm. long, upper lip emarginate, lower lip trilobed, lobes rounded to somewhat emarginate, yellowish- or greenish-white. Filaments 2.0–2.8 mm. long, slender. Ovary 1.5 mm. long, 0.75 mm. in diam., narrowly conical. Style 1.5 mm. long, somewhat arcuate at apex. Capsule about 5.5 mm. long, narrowly oblong in outline, 4-furrowed.

Zambia. C: W. Bank of Luangwa R., Luangwa South Game Res., fl. & fr. 16.vi.1972, *Abel* 672 (SRGH). S: Kafue Flats, Mazabuka, 975 m., fl., 18.vi.1955, *Robinson* 1308 (K; PRE; SRGH). **Mozambique**. T: Boroma (Boruma), fr. v.1891, *Menyhart* 1878 (K).

Also in West Africa, Sudan, Ethiopia and Tanzania. Recorded from Madagascar and India. Grasslands; up to c. 1000 m.

Tab. 16. STEMODIA SERRATA. 1, habit (× ⅔); 2, flower (× 6); 3, flower, longitudinal section showing gynoecium (× 6); 4, stamens (× 12); 5, dehiscing fruit (× 6), all from *Robinson* 1308.

16. STEMODIOPSIS Engl.

Stemodiopsis Engl. in Ann. Ist. Bot. Roma **7**: 25 (1897).

Herbs, perennial, often somewhat shrubby, stems much branched, branches very leafy, procumbent. Leaves opposite, simple, serrate-dentate, crenate or entire, petiolate. Flowers solitary or in 2–3-flowered cymes, pedicels bibracteolate. Calyx 5-lobed, lobes longer than tube, linear-lanceolate to lanceolate, subequal. Corolla bilabiate, subcampanulate; tube cylindric to somewhat dilated at mouth; upper lip shortly bilobed, lobes broadly triangular to rounded; lower lip trilobed, subcircular, lobes broadly triangular, oblong to semi-circular, pilose to papillose within. Stamens 4 perfect, didynamous, included; filaments of longer stamens markedly curved or twisted at or about middle; anthers bithecal, cells ovoid, equal, confluent at apex; staminode where present, filiform, arising near base of corolla tube. Ovary elongate-conical, bilocular, attenuate into slender style; ovules numerous. Capsule ellipsoid-conical, sharply rostrate, downwardly directed in fruit; valves coherent or connivent at apex after dehiscence.

A genus of about 10 species occurring in Africa and Madagascar.

1. Leaves coarsely serrate   -   -   -   -   -   -   -   -   -   -   -   -   2
 - Leaves entire or crenate   -   -   -   -   -   -   -   -   -   -   -   -   4
2. Stems glabrous, or with scattered fine hairs on younger parts and leaves; rest of plant, including fruit, glabrous   -   -   -   -   -   -   -   -   -   1. *buchananii* var. *buchananii*
 - Stems pubescent   -   -   -   -   -   -   -   -   -   -   -   -   -   -   3
3. Plant, including fruit, densely and minutely pubescent   -   -   -   -   -   2. *rivae*
 - Plant glabrous, except for densely and minutely pubescent stem   1. *buchananii* var. *pubescens*
4. Leaves entire, spathulate-ovate; plant, including fruit, minutely pubescent, eglandular   3. *eylesii*
 - Leaves crenate, elliptic to lanceolate-elliptic; stems, leaves and calyx lightly stipitate-glandular; fruit glabrous   -   -   -   -   -   -   -   -   -   -   -   4. *glandulosa*

1. **Stemodiopsis buchananii** Skan in F.T.A. **4**, 2: 315 (1906). Type: Malawi, without locality, *Buchanan* 365 (K, lectotype, chosen here).

Perennial herb (3)6–20 cm. tall, glabrous except occasionally with scattered fine hairs on leaves and young stems, or finely, densely, minutely pubescent on stems only; stems procumbent, much branched, subterete. Leaves opposite, ovate to lanceolate-ovate; lamina (4.5)7–20(37) × (3)4–10(22) mm., acute, subcuneate at base, thin, coarsely serrate-dentate; petiole about equalling lamina. Flowers solitary or in 2(3)-flowered cymes. Pedicels 3–5(7) mm. long, filiform, decurved in fruit. Bracteoles 0.5–1(2) mm. long, narrowly linear. Calyx 3–5.5 mm. long, lobes 2.5–5 mm. long, narrowly linear-triangular, subequal, mid-nerve prominent. Corolla white with lip variously marked in pink to purple, or mauve, 4.5–9.5 mm. long; tube c. 2.5–5(6) mm. long, c. 1.5 mm. broad at base, becoming broader above; upper lip broadly ovate, lobes subtriangular, rounded; lower lip subcircular to broadly obovate, lobes rounded. Capsule 4.5–6(7.5) × (1.5)2.5–3(4.5) mm., ellipsoid-conical including beak 1.5–2.25 mm. long, glabrous, pale yellow-brown.

Var. **buchananii**

Plant totally glabrous except occasionally with scattered fine hairs on leaves and young stems. Corolla white with various pink to purple markings on lower lip.

Zambia. N: near Inono R. by bridge on Mpulungu-Mbala road, 914 m., fl. & fr. 9.iv.1959, *McCallum-Webster* 604 (K). E: Petauke Distr., Lukusuzi Nat. Park, 800 m., fl. & fr. 12.iv.1971, *Sayer* 1141 (K; SRGH). **Malawi.** N: Nkhata Bay Distr., Chikla Beach, 487 m., fl. & fr. 19.vii.1970, *Pawek* 3600 (K). C: Dedza, Mphunzi Hill, fr. 8.iii.1967, *Salubeni* 577 (K; SRGH). S: Mangochi Distr., c. 10 km. S. of Monkey Bay on road to Mangochi, fl. & fr. 1.iii.1977, *Grosvenor & Renz* 994 (K; SRGH).
Also known from Sudan, Kenya and Tanzania. Almost exclusively restricted to rock crevices; up to 1800 m.

Var. **pubescens** Philcox in Kew Bull. **40**: 606 (1985). Type: Zimbabwe, Chimanimani (Melsetter) Distr., Chimanimani Mts., *Grosvenor* 327 (BR; K; LISC, holotype; PRE).

Plant glabrous except for stems being totally finely, densely and minutely pubescent. Corolla mauve.

Zimbabwe. E: Chimanimani Mts., c. 1280 m., fl. & fr. 6.iv.1967, *Grosvenor* 327 (BR; K; PRE; LISC).

Tab. 17. STEMODIOPSIS RIVAE. 1, flowering branch (×⅔); 2, flower (× 4); 3, flower, longitudinal section showing gynoecium (× 8); 4, transverse section of corolla showing hairs (× 36); 5, ripe capsules (× 4), all from *Ngoni* 177.

Mozambique. MS: Chimoio Distr., Monte Chicama, fl. & fr. 24.iv.1948, *Andrada* 1171 (LISC). Not known from elsewhere. Rock crevices up to 1800 m.

2. **Stemodiopsis rivae** Engl. in Ann. Ist. Bot. Roma **7**: 25 (1897); Bot. Jahrb. **23**: 497, t.7, figs. A–F (1897). —Skan in F.T.A. **4**, 2: 315 (1906). TAB. **17**. Type from Ethiopia.
  Stemodiopsis humilis Skan in F.T.A. 4, **2**: 316 (1906). Type: Malawi, near Mt. Chiradzulu, *Cameron* 182 (K).

Perennial herb, somewhat shrubby, 2.5–15 cm. tall, procumbent, much branched with branches 4-angled, reaching 35 cm. long, vegetative parts and calyx densely covered with short, stiff, patent, white hairs. Leaves petiolate; lamina ovate 5–18 × 2–18 mm., acute, cuneate at base, coarsely serrate-dentate; petiole 3–12 mm. long. Flowers solitary. Pedicels 4–9 mm. long, filiform, decurved in fruit. Bracteoles 1.2–1.5 mm. long, subulate. Calyx (2)3–4.5 mm. long, lobes (1.8)2.3–3.2 mm. long, narrowly triangular to linear-triangular, subequal. Corolla white with pink to purple throat, up to 11 mm. long; tube 5.5–6.5 mm. long, c. 1.5 mm. wide at base widening to c. 3.5–4 mm. in diam. at mouth, sparsely pilose to subglabrous without; upper lip c. 3 mm. long, triangular-ovate, lower lip c. 4 × 4.5 mm., shortly trilobed, minutely pubescent within. Capsule 4–5(9) mm. long including beak, pale brown, densely minutely pubescent to pubescent only in upper third.

Zambia. N: Chinsali Distr., 21 km. W. of Shiwa Ngandu, 1425 m., fr. 28.v.1972, *Kornaś* 1860 (K). **Zimbabwe.** N: Guruve (Sipolilo) Distr., c. 11 km. N. of Impinge (Mpingi) Pass, 1370 m., fr. 18.v.1962, *Wild* 5797 (K; SRGH). W: Matobo Distr., hills near Maleme Dam, fl. & fr. 30.i.1973, *Ngoni* 177 (K; P; Mutare Distr., Maranke Res., c. 820 m., fl. 27.ii.1953, *Chase* 4804 (BM; BR; SRGH). S: Beitbridge Distr., Umzingwane R., Fulton's Drift, fl. & fr. 26.ii.1961, *Wild* 5432 (K; SRGH). **Malawi.** S: Midima Hill, c. 24 km. E. of Blantyre, c. 1060 m., fl. & fr. 5.ii.1967, *Hilliard & Burtt* 4672 (E; K; SRGH). **Mozambique.** MS: Beira, fl. & fr. Feb. 1912, *Rogers* 5955 (K).
  Also known from Sudan, Uganda, Kenya, Tanzania and South Africa (Transvaal). Apparently restricted to rock crevices; up to c. 1450 m.

3. **Stemodiopsis eylesii** S. Moore in Journ. Bot. **46**: 71 (1908). Type: Zimbabwe, Mazowe (Mazoe), Iron Mask Hill, c. 1500 m., *Eyles* 252 (BM, holotype; K; SRGH, isotypes).

Subshrubby perennial herb c. 8 cm tall, totally, finely pubescent except for flowers; stems procumbent with main branches quadrangular, up to at least 24 cm. long, much branched. Leaves spathulate-ovate, obtuse, cuneate at base; lamina 5–10 × 3–5 mm., entire to rarely obscurely crenate, margin somewhat revolute, appearing thickened, indumentum much denser beneath. Petiole up to 5 mm. long, narrowly winged. Flowers solitary. Pedicels c. 5 mm. long, reflexed in fruit. Bracteoles c. 1 mm. long, subulate. Calyx 2–3 mm. long, lobes 1.8–2.6 mm. long, narrowly triangular, subequal. Corolla white with purple markings, up to 9.4 mm. long; tube 5–5.5 mm. long, c. 2 mm. wide at base widening to c. 2.5 mm. in diam. at mouth, sparsely pubescent without; upper lip c. 1.7–2 mm. long, lobes triangular, acute; lower lip 3.5–4 mm. long, lobes subcircular to oblong-ovate densely minutely papillose within. Capsule 4.5–5.5 mm. long including beak, minutely pubescent.

Zimbabwe. N: Mazowe (Mazoe), Iron Mask Hill, c. 1500 m., *Eyles* 252 (BM; K; SRGH).
  Not found elsewhere and known only from the type specimen.
  The entire leaves with recurved margins and overall fine pubescence, clearly separate this as a distinct species. Its recorded habitat being in "cliff crevices" compares with the lithophytic habitat of the other species in the genus.

4. **Stemodiopsis glandulosa** Philcox sp. nov.* Type: Zambia, Serenje, fl. & fr. 18.ii.1955, *Fanshawe* 2084 (K, holotype).

Perennial herb 4–10 cm. tall with trailing branches up to 18 cm. long; stems tufted, quadrangular, much branched, moderately to densely short stipitate-glandular; branches terete. Leaves petiolate, opposite; lamina elliptic to lanceolate-elliptic; 5–18 × 2–6 mm., obtuse apex, cuneate at base into the short petiole somewhat densely glandular, crenate, often obscurely so; petiole 1–1.5(2) mm. long. Flowers solitary. Pedicels 4.5–7.5 mm. long, glandular, recurved in fruit. Bracteoles 2 mm. long, linear, glandular. Calyx 3–3.8 mm. long, lobes 2–3.2 mm. long, lanceolate, 1–1.3 mm. wide at base, glandular, subequal. Corolla white, 6–7.8 mm. long; tube 4–5.3 mm. long, subglabrous or with few short,

---

* Affinis *Stemodiopsis eylesii* S. Moore sed planta tota stipitato-glandulosa, capsulis solis exceptis.

glandular, hairs; upper lip 1.5 mm. long, lobes 0.45 × 0.4 mm., oblong; lower lip c. 2 mm. long, lateral lobes c. 1 mm. long broadly triangular, middle lobe c. 1 mm. long, rounded, densely papillose within. Filaments of longer stamens 3.5 mm. long, curved not twisted; filaments of shorter stamens 1.8 mm. long, strict; staminode arising at base of corolla tube, c. 1.4 mm. long. Capsule 5–7.5 mm. long including beak, up to c. 3.5 mm. broad, broadly ovoid, glabrous, brown, somewhat glossy.

Zambia. C: Kapiri Mposhi Hills, c. 1280 m., fl. Feb. 1907, *Allen* 447 (K; SRGH).
Known only from a small area of Zambia.
As in other members of the genus, this plant inhabits rock faces and crevices, but is characterised by being almost totally covered by small glandular hairs, a feature hitherto not met with in the genus.

17. LIMNOPHILA R. Br.

Limnophila R. Br., Prodr.: 442 (1810) *nom. cons.* —Philcox in Kew Bull. **24**: 101–170 (1970).

Annual or perennial, aquatic or marsh herbs, frequently aromatic when bruised, glabrous, pubescent or glandular. Stems erect, prostrate or creeping, rooting at nodes, simple or more usually, branched. Leaves, when submerged in truly aquatic species, laciniate, pinnatifid to capillary-multifid, glabrous; aerial leaves opposite to verticillate, sessile or petiolate, entire to serrate, laciniate or pinnately divided, where undivided, pinnately or parallel nerved; punctate. Flowers sessile or pedicellate, solitary-axillary or in lax or compact, terminal or axillary spikes or racemes; bracteoles 0 or 2. Calyx tubular, 5-lobed; lobes sub-equal or with adaxial lobe enlarged; tube terete with 0-5 prominent nerves or striate with more than 10 prominent nerves present at maturity. Corolla tubular or infundibuliform, 5-lobed, bilabiate; adaxial lip outside in bud, entire or bilobed; abaxial lip three-lobed, erect or spreading. Stamens 4, included, didynamous, posterior pair shorter; anthers free, loculi stipitate. Ovary glabrous. Stigma bilamellate. Style filiform, deflexed at apex. Capsule ellipsoid to globose, septicidally 4-valved; valves bifid. Seeds small, numerous.

A genus of some 35 species from the Old World tropics and subtropics.

1. Finely divided submerged leaves present - - - - - - - - - - 2
  – Finely divided submerged leaves absent - - - - - - - - - 8
2. Flowers sessile - - - - - - - - - - - - 3
  – Flowers pedicellate - - - - - - - - - - 1. *indica*
3. Upper aerial leaves on flowering stems finely divided or deeply lobed almost to base 4
  – Upper aerial leaves on flowering stems entire or shallowly toothed, not finely divided 5
4. Stamens 4 in the chasmogamous flowers and 2 in the cleistogamous flowers; anther thecae divergent, attached by end to greatly inflated filament connective; stigma hooked; fruits from cleistogamous flowers dark brown, opaque, ovoid, truncate-emarginate 2. *ceratophylloides*
  – Stamens 4 in both the chasmogamous and cleistogamous flowers; anther thecae almost parallel, median attached to slightly swollen filament connective; stigma truncate; fruits from cleistogamous flowers pale brown, translucent, subspherical - - - 3. *fluviatilis*
5. Bracteoles present - - - - - - - - - - - 6
  – Bracteoles absent - - - - - - - - - - 6. *dasyantha*
6. Flowers 5–14 mm. long - - - - - - - - - - 7
  – Flowers 3–4 mm. long - - - - - - - - - - 5. *barteri*
7. Stems, and leaves on flowering stems, glabrous to minutely sessile-glandular; inflorescence laxly flowered; flowers 6–12 mm. long - - - - - - - - 1. *indica*
  – Stems, and leaves on flowering stems, densely crisped glandular-hirsute; inflorescence densely flowered; flowers 5–14 mm. long - - - - - - - 4. *bangweolensis*
8. Upper leaves on aerial flowering stem divided - - - - - - - 9
  – Upper leaves on aerial flowering stem undivided - - - - - - 11
9. Flowers sessile - - - - - - - - - - - - 10
  – Flowers pedicellate - - - - - - - - - - 1. *indica*
10. Corolla 10–12 mm. long, tube villous within; anther thecae divergent, attached by end to greatly inflated filament connective - - - - - - - 2. *ceratophylloides*
  – Corolla 4–5 mm. long, tube glabrous to subglabrous within; anther thecae subparallel, median attached to slightly swollen filament connective - - - - - 3. *fluviatilis*

11. Flowers pedicellate	- - - - - - - - - - - -	1. *indica*
– Flowers sessile	- - - - - - - - - - - -	12
12. Bracteoles present	- - - - - - - - - - - -	13
– Bracteoles absent	- - - - - - - - - - - -	6. *dasyantha*
13. Flowers 5–14 mm. long	- - - - - - - - - - - -	14
– Flowers 3–4 mm. long	- - - - - - - - - - - -	5. *barteri*
14. Aerial stem, leaves and calyx densely white crisped glandular-hirsute		4. *bangweolensis*
– Stem and leaves sparsely hirsute, minutely yellow glandular; calyx glabrous		7. *crassifolia*

1. **Limnophila indica** (L.) Druce in Rep. Bot. Exch. Club Brit. Isles 1913, **3**: 420 (1914). —Schlechter in Engl., Bot. Jahrb. **56**: 571 (1921). —Hepper in F.W.T.A. ed. 2, **2**: 357 (1963). —Philcox in Kew Bull. **24**: 115 (1970) excl. *Ambulia baumii* Engl. & Gilg. Type from India.

   Hottonia indica L., Sp. Pl., ed. 2: 208 (1762); in Syst. Nat., ed. 10: 919 (1759). —Willd., Sp. Pl. **1**: 813 (1798). Type as above.

   Limnophila gratioloides R.Br., Prodr. Fl. Nov. Holl.: 442 (1810). —Hook. f. in Hook., Niger Fl.: 474 (1849). —Skan in F.T.A., **4**, 2: 319 (1906). —Eyles in Trans. Roy. Soc. S. Afr. **5**: 472 (1916). —Hutch. & Dalz. in F.W.T.A. **2**, 1: 223 (1931). Type from Australia.

   Ambulia gratioloides (R.Br.) Baillon ex Wettst. in Engl. & Prantl, Nat. Pflanzenfam. IV. **3B**: 73 (1891). —Engl., Pflanzenw. Ost-Afr. **C**: 357 (1895). —A. Chev., Expl. Bot. Afr. Occ. Fr. **1**: 470 (1921). Type as above.

   Ambulia baumii Engl. & Gilg in Warb., Kunene-Samb.-Exped. Baum: 361, t.7, figs. F & G (1903). Type from Angola.

   Limnophila gratioloides var. *nana* Skan in F.T.A. **4**, 2: 319 (1906). —Hutch. & Dalz. in F.W.T.A. **2**, 1: 233 (1931). Type from Nigeria.

Amphibious perennial herb. Stems: aerial part 2.5–14 cm. tall, simple to much branched, slender, with sessile or stipitate glands above becoming sub-glabrous; submerged stem up to 1 m. long, much branched, glabrous. Leaves on aerial stem usually all verticillate and variously dissected, (2.5)4–12(22) mm. long, sometimes 2–3 pairs of opposite, undissected, crenate-serrate to lacerate, 1–3 nerved leaves towards apex, up to 15 × 4 mm., sessile-glandular to sub-glandular, rarely all aerial leaves undissected; submerged leaves up to 30 mm. long, verticillate in whorls of 6–12, pinnatisect with lobes flattened or capillary. Flowers solitary, axillary, slender pedicellate. Pedicels (1)3.5–10(15) mm. long, sessile glandular to stipitate glandular, usually, though not always, longer than subtending leaves. Bracteoles 2, (1.5)3–4 mm. long, linear to linear-oblong to obovate-lanceolate, acute, entire to irregularly remotely serrate-dentate to occasionally deeply incised, glandular to subglabrous. Calyx 3.5–6 mm. long, sessile glandular, rarely sparsely hirsute, not striate at maturity; lobes 2–3 mm. long, broadly ovate to lanceolate, shortly acuminate, occasionally ciliate. Corolla (6)8–12 mm. long, white to pale yellow, or yellow at base of tube, mauve-pink above, glabrous without; lobes all entire. Stamens with anthers contiguous; posterior filaments 2 mm. long, anterior 4 mm. long, all glabrous. Style up to 4.5 mm. long with two lateral processes about 0.2 mm. wide at apex and below stigma. Capsule about 3.5 mm. long, compressed ellipsoid to sub-globose, dark brown.

Zambia. B: Masese, fl. 19.vi.1960, *Fanshawe* 5748 (K; SRGH). C: Munali, E. of Lusaka, fl. 17.vii.1955, *King* 66(K). S: Pemba, fl. 19.ii.1929, *Greenway* 1496 (EA; K). **Zimbabwe**. N: Hurungwe (Urungwe), fl. 13.v.1952, *Whellan* 661 (K; SRGH). C: Marondera (Marandellas), c. 1500 m., fl. Dec. 1927, *King* 5153 (K). **Malawi**. S: Mangochi, fl. 2.vi.1955, *Banda* 106 (K; SRGH). **Mozambique**. N: Nampula, fl. & fr. 3.v.1937, *Torre* 1460 (COI). M: Namaacha, Rio Mazeminhoma, 40 km. from Catvane towards Porto Henrique, fl. & fr. 15.v.1964, *Moura* 71 (COI).

   Common throughout the Old World tropics. Usually in rice-fields, river flood plains and marshes.

   This species varies widely from a slender plant about 5 cm. tall, when growing on fairly firm drying mud of river-banks and rice-fields to a luxuriant submerged aquatic with much-branched stems up to 1 m. long. In the terrestrial or semi terrestrial state, the leaves are usually all whorled and deeply pinnatisect or lacerate, whereas in the submerged plant, although the submerged leaves remain whorled and divided, the leaves on the aerial stems can vary considerably, perhaps becoming opposite and crenate-dentate or even entire. Floral characters in general appear to be unaffected by these variations in habit except that some variation is shown in the size of the lateral processes or swellings at the top of the style - those appearing to be far more pronounced in the aquatic state of the species.

   Another character which also varies to some degree is the form of the submerged leaves. In general the upper submerged leaves have flat lobes, but rarely do they all develop into very fine capillary segments. The form with broad and deeply lacerate upper submerged leaves has developed from normal plants in cultivation under conditions of constant shallow standing water with normal day length and low light intensity.

2. **Limnophila ceratophylloides** (Hiern) Skan in F.T.A. **4**, 2: 317 (1906). —Eyles in Trans. Roy. Soc. S. Afr. **5**: 472 (1916). —Peter, Wasserpf. Deutsch. Öst-Afr.: 127 (1928). —Philcox in Kew Bull. **24**: 122 (1970) pro parte. —Raynal & Philcox in Adansonia, sér. 2, **15**, 2: 234 (1975). Type from Angola.

Stemodiacra ceratophylloides Hiern, Cat. Afr. Pl. Welw. **1**: 759 (1898). Type as above.
Stemodia ceratophylloides (Hiern) K. Schum. in Just, Jahresber. **26**, 1: 395 (1900). Type as above.
Ambulia ceratophylloides (Hiern) Engl. & Gilg in Warb., Kunene-Samb.-Exped. Baum: 362 (1903). Type as above.
Ambulia baumii Engl. & Gilg, tom. cit.: 361. Type from Angola.
Stemodiacra sessiliflora auct. non (Vahl) Hiern— Hiern, op. cit.: 758, quoad pl. afric. non Hottonia sessiliflora Vahl 1791.

Amphibious perennial. Stems: aerial stems to 20 cm. tall, simple or branching, glabrous to laxly white hirsute or covered lightly with small sessile yellow glands; submerged stems to 60 cm. long, simple or branching, glabrous. Leaves on aerial stems verticillate to opposite, irregularly pinnatisect to lacerate, 5–8 × 1–2 mm., densely punctate, glabrous, hirsute or yellow-glandular; submerged leaves to 2.5 cm. long, pinnatisect-multifid, segments capillary or more usually flattened, glabrous. Flowers solitary axillary, sessile, cleistogamous flowers present on submerged stems. Chasmogamous flowers: bracteoles 1.5–4 mm. long, narrowly linear, glabrous to very shortly hirsute; calyx 3–5 mm. long, glabrous or hirsute, yellow-glandular; corolla 6–10 mm. long, mauve to lilac with darker throat, externally glabrous, densely villous within tube mainly on the posticous side; stamens four with contiguous anthers at anthesis, filaments 0.75–3 mm. long, anthers attached by one pole to largely inflated connective, thecae divergent; stigma unequally bilobed with one lobe somewhat extended, narrowly deltoid, perpendicular to style, style 1–4 mm. long; capsule 2.5–3.5 mm. long, dark brown, emarginate, broadly ovoid. Cleistogamous flowers: bracteoles 2.5–4 mm. long, narrowly linear, glabrous; calyx 3–3.5 mm. long, glabrous; corolla 3.5–4 mm. long, villous within tube; stamens generally two, filaments c. 0.75–1 mm. long; style c. 1.5 mm. long; capsule 2.5 mm. long, dark brown to light brown, opaque, broadly ovoid.

Botswana. N: Thamalakane R., near Maun, 900 m., fl. 13.iii.1961, *Richards* 14686 (K). **Zambia.** W: Mwinilunga, S. of Dobeka, fl. 17.xii.1937, *Milne-Redhead* 3709 (BR; K). S: Kafue Gorge, fl. 10.x.1957, *West* 3541 (K; SRGH). **Zimbabwe.** N: Mazowe (Mazoe) Distr., 1200 m., fl. 7.v.1952, *Wild* 3881 (K; SRGH). W: Matobo Distr., fl., *Miller* 4411 (BR). C: Harare Distr., Cleveland Dam, fl., 4.v.1946, *Wild* 1078 (K; SRGH).

Known also from tropical Central Africa and Namibia. In wet marshy areas usually bordering rivers, and river flood-plains.

3. **Limnophila fluviatilis** A. Chev. in Bull. Mus. Nat. Hist. Nat. Paris, sér. 2, **4**: 587 (1932). —Raynal in Adansonia **7**, 3: 351 (1967). —Raynal & Philcox in Adansonia, sér. 2, **15**, 2: 236 (1975). Type from Mali.

Limnophila fluviatilis forma fluviatilis A. Chev. loc. cit. (1932).
Limnophila fluviatilis forma terrestris A. Chev., loc. cit.: 588 (1932). Type from Mali.

Amphibious perennial. Stems: aerial stems to 10 cm. tall, simple or rarely branching except at base, glabrous to sparsely hirsute, particularly above; submerged stems to 35 cm. long, branching, glabrous. Leaves on aerial stems verticillate, irregularly pinnatisect to lacerate, 4–18 × 1–6 mm., glabrous, densely punctate, frequently with sessile yellow glands; submerged leaves to 2.5 cm. long, pinnatisect-multifid, segments capillary or more usually flattened, glabrous. Flowers solitary, axillary, sessile on submerged stems, solitary or clustered on very short axillary few-flowered racemes on aerial stems; cleistogamous flowers usually present on submerged stems. Chasmogamous flowers: bracteoles 2–3.5 mm. long, narrowly linear, glabrous to sparsely hirsute; calyx 3–4.5 mm. long, glabrous, frequently with yellow glands; corolla 4–6.5 mm. long, white to lilac, throat yellow, externally glabrous, occasionally very sparsely villous within tube; stamens four with contiguous anthers at anthesis, filaments 0.5–2 mm. long, anthers median attached to slightly swollen connective, thecae almost parallel; stigma truncate to emarginate, style 1–2 mm. long; capsule 3–4 mm. long, dark brown, subtruncate to emarginate, somewhat flattened, ovoid. Cleistogamous flowers; bracteoles 2.5–6 mm. long, narrowly linear, glabrous; calyx 3–6 mm. long, glabrous; corolla 3.5–4.75 mm. long, glabrous to sparsely villous within tube; stamens four with contiguous anthers, filaments c. 0.25 mm. long, anthers as above; style 0.75–1 mm. long; capsule 2.25–4 mm. long, light brown, translucent, broadly ovoid, emarginate to subspherical.

46

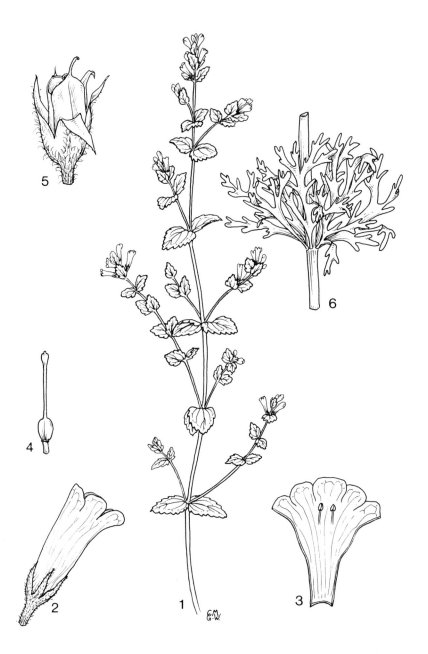

Tab. 18. LIMNOPHILA BANGWEOLENSIS. 1, flowering branch (×⅔), from *Sanane* 875; 2, flower (× 4); 3, corolla opened out showing androecium (× 4); 4, gynoecium (× 6); 5, dehiscing fruit (× 4), 2–5 from *Fanshawe* 3424; 6, submerged node (× 2), from *Fanshawe* 2554.

Zambia. N: Mbala Distr., Lumi R., 1680 m., fl., 31.v.1957, *Richards* 9930 (BR; K). **Zimbabwe**. S: Matobo Distr., c. 1600 m., fl. September 1957, *Miller* 4548 (K; SRGH).
Known also from tropical West and Central Africa. In similar areas to the preceding species.

4. **Limnophila bangweolensis** (R.E. Fries) Verdc. in Kew Bull. **5**: 379 (1951). —Philcox in Kew Bull. **24**: 126 (1970). TAB. **18**. Type: Zambia, Lake Bangweulu, *Fries* 895 (UPS; K, photo).
  Ambulia bangweolensis R.E. Fries, Wiss. Ergebn. Schwed. Rhod.-Kongo-Exped. 1911–1912, **1**: 288 (1916). Type as above.

Amphibious annual. Stems to 50 cm. tall, simple or branched towards apex, aerial part densely white crisped glandular-hirsute, submerged part glabrous. Leaves on aerial stem opposite becoming verticillate, sessile, 6–13 × 5–11 mm., broadly ovate, subcircular or oblong, densely glandular-hirsute, becoming glabrous below dependent on degree of immersion, serrate-dentate with thickened margin, somewhat amplexicaul, revolute, with 3–5 parallel nerves; submerged leaves at base of aerial stem becoming laciniate to occasionally capillary-multifid; finely divided leaves with capillary segments, usually borne on sterile totally submerged stems, dying off before maturity of aerial stem. Flowers sessile to sub-sessile, bracteate, in densely flowered spikes, or occasionally solitary-axillary; bracts reduced, leaflike; bracteoles two, c. 3.5 mm. long, linear, glandular-hirsute. Calyx 5–6 mm. long, glandular-hirsute, not striate at maturity; lobes 3 mm. long, narrowly lanceolate-acuminate. Corolla 5–14 mm. long, white, pale pink, lilac to mauve-pink; adaxial lip broad emarginate; abaxial lip trilobed, lobes broadly ovate; tube with finely clavellate papillae within, glabrous without. Stamens with posterior filaments 2 mm. long, anterior, 3.75–4 mm. long, glabrous. Capsule about 3 mm. long, compressed ovoid, black.

Botswana. N: above Kwando, fl. 13.xi.1974, *Smith* 1184 (BR). **Zambia**. N: W. shores of Lake Young, Shiwa Ngandu, 1470 m. fl., 20.ix.1938, *Greenway* 5747 (EA; K; M). C: Chiwefwe, fl. 7.viii.1957, *Fanshawe* 3424 (BR; EA; K; SRGH).
Also in Tanzania. Aquatic or amphibious in shallow water or mud of bogs, swamps or riversides; 1400–1750 m.

5. **Limnophila barteri** Skan in F.T.A. **4**, 2: 317 (1906). —A. Chev., Expl. Bot. Afr. Occ. Fr. **1**: 470 (1920). —Hutch. & Dalz. in F.W.T.A. **2**: 223 (1931). —Hepper in Hutch. & Dalz. in F.W.T.A. ed. 2, **2**: 357 (1963) excl. *L. fluviatila* Chev. et formae. —Philcox in Kew Bull. **24**: 127 (1970). Type from Nigeria.

Semi-aquatic annual. Stems: aerial stem to 40 cm. long, sparsely branched, densely short, white, patent hirsute above to glabrescent below; submerged stem up to 20 cm. long, sparsely branched, glabrous, usually sterile. Leaves on aerial stem verticillate near base to opposite above, 10–18 mm. long, 6–9 mm., ovate-elliptic, sessile, somewhat amplexicaul, crenate-serrate, shortly hirsute, densely, minutely punctate, with 3–7 parallel nerves; submerged leaves up to 18 mm. long, capillary-pinnatifid, glabrous. Flowers either sessile in axils of leaf-like bracts or in slender, distantly flowered spikes; bracteoles two, about 2 mm. long, linear, sub-acute, densely pilose. Calyx 2.5–3 mm. long, densely pilose and with round yellowish glands, not striate at maturity; lobes about 1.25 mm. long, ovate-lanceolate, acuminate, divergent. Corolla 3–4 mm. long, tube pale pink to white, limb white; totally glabrous; abaxial lip trilobed, lobes entire, adaxial lip erect, entire. Stamens with posterior filaments about 0.4 mm. long, anterior 1 mm. long, glabrous. Capsule about 2 mm. long, broadly ovoid, pale brown.

Zambia. B: NE. of Sesheke, Masese, fl., 20.vii.1962, *Fanshawe* 6949 (K; LISC; SRGH).
Also in tropical West and Central Africa. Wet muddy, grassy areas.

6. **Limnophila dasyantha** (Engl. & Gilg) Skan in F.T.A. **4**, 2: 318 (1906). —A. Chev., Expl. Bot. Afr. Occ. Fr. **1**: 470 (1920). —Hutch. & Dalz. in F.W.T.A. **2**: 223 (1931). —Hepper in Hutch. & Dalz. in F.W.T.A. ed. 2, **2**: 358 (1963). —Philcox in Kew Bull. **24**, 1: 128 (1970). Type from Angola.
  Ambulia dasyantha Engl. & Gilg in Warb., Kunene-Samb.-Exped. Baum: 362; t.7, figs A–E (1903). Type as above.

Semi-aquatic annual. Stems up to 1 m. long, glabrous and leaflets in extreme submerged part, becoming densely white villous above then glabrous on emergence. Leaves on aerial stem opposite, 10–15 × 3–5 mm., elliptic to narrowly ovate to obovate becoming divided below, sessile, entire to crenate or occasionally obscurely serrate, hirsute beneath particularly upper leaves, glabrous above, with 3–5 parallel nerves, sparsely punctate; lower submerged leaves multipartite, segments capillary. Flowers sessile, clustered together into short compact terminal spike, each flower subtended by much reduced leaf-like bract, or occasionally solitary-axillary; ebracteolate; occasionally cleistogamous flowers present on submerged stems. Calyx about 3.5 mm. long, glabrous, not striate at maturity; lobes 1.5 mm. long, ovate, sub-obtuse. Corolla about 14 mm. long, yellow, shortly pilose without, denser within; adaxial lip shallowly bilobed, abaxial lip with entire lobes. Stamens with posterior filaments about 2 mm. long, anthers 4.5 mm. long, all glabrous; anthers contiguous. Capsule 3 × 2 mm., compressed broad-ellipsoid, brown.

Zambia. N: Lake Bangweulu, fl., 27.v.1964, *Fanshawe* 8695 (K).
Also in West Africa, Portuguese Guinea and Angola. Aquatic in slow-running or standing water or mud fringes to pools; above 80 m.

7. **Limnophila crassifolia** Philcox in Kew Bull. **21**: 157 (1967); op. cit. **24**: 134 (1970). Type: Zambia, Mwinilunga, fl., 14.x.1953, *Fanshawe* 412 (K, holotype).

Annual up to about 6 cm. tall. Stems procumbent and rooting at nodes, to erect, simple, sparsely hirsute with minute sessile yellow glands above to subglabrous and eglandular below. Leaves all opposite, 2–6 × 1.5–2.5 mm., undivided, sessile, thick, sub-rotund below to ovate-elliptic above, serrate-dentate with few small teeth, particularly in the upper third, minutely yellow glandular becoming glabrous, with 1–5 parallel nerves; apex obtuse; base semi-amplexicaul. Flowers sessile, solitary in upper leaf-axils; bracteoles about 2.5 mm. long, glabrous. Calyx 4.5 mm. long, glabrous, not striate at maturity; lobes 2 mm. long, lanceolate-acuminate. Corolla 7.5–9 mm. long, purple, densely villous within. Stamens with anthers shortly stipitate; posterior filaments 0.5 mm. long, anterior 1 mm. long, all glabrous. Style 4 mm. long. Capsule unknown.

Zambia. W: Mwinilunga, fl., 14.x.1953, *Fanshawe* 412 (K).
Muddy river-sides.
This small plant, at present known only from one gathering, appears to be quite distinct by reason of its thick undivided leaves, comparatively stout simple stems and solitary flowers.

18. BACOPA Aubl.

Bacopa Aubl., Pl. Guian. **1**: 128, t. 49 (1775) *nom. cons.*

Annual herbs, erect, prostrate or decumbent, or floating, mostly glabrous. Leaves opposite, entire or variously dentate to crenate, or capillary-divided in aquatic species. Flowers axillary, solitary, pedicellate or sessile, bracteate or ebracteate. Calyx 5-partite; sepals imbricate, posterior broadest, lateral usually very narrow. Corolla bilabiate, upper lip bilobed or emarginate, lower trilobed. Stamens four, didynamous, perfect, included. Style apically dilated, entire or bilobed. Ovules many. Capsule ovoid to globose, 2 or 4 loculicidal or septicidal. Seeds small, numerous.

A genus of about 100 species from tropical and subtropical regions throughout the world.

1. Plant creeping or decumbent   -   -   -   -   -   -   -   -   -   -   -   2
  – Plant erect   -   -   -   -   -   -   -   -   -   -   -   -   -   -   3
2. Plant creeping and frequently rooting at nodes; leaves spathulate to obovate, obtuse; fruiting pedicels exceeding to much exceeding leaves, not winged below calyx; sepals not markedly prominent nerved   -   -   -   -   -   -   -   -   -   -   1. *monnieri*
  – Plant decumbent or ascending, not rooting at nodes; leaves lanceolate to ovate, obtuse or subacute; fruiting pedicels shorter than leaves, winged below calyx; sepals strongly reticulate-nerved   -   -   -   -   -   -   -   -   -   -   -   -   - 2. *crenata*
3. Flowers sessile or subsessile, pedicels rarely more than 0.5 mm. long, glabrous   3. *hamiltoniana*
  – Flowers pedicellate, pedicels 4–5 mm. long, hispid   -   -   -   -   4. *floribunda*

1. **Bacopa monnieri** (L.) Pennell in Proc. Acad. Nat. Sci. Philad. **98**: 94 (1946). —Philcox in Kew Bull. **33**: 679 (1979). Type from South America.
    Lysimachia monnieri L., Cent. Plant. **2**: 9 (1756). Type as above.
    Monniera cuneifolia Michaux, Fl. Bor. Amer. **2**: 22 (1803). —Hiern in F.C. **4**, 2: 355 (1904).
    —Hemsl. & Skan in F.T.A. **4**, 2: 320 (1906) [as "*Moniera*"] pro parte excl. *Herpestis crenata* Beauv. Type from North America (Carolina).
    Monniera africana Pers. Syn. **2**: 166 (1806). Type from Africa.

Herpestis monnieria (L.) Kunth, Nov. Gen. & Sp. **2**: 366 (1818). —Benth. in Hook, Comp. Bot. Mag. **2**: 58 (1836) [as "*monniera*"]; in DC., Prodr. **10**: 400 (1846). —Klotzsch in Peters, Reise Mossamb. Bot.: 223 (1861) nom. illegit.

Herpestis africana (Pers.) Steud., Nom. Bot. ed. 1: 402 (1821). Type as above.

Bacopa monnieria (L.) Wettst. in Engl. & Prantl, Nat. Pflanzenfam. IV, **3B**: 77 (1891) [as "*monniera*"]. —Engl., Pflanzenw. Ost-Afr. **C**: 357 (1895) nom. illegit.

Creeping, glandular-punctate herb often rooting at nodes, glabrous; stems simple, rarely branched. Leaves 10–15 × 5–7 mm., obovate-cuneate to spathulate-oblong, entire or occasionally subcrenulate particularly towards apex, obtuse, glandular-punctate. Flowers solitary, axillary, blue, pedicellate, bracteate. Pedicels up to 25 mm. long at fruiting, always exceeding leaves. Bracts 2–3 mm. long, oblong. Calyx with lateral segments narrow, oblong; posterior 4–6 × 3–4 mm., broadly ovate; anterior c. 5 × 2.5 mm., ovate; no segments markedly prominently nerved. Corolla 6–7.5 mm. long, subequally 5-lobed. Capsule 2–2.5 mm. long, ovoid.

Mozambique. N: Mozambique Is., *Peters* s.n. (K).
Widespread throughout much of the tropics and subtropics. Creeping, at times aquatic, marsh plant.

2. **Bacopa crenata** (Beauv.) Hepper in Kew Bull. **14**: 407 (1960); in F.W.T.A. ed. 2, **2**: 359, fig. 286 (1963). Type from Nigeria.

Herpestis crenata Beauv., Fl. Oware & Benin **2**: 83, t. 112 (1819). Type as above.

Herpestis calycina Benth. in Hook., Comp. Bot. Mag. **2**: 5 (1836). Type from West Africa.

Moniera calycina (Benth.) Hiern, Cat. Afr. Fl. Welw. **1**: 760 (1898). —Hemsl. & Skan in F.T.A. **4**, 2: 320 (1906). Type as above.

Bacopa calycina (Benth.) Engl. ex De Wild. in Bull. Herb. Boiss. **1**: 832 (1901).

Decumbent or ascending, glandular-punctate herb up to 30 cm. tall, subglabrous; stems simple or branched. Leaves 9–33.5(50) × 3–12(18) mm., lanceolate to ovate, obtuse or subacute, cuneate at base into short, broad petiole or subsessile, crenulate to crenate-serrate, glandular-punctate, markedly so beneath. Flowers solitary-axillary, white or occasionally pale violet, bibracteate. Pedicels 5–9 mm. long, winged just below calyx. Bracts 1–1.5 mm. long, narrowly linear. Calyx with lateral lobes ovate-lanceolate, narrow, keeled, ciliate; posterior lobe 6–7 mm. long, 4–4.5 mm. wide, up to 9 mm. long, 6 mm. wide in fruit; anterior similar to but somewhat narrower than posterior; all lobes strongly reticulate-nerved. Capsule 4 × 2.5 mm., ellipsoid, foveolate, somewhat densely minutely glandular.

Zambia. N: Mporokoso Distr., Pakefupa, near Bulaya, c. 1000 m., fl. & fr. 13.viii.1962, *Tyler* 443 (BM; SRGH). **Malawi**. S: Mulanje, fl. & fr. 25.ix.1932, *Young* 924 (BM).
Known also from West Africa to Angola, the Sudan and Madagascar; Kenya and Tanzania. In swamps and ditches.

3. **Bacopa hamiltoniana** (Benth.) Wettst. in Engl. & Prantl, Nat. Pflanzenfam. IV, **3B**: 77 (1891). —Hepper in F.W.T.A. ed. 2, **2**: 358 (1963). TAB. **19**, fig. B. Type from India.

Herpestis hamiltoniana Benth., Scroph. Ind.: 30 (1835); in DC., Prodr. **10**: 400 (1846). Type as above.

Moniera hamiltoniana (Benth.) T. Cooke, Fl. Pres. Bombay **2**: 286 (1905). —Hemsl. & Skan in F.T.A. **4**, 2: 323 (1906). Type as above.

Erect herb up to 40 cm. tall, glabrous; stem simple or branched, smooth to subquadrangular especially towards base. Leaves 15–40 × 1.5–5(8) mm., sessile, elliptic-lanceolate, acute, cuneate into narrow, petiole-like, somewhat amplexicaul base, entire to obscurely, remotely denticulate or evenly crenate-dentate, glandular-punctate. Flowers axillary, solitary, pale mauve or pale blue, shortly pedicellate, bibracteate. Pedicels c. 0.5 mm. long, stout, glabrous. Bracts 1–1.5 mm. long, very narrowly-linear. Calyx minutely ciliate, glandular-punctate; lateral sepals 3–3.5 mm. long, linear-lanceolate, acuminate; posterior 4.5–5 × 2.5–3 mm., broadly ovate, reticulate veined; anterior c. 3 mm. long. Corolla c. 4 mm. long, upper lip, truncate, emarginate; lower lip equally trilobed. Capsule 3–3.5 mm. long, broadly ovoid to subglobose.

Zambia. W: 7 km. E. of Chizela (Chizera), fl. & fr. 27.iii.1961, *Drummond & Rutherford-Smith* 7431 (BR; K; LISC; SRGH). **Mozambique**. N: Nampula, to the north, fr. 16.viii.1936, *Torre* 606 (LISC).
Known also from West Africa where it is widely spread, and India. Damp places.
A shortly scabrid-pubescent plant with pedicels 3–6.5 mm. long which may eventually prove to be distinct, occurs in West Africa along with the common form.

50

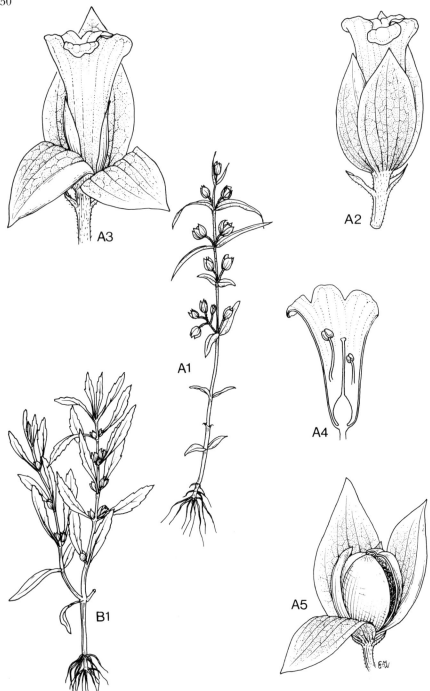

Tab. 19. A.—BACOPA FLORIBUNDA. A1, habit (× $\frac{2}{3}$); A2, flower (× 6); A3, flower opened displaying unequal calyx lobes (×6); A4, flower, longitudinal section showing gynoecium (×6); A5, dehiscing capsule (× 6), A1–5 from *Robinson* 2838. B.—BACOPA HAMILTONIANA. B1, habit (× $\frac{2}{3}$), from *Drummond & Rutherford-Smith* 7431.

4. **Bacopa floribunda** (R.Br.) Wettst. in Engl. & Prantl, Nat. Pflanzenfam. IV, **3**B: 77 (1891). —Engl., Pflanzenw. Ost-Afr. **C**: 357 (1895). TAB. **19**, fig. A. Type from Australia.

   Herpestis floribunda R.Br., Prodr.: 442 (1810). —Benth. in Hook. f., Comp. Bot. Mag. **2**: 57 (1836); in DC., Prodr. **10**: 400 (1846). Type as above.

   Moniera floribunda (R.Br.) T. Cooke, Fl. Pres. Bombay **2**: 286 (1905). —Hemsl. & Skan in F.T.A. **4**, 2: 322 (1906). Type as above.

   Moniera pubescens Skan in F.T.A. **4**, 2: 322 (1906). Type from Sierra Leone.

   Bacopa pubescens (Skan) Hutch. & Dalz. in F.W.T.A., ed. 1, **2**: 222 (1931). Type as above.

Erect, slender herb 8–14(35) cm. tall, scabrid, pubescent to subglabrous; stem simple to more frequently much branched, terete to channelled above, tetragonal below, occasionally subulate towards base. Leaves (10)15–24(40) × (1.5)2–5 mm., sessile, linear to linear-lanceolate, barely acute, narrowed shortly or at length to subamplexicaul base, entire, shortly scabrid to subglabrous, glandular-punctate beneath. Flowers axillary, usually solitary, white, through pink to blue or violet, pedicellate, bibracteate. Pedicels (2)4–5 mm. long, hispid. Bracts 1–1.5 mm. long, setaceous, subpatent, arising c. 0.5 mm. below base of calyx. Calyx minutely pubescent to subglabrous; lateral sepals c. 3.5 mm. long, 1–1.5 mm. wide, lanceolate, midrib somewhat keeled, shortly hispid; posterior 6.5 × 6 mm., broadly ovate; anterior c. 6–6.4 × 4.3–5 mm., broadly ovate, all segments prominently nerved though laterals less so. Corolla 4–6 mm. long; upper lip c. 2 mm. wide, emarginate; lower lip 3-lobed, lobes equal. Capsule 3–3.25 mm. long, 2.5 mm. in diam., somewhat flattened-ellipsoid, glabrous.

Zambia. B: Masese, fl. & fr. 3.v.1961, *Fanshawe* 6536 (SRGH). S: Kabulamwanda Dam, 128 km. N. of Choma, c. 1000 m., fl. & fr. 24.iv.1954, *Robinson* 744 (K; SRGH). **Zimbabwe**. N: Kariba Distr., Charare Fish Camp, fl. & fr. 12.xi.1964, *Jarman* 69 (K; LISC; SRGH). C: Chegutu (Hartley) Distr., Poole Farm, fr. 3.iv.1946, *Wild* 1036 (SRGH). E: Nyamkwarara (Nyumguarara) Valley, 915 m., fl. ii.1935, *Gilliland* 1560 (K). S: Mwenezi (Nuanetsi) Distr., Malangwe R., SW. Mateke Hills, 620 m., fl. & fr. 6.v.1958, *Drummond* 5588 (BR; K; LISC; SRGH). **Mozambique**. N: Larde, fl. & fr. 15.vii.1948, *Pedro & Pedrogão* 4556 (SRGH). Z: Mocuba Distr., Namagoa Estate, fl. & fr. vi.1947, *Faulkner* 158 (BR; K; SRGH).

Widespread throughout much of tropical Africa, Asia and Australia. Wet grasslands and swamps; up to 1000 m.

19. **DOPATRIUM** Buch.-Ham. ex Benth.

Dopatrium Buch.-Ham. ex Benth., Scroph. Ind.: 30 (1835). —Benth. & Hook. f., Gen. Pl. **2**: 953 (1876).

Annual herbs, erect, mostly glabrous. Leaves opposite, mostly small, closely arranged at base of stem; cauline leaves much reduced in size, few, distant. Flowers solitary in axils of upper leaves, occasionally cleistogamous. Calyx 5-lobed. Corolla bilabiate; tube short to long, often dilated at throat; upper lip bilobed, lower trilobed, all lobes spreading. Stamens: 2 posterior fertile, perfect, included, filaments filiform; anther cells parallel, equal; 2 anterior reduced to minute staminodes. Style short, often totally, or in part, persistent. Ovules many. Capsule ovoid or globose, small, loculicidally dehiscent, with 2 entire or bifid valves. Seeds small, numerous.

A genus of some 20 species, represented in tropical Africa, Asia and Australia.

Corolla tube c. 2.5 mm. long, campanulate; pedicels filiform; stem branched mainly from base - - - - - - - - - - - - - - - - 1. *junceum*

Corolla tube 11–14 mm. long, slender-cylindric; pedicels slender; stem branched only within inflorescence - - - - - - - - - - - - 2. *stachytarphetioides*

1. **Dopatrium junceum** (Roxb.) Buch.-Ham. ex Benth., Scroph. Ind.: 31 (1835). TAB. **20**, fig. A. Type from India.

   Gratiola juncea Roxb., Pl. Corom. **2**: 16, t. 129 (1799). Type as above.

Erect annual, (4)12–18(24) cm. tall, glabrous, somewhat fleshy, branching mainly at base. Lower and radical leaves 6.5–22 × 2–4 mm., oblong, obtuse, entire, narrowing at base, becoming smaller above; upper leaves 2.5–3 × 1–1.5 mm., sessile, ovate, few, distant. Flowers in axils of upper leaves, pink to mauve or blue, 4–5.25 mm. long, pedicellate, chasmogamous, or lower flowers sessile to subsessile, cleistogamous. Pedicels (0)4–15 mm.

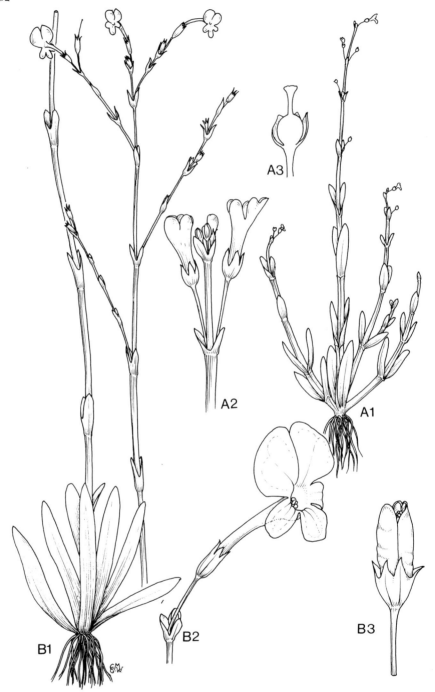

Tab. 20. A.—DOPATRIUM JUNCEUM. A1, habit (× ⅔); A2, part of flowering stem (× 4); A3, gynoecium (× 8), A1–3 from *Richards* 10902. B.—DOPATRIUM STACHYTARPHETIOIDES. B1, habit (× ⅔); B2, flower (× 3); B3, dehiscing fruit (× 6), B1–3 from *Drummond & Rutherford-Smith* 7427.

long, filiform. Calyx c. 2 mm. long; lobes c. 0.5 mm. wide, connate to base from about midway. Corolla tube c. 2.5 mm. long, campanulate, mouth wide; upper lip bilobed; lower lip very broadly trilobed, 2–2.5 mm. long. Capsule c. 2 mm. in diam., spherical, bearing remains of persistent style.

Zambia. W: Kasempa Distr., 7 km. E. of Chizela (Chizera) fl. & fr. 27.iii.1961, *Drummond* 7432 (K; SRGH). C: Luangwa Valley, Lubi R. to Mfuwe, c. 700 m., fl. & fr. 26.iii.1969, *Astle* 5670 (K; SRGH). S: Choma Distr., Mapanza, c. 1050 m., fl. & fr. 16.iii.1958, *Robinson* 2795 (K). Zimbabwe. N: Hurungwe (Urungwe) Distr., c. 6.5 km. N. of Hanora, fl. & fr. 1.iii.1958, *Phipps* 1004 (BR; SRGH). C: Harare Distr., Cranborne, fl. 27.iii.1944, *Greatrex* in GHS 18414 (SRGH). E: Nyamkwarara (Nyumguarara) Valley, fl. & fr. ii.1935, *Gilliland* K1557 (K). S: Mwenezi Distr., Mwenezi Gorge, fr. 28.iv.1962, *Thompson* s.n. (K; SRGH). Malawi. C: Kongwe Mt., near Dowa, 1525 m., fl. & fr. 18.ii.1959, *Robson* 1653 (BM; K; LISC; SRGH).

Widespread in tropical Africa and Asia and extending to Australia (vide Bentham, Fl. Austral.). Seasonal semi-aquatic, in or around pools either on granitic rock or in grassland; from 600–1500 m.

2. **Dopatrium stachytarphetioides** Engl. & Gilg in Warb., Kunene-Samb. Exped. Baum: 362 (1903). —Hemsl. & Skan in F.T.A. 4, 2: 326 (1906). TAB. 20, fig. B. Type from Angola.

Erect, glabrous herb, 10–35(50) cm. tall; stem simple or only branched within inflorescence, somewhat fleshy. Lower and basal leaves 9–35(58) × 1.5–4.5(7) mm., oblong, obtuse, sessile, entire; cauline leaves 2–8 × 1–2.5 mm., becoming small above, distant, broadly ovate, connate at base up to midway, in some cases forming distinct sheath. Flowers solitary in axils of inflorescence bracts, mauve, violet-blue to purple, forming terminal, few-branched, panicle. Bracts like upper cauline leaves but smaller. Pedicels 3–7 mm. long, slender. Calyx 3.5–5(6.5) mm. long; lobes 1–1.5(2.5) mm. long, broadly lanceolate. Corolla tube 11–14 mm. long, slender, cylindric, gibbous-geniculate at apex; upper lip bilobed; lower lip broadly trilobed, 9–12 mm. long. Capsule 4–5 mm. long, 2–3 mm. in diam., ellipsoid.

Zambia. N: Kasama Distr., 13 km. NE. of Kasama, fl. & fr. 27.iv.1961, *Robinson* 4623 (BR; K; SRGH). W: Mwinilunga Distr., Zambezi Rapids, 6 km. N. of Kalene Hill, fl. & fr. 22.ii.1975, *Williamson & Gassner* 2466 (K; SRGH). E: Chipangali Area, N. of Chipata, fl. & fr. 23.ii.1971, *Anton-Smith* in GHS 213276 (SRGH).

Also in Angola. In or by seasonal pools on granitic outcrops or poorly drained soils; up to 1150 m.

There is one specimen in the Kew Herbarium, (*Rogers* 6025 from Victoria Falls in Zimbabwe), which could possibly be related to the Angolan *Dopatrium caespitosum* P. Taylor. More precise determination is not possible until further similar material is collected from the area, and I have chosen not to include this taxon here.

20. CRATEROSTIGMA Hochst.

By F.N. Hepper

Craterostigma Hochst. in Flora **1841**: 668 (1841).

Rosette perennial herbs, stem very short often encased in old leaf-bases, hirsute to subglabrous. Roots fibrous, sometimes coloured yellow, orange or red. Leaves sessile or shortly petioled, linear-lanceolate to broadly ovate. Inflorescence capitate or shortly racemose, usually pedunculate; flowers subsessile to rather long-pedicellate. Calyx tubular, 5-ribbed or slightly winged, teeth 5, subequal. Corolla blue, purple or white, upper lip hooded, entire or emarginate, lower lip with 3 broad lobes. Stamens 4; upper pair included under the hood with broad straight filaments; lower pair inserted in the throat, each filament with a yellow appendage at the base and sharply angled about the middle, the thecae divaricate. Style filiform, dilated and very shortly bilobed. Fruit an oblong capsule, about as long as the calyx; seeds numerous.

About 15 species mainly in tropical and S. Africa, extending to Madagascar, Arabia and India. In the concept of the genus accepted here the species with slender erect stems hitherto included in *Craterostigma* are now placed in *Torenia*.

1. Leaves ± pubescent only on the nerves of the lower surface; flowers racemose   1. *lanceolatum*
 – Leaves pubescent all over lower surface   -   -   -   -   -   -   -   -   -   2

2. Pedicels 1.5–3 cm. long   -   -   -   -   -   -   -   -   -   -   2. *pumilum*
- Pedicels 1 cm. long or less   -   -   -   -   -   -   -   -   -   -   -   3
3. Leaves with upper surface and the rest of the plant hirsute   -   -   -   3. *hirsutum*
- Leaves with upper surface ± glabrous   -   -   -   -   -   -   -   4. *plantagineum*

1. **Craterostigma lanceolatum** (Engl.) Skan in F.T.A. **4**, 2: 331 (1906). Type: Malawi, *Buchanan* 796 (K, isotype).
    Craterostigma nanum var. *lanceolatum* Engl., Pflanzenw. Ost-Afr. **C**: 357 (1895). Type as for species above.

Rosette, perennial herb; rhizome stout, horizontal, with numerous roots. Leaves held erect, lanceolate, 3–7 ×× 1–2.5 cm., narrowed at the base, subacute, entire, 3–5 nerves prominent beneath, glabrous and shining above, pubescent on the nerves beneath; petiole up to half as long as lamina, broad. Inflorescence 3–12 cm. long, sparsely pilose, usually with several opposite, ovate flowerless bracts well below inflorescence, flowers shortly racemose in upper third or half, subsessile; bracts ovate. Calyx c. 8 mm. long, shortly and acutely toothed, sparsely hispid on nerves. Corolla white with purple markings; c. 1.5 cm. long, upper lip nearly as long as lower, lobes very obtuse. Capsule no longer than calyx.

Zambia. N: Uningi Pans, Mbala, 1500 m., fl. 1.i.1968, *Richards* 22844 (K). **Zimbabwe**. E: Mutare Distr., Engwa, fl. 8.ii.1955, *Exell, Mendonça & Wild* 292 (BM). **Malawi**. C: Dedza Distr., Ciwas Hill, Chongoni, fl. 19.i.1959, *Robson* 1261 (BM; K; MO). S: Zomba Distr., Zomba Plateau, *Buchanan* 246 (K).
Also in Kenya. Seasonally wet shallow soils over rock; 1740–1830 m.

2. **Craterostigma pumilum** Hochst. in Flora **1841**: 67 (1841). —Skan in F.T.A. **4**, 2: 330 (1906). Type from Ethiopia.

Perennial herb with short rhizome, red-orange roots and rosette of leaves. Leaves obovate 2–4 × 1–2 cm., obtuse, subacute or shortly acuminate, densely pubescent beneath closely ciliate along the entire margin, glabrous and smooth above or at most with a few cilia towards the apex; petiole short and broad. Flowers arising from the centre of the rosette on a ± sessile inflorescence; bracts lanceolate, c. 1 cm. long, ciliate. Pedicels 2–3 cm. long, pilose. Calyx 6 mm. long, up to 9 mm. long in fruit, teeth acuminate, pubescent along the median nerves. Corolla c. 15 mm. long, variable from purple or blue to pink wtih white and yellowish throat, upper lip entire, lower lip trilobed, 10 mm. long. Capsule as long as or slightly longer than the calyx, broadly ovoid.

Zambia. C: Lusaka, Stewart Park, 27.xi.1961, *Lasaka Nat. Hist. Club* 69 (K). S: Choma Distr., Mapanza, 1036 m., 23.xi.1958, *Robinson* 2931 (K); Siamambo, 21.xii.1963, *Mutimushi* 523 (K, NDO). **Zimbabwe**. N: Murehwa Distr., Nyadiri R., 5.xii.1968, *Müller & Burrows* 980 (K; SRGH). W: Bulawayo Distr., 20.xii.1920, *Borle* 28 (K; PRE). C: Chegutu (Hartley) Distr., Poole Farm, 7.xii.1962, *Hornby* in GHS 139203 (K; SRGH). S: Mwenezi Distr., Rutenga, 3.ii.1973, *Ngoni* 188 (K; SRGH).
Also in Kenya, Uganda, Somalia and Ethiopia. Open grassland in moist shallow soil overlying rocks; 2000–2600 m.

3. **Craterostigma hirsutum** S. Moore in Journ. Bot. **38**: 461 (1900). —Skan in F.T.A. **4**, 2: 330 (1906). TAB. **21**. Type from Kenya.
    Craterostigma ndassekerense Engl., Bot. Jahrb. **57**: 611 (1922). Type from Tanzania.

Perennial, hairy herb with small rosettes of leaves; roots fibrous; stems very short, covered by fibres or persistent leaf nerves. Leaves oblong-spathulate, 1.5–3(4.5) × 0.5–1(2) cm., entire or undulate, denesly pilose on both sides, parallel 5-nerved. Inflorescence scape 3–5(9) cm. long, pilose, flowers subsessile or with pedicels up to 8 mm. long, in dense apical raceme c. 1 cm. long, sometimes laxer. Bracts ovate to ovate-lanceolate, 4–8 mm. long. Calyx 4–5 mm. long, in flower, 6 mm. in fruit, pilose especially on the nerves, shortly 5-toothed. Corolla c. 9 cm. long, glabrous, white with violet patches on each of three lower lobes, 2 yellow protruberances in throat, upper lip hooded, shortly bilobed. Ovary ovoid-cylindric, apiculate, little longer than the calyx, glabrous.

Zimbabwe. W: Matobo Distr., near Rhodes' grave, 30.xi.1930, *Hill* s.n. (K). **Malawi**. S: Zomba Plateau, Chivunde Peak, 13.ii.1982, *Hepper* 7335 (K).
Also in Tanzania, Kenya and Uganda. Shallow soil over rock or impervious pan where water runs or collects during rains, plants rapidly flower after remaining dormant during dry season; 350–2130 m.

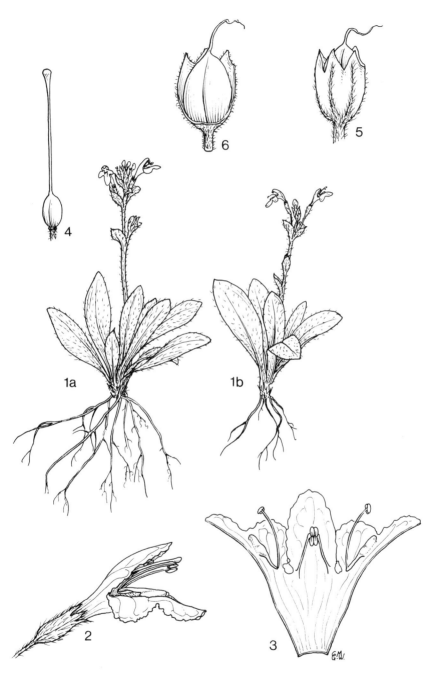

Tab. 21. CRATEROSTIGMA HIRSUTUM. 1a & b, habit (× ⅔), from *Mavi* 143; 2, flower (× 3); 3, corolla opened out showing androecium (× 3); 4, gynoecium (× 3), 2–4 from *Hanger* 175–32; 5, fruit with accrescent calyx (× 3); 6, two calyx lobes removed to show capsule (× 3), 5–6 from *Hill* s.n.

4. **Craterostigma plantagineum** Hochst. in Flora **1841**: 669 (1841). —Skan in F.T.A. **4**, 2: 329 (1906).
Type from Ethiopia.
> *Torenia plantagineum* (Hochst.) Benth. in DC. Prodr. **10**: 411 (excl. syn. *Dunalia acaulis*). Type
> as above.

A rosette perennial with short stout rhizome and numerous orange-yellow roots.
Leaves petiolate; lamina broadly ovate to lanceolate, 1.5–5 × 1–3 cm., entire, broadly
cuneate at base, subacute at apex, glabrous above, densely pubescent beneath, numerous
parallel nerves obscure or indented above, prominent beneath, often purple beneath;
petiole short and broad. Inflorescence 4–8 cm. long, with flowers racemose in upper
third, pubescent; bracts opposite, lanceolate 4–10 mm. long, shorter than the pedicels at
least in fruit. Calyx c. 5 mm. long up to 9 mm. in fruit, shortly toothed, pubescent mainly on
the nerves. Corolla purple and white, up to 18 mm., long, upper lip short bilobed, lower lip
obtusely 3-lobed the central very broad. Capsule ovoid-cylindric, glabrous, slightly longer
than calyx.

Botswana. N: frontier near Plumtree, 900 m., 7.iii.1961, *Richards* 14554 (K). SE: Gaborone Distr.,
Aedume Park, 1050 m., 18.xii.1977, *Hansen* 3306 (C; GAB; K; PRE; SRGH; WAG).
Zambia. B: Seshehe Distr., Machili, 20.xii.1952, *Angus* 974 (FHO; K). N: Mbala Distr., Uningi Pans,
1500 m., 19.xii.1959, *Richards* 12011 (K). **Zimbabwe.** S: Mberengwa Distr., Mt. Buhwa, 20.iv.1973,
Pope 934 (LISC). W: Matopos, iii.1918, *Eyles* 965 (BM). E: Mutare Distr., Kellys Park, 26.xi.1948,
Chase 1561 (BM).
Also in Sudan, Ethiopia, Uganda, Kenya, Tanzania, Angola and S. Africa (Transvaal). Shallow soil
over rock; 900–2200 m.

21. TORENIA L.

Torenia L., Gen. Pl., ed. 5: 270 (1754).

Annual or perennial, erect, ascending or decumbent, glabrous, pubescent or hirsute
herbs; stems quandrangular or subquadrangular. Leaves opposite, sessile or petiolate,
entire or variously toothed. Flowers solitary, axillary, or either in terminal or axillary
well-developed racemes or with few flowers on much shortened branches appearing
somewhat fasciculate. Pedicels ebracteolate. Calyx tubular, 3–5-lobed or bilabiate,
shallowly or deeply 3–5-winged, sometimes appearing plicate. Corolla tubular; tube
cylindric; bilabiate with upper lip bilobed emarginate or subentire, lower lip larger with 3
subequal lobes. Stamens 4, perfect; posterior pair included, filaments filiform; anterior
attached to throat of corolla tube, filaments arched, connivent under upper lip, each with
small appendage at base; anthers usually cohering in pairs. Ovules many. Capsule equal
to or slightly shorter than calyx, septicidal. Seeds small, numerous.

A genus of about 60 pantropical species.
In his treatment of the genus *Craterostigma* for the Flora of Tropical East Africa, F.N. Hepper
decided to restrict the limits of the genus to include the acaulescent species, meanwhile transferring
those caulescent taxa to the closely allied genus *Torenia*. I follow Hepper and include those
caulescent species from our area, hitherto in *Craterostigma*, in a broader concept of *Torenia*. It may be
found at a later date that *Torenia involucrata*, bearing all the characters of an acaulescent
Craterostigma except for the apparently normal basal rosette supported on a leafless stem, may be a
species linking the two groups within *Craterostigma* and could be a future reason for returning them
to that genus.

1. Leaves petiolate; flowers distinctly pedicellate, pedicels 4–22 mm. long   1. *thouarsii*
 – Leaves sessile; flowers very shortly pedicellate, pedicels 1–5 mm. long   -   -   - 2
2. Flowers in lax, spike-like racemes   -   -   -   -   -   -   -   -   -   - 3
 – Flowers in capitate inflorescences   -   -   -   -   -   -   -   -   -   - 5
3. Stems simple, tufted, not markedly quadrangular; basal leaves many, linear; cauline leaves not
    exceeding 0.5 mm. wide   -   -   -   -   -   -   -   -   -   -   - 2. *monroi*
 – Stems branched, not tufted, quadrangular; basal leaves absent; cauline leaves 1.5–20 mm.
    wide   -   -   -   -   -   -   -   -   -   -   -   -   -   - 4
4. Plant erect to 15(25) cm. tall; corolla 4–5 mm. long; calyx 5-lobed, bilabiate   3. *spicata*
 – Plant creeping or ascending to 65 cm. tall; corolla 10.5–13 mm. long; calyx 5-lobed, lobes
    subequal   -   -   -   -   -   -   -   -   -   -   -   -   - 4. *goetzei*
5. Bracts 6–10 mm. wide, broadly ovate to subreniform   -   -   -   -   -   - 6
 – Bracts never more than 2 mm. wide, subulate to ovate lanceolate   -   -   -   - 7

6. Stem leafy throughout, but sparsely so; branched throughout
   or simple   -   -   -   -   -   -   -   -   -   -   5. *latibracteata* subsp. *parviflora*
 - Stem leafless except for basal pair; branched only from base or simple   6. *involucrata*
7. Stout perennial up to 70 cm. tall; corolla 11–12 mm. long, pubescent without   7. *schweinfurthii*
 - Slender annual, rarely exceeding 20 cm. tall; corolla 4–10 mm. long, glabrous without   8
8. Stem very slender; leaves narrowly linear, never exceeding 1.4 mm. wide; calyx lobes longer than
   tube   -   -   -   -   -   -   -   -   -   -   -   -   -   -   -   8. *tenuifolia*
 - Stem slender, leaves ovate-lanceolate, up to 4.5 mm. wide; calyx lobes shorter
   than tube   -   -   -   -   -   -   -   -   -   -   -   -   9. *ledermannii*

1. **Torenia thouarsii** (Cham. & Schlechtend.) Kuntze, Rev. Gen. Pl. **2**: 468 (1891). —Hepper in
   F.W.T.A., ed. 2, **2**: 363 (1963). TAB. **22**, fig. A. Type from Mauritius.
   Nortenia thouarsii Cham. & Schlechtend. in Linnaea **3**: 18 (1828). Type as above.
   Torenia parviflora Buch.-Ham. ex Benth., Scroph. Ind.: 39 (1835). —Benth. in DC., Prodr. **10**:
   410 (1846). —Hook. f., Fl. Brit. Ind. **4**: 278 (1884). —Engl., Pflanzenw. Ost-Afr. **C**: 357 (1895).
   —Hiern, Cat. Afr. Fl. Welw. **1**: 762 (1898). —Engl. in Schlechter, Westafr. Kautsch.-Exped.: 313
   (as *Torenia*) (1901). —Hemsl. & Skan in F.T.A. **4**, 2: 335 (1906) excl. *Lindernia senegalensis* nom.
   illegit., superfl. based on *Nortenia thouarsii*.
   Torenia ramosissima Vatke in Oest. Bot. Zeitschr. **25**: 10 (1875). Type from Zanzibar.

Slender, straggling, procumbent or weakly ascending; stems simple or branched up to
40 cm. or more long, subpilose or often glabrous, quadrangular, sometimes rooting at
nodes. Leaves 12–20(30) × 8–15 mm., broadly lanceolate to ovate-lanceolate, shortly
patent hirsute, acute or obtuse at apex, rounded, subtruncate or shortly cuneate at base,
serrate to serrate-crenate, margin occasionally thickened. Petiole 0.5–3 mm. long, hirsute.
Flowers white or pink, through blue to purple or mauve, solitary axillary or more usually
2–6 together, fascicled or on very small leafless branches. Bracts 1.25–4 mm. long,
narrowly linear. Pedicels 4–22 mm. long, hirsute, somewhat reflexed in fruit. Calyx 5–6
mm. long, up to 10 mm. long at fruiting, sparingly hirsute to subglabrous, narrowly
winged, markedly so at fruiting. Corolla 10–12 mm. long. Capsule 7–9.5 mm. long,
2.5–3.25 mm. in diam., narrowly oblong-ellipsoid. Seeds pale yellow.

Botswana. N: Ng-Gokha (Ngokha) R., downstream from Xaenga, fl. & fr. 2.x.1975, *Smith* 1462
(BR; K; SRGH). **Zambia**. B: 16 km. N. of Senanga, c. 1030 m., fl. & fr. 31.vii.1952, *Codd* 7307 (BM;
K). N: Mbereshi, Luapula R. swamp, 1050 m., fl. & fr. 11.i.1960, *Richards* 12320 (K). W: Mwinilunga
Distr., Matonchi R. banks, fl. & fr. 23.x.1937, *Milne-Redhead* 2917 (K). **Zimbabwe**. E: Chipinge Distr.,
Musirizwe R., road to Mt. Selinda, 500 m., fl. & fr. 28.i.1975, *Gibbs-Russell* 2629 (BR; K; SRGH).
S: Chiredzi Distr., Gonarezhou, Save-Runde Rivers junction, Save R., fl. 31.v.1971, *Grosvenor* 580
(SRGH). **Malawi**. N: Mzimba Distr., Mzuzu, Marymount, c. 1400 m., fl. 14.ii.1974, *Pawek* 8109
(SRGH). S: Mulanje, fl. & fr. 24.ix.1932, *Young* 881 (BM). **Mozambique**. N: Moginqual, fl.
15.xi.1936, *Torre* 1035 (LISC). Z: Mocuba, Namagoa, 200 km. inland from Quelimane (Quilimane),
60–120 m., fl. ix.1943, *Faulkner* 86 (K). MS: Beira, 15 m., fl. 25.xii.1906, *Swynnerton* 1979
(BM). M: Inhaca Island, 36 km. E. of Maputo (Lourenco Marques), fl. 17.vii.1959, *Mogg* 29400 (K;
SRGH).
Widespread from West Africa, through Central and East Africa, Sudan, Angola and the Mascarene
Islands. In the New World, it is recorded from Costa Rica and Trinidad and through South America
from Venezuela to Bolivia and southern Brazil. Grasslands, marshes, swamps, lake- and riversides;
from c. sea-level to 1750 m.

2. **Torenia monroi** (S. Moore) Philcox in Bol. Soc. Brot., sér. 2, **60**: 267 (1987). Type: Zimbabwe,
   Masvingo (Victoria), 1909, *Monro* 788 (BM, holotype).
   Craterostigma monroi S. Moore in Journ. Bot. **57**: 214 (1919). Type as above.

Erect annual herb, 3.5–13 cm. tall, fibrous rooted; stems simple, glabrous, tufted, not
markedly quadrangular. Leaves: basal, alternate, crowded, 3–10(15) × c. 1 mm. above, to
2.5 mm. below at sheathing base, linear, subsucculent; cauline 5–15(20) × 0.2–0.5 mm.,
opposite, narrowly linear, acute, somewhat involute, especially so at base, appearing
sheathed, somewhat amplexicaul. Flowers few in axils of upper leaves, pedicillate.
Pedicels 5–10 mm. long, slender. Calyx 2.25–2.75(2.9) mm. long, glabrous, main nerves
slightly prominent, not winged; lobes 0.6–0.8(1) mm. long, deltoid, obtuse or acute.
Corolla pale mauve, pale pink or white with yellow palate, 10–11 mm. long,
infundibuliform, glabrous without, upper lip up to 5 mm. long, ovate, bifid, lower lip c. 3
mm. long, subobovate. Anterior filaments c. 4 mm. long, appendage at base c. 1 mm. long,
0.7 mm. broad, oblong-ovate to subreniform in outline, minutely, densely papillose.
Capsule 3.5 × 2 mm., ellipsoid.

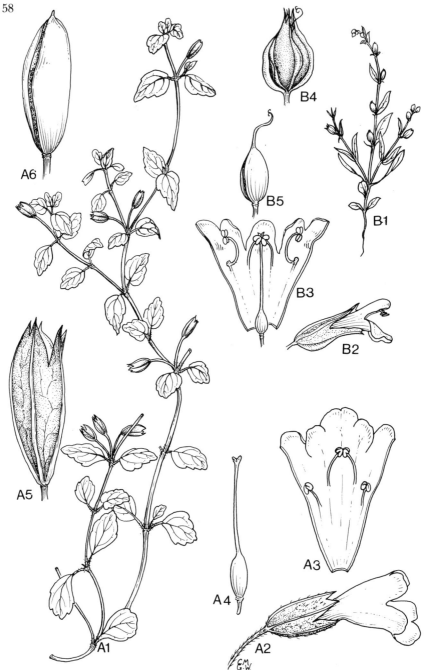

Tab. 22. A.—TORENIA THOUARSII. A1, fruiting branch (× ⅔), from *Goldsmith* 82/69; A2, flower (× 4); A3, corolla opened showing androecium (× 4); A4, gynoecium × 4); A2–4 from *Smith* 1462; A5, fruiting calyx (× 4); A6, fruit with calyx removed (× 4), A5–6 from *Goldsmith* 82/69. B.—TORENIA SPICATA. B1, habit (× ⅔); B2, flower (× 4); B3, corolla opened showing androecium and gynoecium (× 4); B4, fruiting calyx (× 4); B5, fruit with calyx removed (× 4), B1–5 from *Philcox & Drummond* 9051.

Zimbabwe. N: Mutoko Distr., Nyamahere Hill, fl. & fr. 14.iii.1978, *Pope* 1662 (K; SRGH). W: Bulilima Mangwe Distr., top of Mt. Jim, fl. 10.iv.1974, *Ngoni* 365A (K; SRGH). C: Gweru (Gwelo) Distr., Gwenoro Dam, 1370 m., fl. & fr. 5.ii.1967, *Biegel* 1886 (K; LISC; SRGH). E: Mutare Distr., Sabi Drift, 820 m., fl. & fr. 1.xii.1954, *Wild* 4664 (BR; K; SRGH).

Not known from elsewhere. Rooted in ephemeral rock pools; 800–1400 m.

3. **Torenia spicata** Engl., Bot. Jahrb. **23**: 502, t. 7, G–M (1897). —Hemsl. & Skan in F.T.A. **4**, 2: 334 (1906). —Hepper in F.W.T.A. ed. 2, **2**: 363 (1963). TAB. **22**, fig. B. Type from the Sudan.
    Torenia inaequalifolia Engl., Bot. Jahrb. **23**: 502 (1897). —Hiern, Cat. Afr. Pl. Welw. **1**: 762. Type from Angola.
    Canscora ramosissima Baker in Bull. Misc. Inf., Kew **1898**: 158 (1898). Type: Malawi, Chitipa (Fort Hill), *Whyte* s.n. (BM; K, holotype).

Erect herb, 5–15(25) cm. tall, simple to more usually branched; stems quadrangular, 4-winged, glabrous with few small glands at or near nodes. Leaves sessile (5)10–25(30) × (1.5)2–8 mm., lanceolate to elliptic-lanceolate, obtuse or acute, entire to remotely, minutely serrate, those of upper pairs often very unequal. Flowers blue to occasionally purple or white, solitary, axillary below becoming subracemose above, very shortly pedicellate. Pedicels 0.5–1 mm. long, occasionally minutely stipitate-glandular. Calyx 3.5–4.5 mm. long, becoming 5–6 mm. long in fruit, bilabiate, 5-winged; upper lip trilobed, lower bilabiate. Corolla 4–5 mm. long, lobes rounded. Capsule (2.5)3.5–4 × (1.8)2–2.5 mm., cylindric-ovoid. Seeds reddish.

Botswana. N: Gomoti R., fl. & fr. 30.iii.1975, *Smith* 1335 (SRGH). SE: Gaborone Distr., Aedume Park, 1050 m., fl. & fr. 1.v.1978, *Hansen* 3417 (K; SRGH). **Zambia**. B: Sioma Falls, 76 km. S. of Senanga, 1050 m., fl. & fr. 1.ii.1975, *Brummitt et al.* 14219 (BR; K; SRGH). N: Kundabwika Falls, near Lake Mwene, c. 1200 m., fl. & fr. v.1931, *Walter* 7 (K). W: 45 km. E. of Mwinilunga, fl. & fr. 15.iv.1960, *Robinson* 3565 (K). C: 55 km. NE. of Serenje, fl. & fr. 2.iii.1962, *Robinson* 4978 (K). E: c. 27 km. from Chipata on Chadiza road, fl. & fr. 25.ii.1971, *Anton-Smith* 828 (SRGH). S: Mazabuka Distr., c. 6 km. from Chirundu Bridge on Lusaka road, fl. & fr. 6.ii.1958, *Drummond* 5500 (SRGH). **Zimbabwe**. N: Goromonzi Distr., Chinamora Res., Ngomakurira, fl. & fr. 14.iii.1965, *Loveridge* 1375 (SRGH). W: Hwange Distr., Kazuma Range, fl. & fr. 9.v.1972, *Gibbs-Russell* 1888 (K; LISC; SRGH). C: Chegutu Distr., Poole Farm, fl. & fr. 3.iii.1948, *Hornby* 2897 (K; SRGH). E: Mutare Distr., Odzani R. Valley, fl. & fr. 1915, *Teague* 456 (K). S: Bikita Distr., Turgwe-Dafana R. confluence, fl. & fr. 5.v.1969, *Biegel* 3026A (SRGH). **Malawi**. N: Mzimba Distr., Katoto, c. 5 km. SW. of Mzuzu, 1370 m., fl. & fr. 17.iv.1973, *Pawek* 6548 (K; P; SRGH). C: Kasungu Game Reserve, fl. & fr. 20.iii.1970, *Hall-Martin* 578 (SRGH). S: Ntcheu Distr., Msasa Escarpment, Dedza-Golomoti road, 1250 m., fl. 19.iii.1955, *Exell et al.* 1022 (BM; LISC; SRGH). **Mozambique**. N: Nampula, fl. 7.v.1937, *Torre* 1405 (LISC).

Known also from West Africa, Zaire, Sudan, Angola and S. Africa (Transvaal). Grasslands, or more rarely semi-open woodlands on well-drained sandy or gravelly soils; up to 1650 m.

4. **Torenia goetzei** (Engl.) Hepper in Bol. Soc. Brot., sér. 2, **60**: 271 (1987). Type from Tanzania.
    Craterostigma goetzei Engl., Bot. Jahrb. **28**: 477 (1900). —Skan in F.T.A. **4**, 2: 332 (1906). Type as above.
    Torenia brevifolia Engl. & Pilger in Engl., Bot. Jahrb. **45**: 214 (1910). Type from Tanzania.

Straggling, creeping or ascending perennial herb, branched from base; stems 10–65 cm. long, quadrangular, glabrous throughout, lower internodes short, lengthening above. Leaves 5–17(28) × 3–15(20) mm., opposite, sessile, 3–5-nerved, ovate to ovate-oblong, obtuse, subcordate at base, subentire, crenate or shallowly obscurely toothed, glabrous, pellucid-punctate beneath. Flowers few in lax, spike-like raceme, in axil of one of opposite pair of bracts, pedicillate. Bracts 1–2.5 × c. 0.5 mm., narrowly triangular, acute, shortly hispid-ciliate, slightly pubescent at base. Pedicels 1–2 mm. long. Calyx 5(7.5) mm. long, tube glabrous, markedly winged along major nerves, intermediate nerves more or less prominent; lobes (0.8)1.25–1.5 mm. long, broadly triangular, shortly ciliate-pubescent. Corolla blue, violet or purple, 10.5–13(16.5) mm. long, shortly glandular-pilose without, upper lip broadly ovate, subentire, lower lip clearly trilobed, lobes 5–6 × c. 4 mm., obovate; throat and palate densely covered with brown hairs. Anterior filaments c. 6 mm. long, appendages at base 1–1.5 mm. long, clavate. Capsule 5 × 2–2.5 mm., ovoid with persistent style.

Zambia. N: Mbala Distr., Zombe plain, Sumbawanga road, 1500 m., fl. & fr. 29.xii.1964, *Richards* 19402 (BR; K; LISC). C: E. of Chiwefwe, fl. 15.vii.1930, *Hutchinson & Gillett* 3694 (K). E: Lundazi Distr., Nyika Plateau, near Govt. Rest House, 2100 m., fl. & fr. 2.i.1959, *Richards* 10405 (BR; K). **Malawi**. N: Nyika Plateau, Lake Kaulime, 2200 m., fl. & fr. 23.x.1958, *Robson & Angus* 281 (BM; K; LISC; SRGH). C: Lilongwe Distr., Dzalanyama For. Res., above Chiunjiza road, c. 5 km. SE. of Chaulongwe Falls, 1260 m., fl. & fr. 22.iii.1970, *Brummitt* 9296 (K; SRGH).

Also known from Zaire, Burundi and Tanzania. Marshes, bogs and streamsides; from 670–2255 m.

5. **Torenia latibracteata** (Skan) Hepper in Bol. Soc. Brot., sér. 2, **60**: 271 (1987).

Subsp. **parviflora** Philcox in Bol. Soc. Brot., sér. 2, **60**: 267 (1987). Type: Zambia, Kawimbe, 1524 m., 20.v.1955, *Richards* 5792 (K, holotype).

Erect herb, 10–36 cm. tall; stem simple or branched, quadrangular, furrowed, pilose in varying densities to glabrous especially above. Leaves similar throughout, 10–20(27) × 2–8(12) mm., opposite, sessile, ovate to ovate-lanceolate, obtuse, shallowly dentate to subentire, hispid-pilose especially on nerves beneath, to subglabrous, shortly ciliate. Flowers few, shortly pedicellate in terminal, capitate inflorescences. Pedicels c. 0.5 mm. long, stout. Bracts up to 7 × 6-8 mm., broadly ovate, subcircular or subreniform, obtuse, entire to occasionally toothed, ciliate. Calyx c. 4.5–5.5 mm. long, densely to sparsely white hairy, major nerves prominent; lobes c. 2 mm. long, triangular, acute, broad at base, ciliate. Corolla blue to purple, 7–8 mm. long, minutely pubescent without, tube c. 4 mm. long; upper lip 3 mm. long, slightly emarginate, lower lip 3-lobed, large brown clavate hairs on palate. Anterior filaments 2–2.5 mm. long, appendage at base 0.35–0.5 mm. long, clavate, minutely papillose. Capsule 3 × 1.75 mm., cylindric-ellipsoid.

Zambia. N: Mbala Distr., Nkali Dambo, 1740 m., fl. & fr. 25.v.1967, *Richards* 22268 (K). **Malawi**. S: Likabula, below Mulanje Mt., 720 m., fl. 15.vi.1962, *Richards* 16699 (K). **Mozambique**. N: Lichinga (Vila Cabral), fl. & fr. x.1933, *Torre* 29 (COI). Z: Mocuba Distr., Namagoa, fl. & fr. 18.v.1948, *Faulkner, Kew* 270 (K).
Known only from the Flora Zambesiaca area. Damp swampy ground; 720–1750 m.
Material as described here differs somewhat from the concept of Skan when he described *Craterostigma latibracteatum* from Zaire. His material and that collected subsequently from that area, shows a plant with a more glabrous stem, smaller cauline leaves and much larger flowers. In all other respects I consider our material shows no further differences and, from the specimens available, am giving it subspecific rank.

6. **Torenia involucrata** Philcox in Bol. Soc. Brot., sér. 2, **60**: 267 (1987). Type: Zambia, Mwinilunga Distr., 25 km. N. of Kabompo Gorge, fl. & fr. 19.iv.1965, *Robinson* 6657 (K, holotype; SRGH, isotype).

Erect annual, 4–11 cm. tall; stems simple or branched from base, glabrous to subglabrous above, becoming sparsely white hirsute below, leafless except for basal pair. Leaves sessile 5–10(20) × 1–9 mm., ovate-lanceolate to obovate, entire to somewhat sinuate, obtuse, narrowed at base, glabrous above, hirsute on nerves beneath. Flowers many, subsessile in globular, large-bracted heads. Bracts several, completely surrounding inflorescence, leaflike up to 25 × 10 mm., broadly ovate to ovate-lanceolate, somewhat amplexicaul, strongly 3–5-nerved, entire to occasionally sparsely toothed. Pedicels up to 0.5 mm. long, stout. Calyx 7.5 mm. long, shaggy pilose, ribs very prominent; lobes 3 mm. long, narrowly lanceolate. Corolla bright blue, 5.5 mm. long; upper lip 3.5 mm. long, bifid. Anterior filaments 3–3.5 mm. long, with appendage at base 0.5 mm. long, clavate. Capsule 5.5 × 2.5–3 mm., ovoid, bearing persistent style.

Zambia. W: Solwezi Distr., Chifubwa R. gorge, 3 km. S. of Solwezi, fl. & fr. 20.iii.1961, *Drummond & Rutherford-Smith* 7138 (K; SRGH).
Not known from elsewhere. Woodlands, roadsides and stony areas in and bordering gorges.

7. **Torenia schweinfurthii** Oliv. in Hook., Ic. Pl. t.1251 (1878). Type from Sudan.
    Craterostigma schweinfurthii (Oliv.) Engl., Bot. Jahrb. **23**: 501 (1897). —Hiern, Cat. Afr. Pl. Welw, **1**: 762 (1898). —Skan in F.T.A. **4**, 2: 332 (1906). —Hepper in F.W.T.A. ed. 2, **2**: 361 (1963). Type as above.

Erect perennial, 20–70 cm. tall; stems simple or sparingly branched, quadrangular, glabrous throughout or occasionally pilose at or towards base. Leaves opposite, sessile, 5–33 × 1–10 mm., sessile; lower elliptic or elliptic lanceolate, obtuse, 3-nerved, entire or sparsely short-toothed, scabrid-ciliate; upper becoming smaller, linear-lanceolate, acute. Inflorescence terminal, capitate, capitula c. 1–1.5 cm. in diam., flowers shortly pedicillate. Bracts (2.5)3.5–5 × 0.25–0.5 mm., subulate to narrowly lanceolate, shorter than calyx, minutely ciliate. Pedicels 1–1.5 mm. long. Calyx 4.5–7 mm. long, glabrous, shallowly winged on major nerves, wings glabrous to shortly ciliate, lobes 1–1.5(2) mm. long, triangular, acute, minutely ciliate on somewhat thickened margin. Corolla 11–12 mm. long, minutely pubescent without, tube 5–7 mm. long, upper lip broadly oblong-ovate, slightly emarginate, lower lip trilobed, lobes appearing crenate, few brown clavate hairs on palate. Anterior filaments 3.5(5) mm. long, somewhat gibbous at base, appendage 0.3–0.5 mm. long, blunt, hardly clavate; anthers connivent. Capsule c. 4.5 × 3.5 mm., broadly ovate, persistent base of style present.

Zambia. N: Mansa Distr., S. end of Lake Bangweulu, fl. 12.ii.1959, *Watmough* 239 (K; LISC; SRGH). W: Mufulira, 1220 m., fl. & fr. 8.vi.1934, *Eyles* 8039 (BM; K; SRGH). C: Serenje Distr., Mulembo R. on Mukaru road, 1230 m., fl. & fr. 6.v.1972, *Kornaś* 1706 (K). **Mozambique**. N: Murrupula, fl. & fr. 19.i.1961, *Carvalho* 449 (K).

Also known from West Africa, Sudan, Tanzania and Angola. Marshes, bogs and swamps; 1050–1525 m.

8. **Torenia tenuifolia** Philcox in Bol. Soc. Brot., sér. 2, **60**: 267 (1987). Type: Zambia, Mbala Distr., Chilongowelo, 1525 m., 15.v.1952, *Richards* 1690 (K, holotype).

Erect, very slender annual herb, 8–35 cm. tall; stems quadrangular, simple or branched, glabrous. Leaves opposite, 4–11(35) × 0.15–1(1.4) mm., narrowly linear, acute, entire or rarely with few minute teeth, glabrous with few minute hairs at point of insertion at stem. Flowers terminal in axils of bract-like, reduced leaves, appearing capitate, pedicellate. Bracts 1.5–3 mm. long, up to c. 0.5 mm. wide, linear to linear-lanceolate, glabrous. Pedicels c. 1.5 mm. long. Calyx 3.5–4(5) mm. long, weakly nerved with major nerves only slightly prominent, subglabrous to minutely stipitate-glandular; lobes (2)2.25–3 mm. long, narrowly lanceolate, much longer than tube, sinus between lobes very broad, rounded, especially in fruit. Corolla blue or mauve, 4–5(13) mm. long, upper lip rounded, entire; lower lip trilobed, lobes subcircular, few pale brown, clavate hairs on palate. Anterior filaments to 4.5 mm. long with appendage at base 0.5–0.65 mm. long, clavate with papillose head c. 0.25 mm. in diam. Capsule c. 4 × 3.5 mm., compressed subglobular.

Zambia. N: Kansanshi Dambo, 55 km. ESE. of Mporokoso, fl. & fr. 20.v.1961, *Robinson* 4665 (K). Only known from northern Zambia. Bogs, swamps and marshy ground; 1250–1550 m.

9. **Torenia ledermannii** Hepper in Bol. Soc. Brot., sér. 2, **60**: 271 (1987). Type from Cameroon.
   Craterostigma gracile Pilger in Engl., Bot. Jahrb. **45**: 213 (1910). —Raynal in Adansonia, sér. 2, **6**: 431, pl. (1966) non *Torenia gracilis* Benth. Type as above.

Erect, slender annual or perennial herb, 5–20(35) mm. tall; stems quadrangular, simple or branched, subglabrous above becoming densely to subdensely minutely white pilose below. Leaves opposite, 2–8 cauline pairs, 2–10(16) × 0.5–2(4.5) mm., ovate-lanceolate, obtuse, subcuneate at base, glabrous, margin somewhat thickened especially towards apex, shortly hispid-ciliate, minutely glandular-punctate. Flowers few, terminal, in axils of alternate, bract-like, reduced leaves, appearing subcapitate, shortly pedicellate. Pedicels c. 1 mm. long or less. Bracts 3–6 mm. long, 1–2 mm. wide, ovate-lanceolate, pilose. Calyx 2.2–6 mm. long, shortly stipitate-glandular or white pilose to subglabrous, 5-nerved with nerves occasionally shallowly raised or slightly winged; lobes 0.8–1.5 mm. long, deltate at base, acuminate, shortly ciliolate, markedly shorter than tube. Corolla purple or violet with white markings, 7–10.5 mm. long, upper lip subtruncate, obscurely, shallowly emarginate, glabrous without; lower lip trilobed, lobes rounded. Anterior filaments 2–3 mm. long with appendage at base c. 1 mm. long, clavate with globular head, papillose c. 0.35 mm. in diam. Capsule up to 3.25 × 1.5 mm., broadly ellipsoid-ovoid.

Zambia. N: Kawambwa, c. 1350 m., fl. & fr. 21.vi.1957, *Robinson* 2330 (K). W: Mwinilunga Distr., Kalenda Dambo, fl. & fr. 8.x.1937, *Milne-Redhead* 2655 (K).

Known also from West Africa and Tanzania. Riverbanks and moist, peaty grasslands; 300–1350 m.

22. LINDERNIA Allioni

Lindernia Allioni, Misc. Taur. **3**: 178, t. 5, fig. 1 (1762?).

Annual or perennial herbs, glabrous, pubescent or subglandular. Stems slender, erect, prostrate or creeping and rooting at nodes, simple or more usually branched, obscurely to distinctly quadrangular. Leaves opposite to almost opposite, simple, sessile or petiolate, entire, crenate or toothed, pinnately nerved or 3–5-nerved with nerves arising at base of lamina. Flowers solitary, axillary or terminal, or in terminal or axillary racemes or clusters, small, pedicellate or occasionally subsessile. Calyx 5-lobed, shallowly lobed with lobes spreading or somewhat connivent when mature, or lobed almost to base **with** lobes lanceolate to linear-lanceolate; tube 5-nerved, each nerve with obscure or distinct rib or minute wing. Corolla tubular, bilabiate; anterior lip trilobed, lobes usually spreading; posterior lip either entire or emarginate to bilobed, suberect. Stamens either 4, all antheriferous and fertile, or 2, with posterior pair fertile and anterior pair reduced to staminodes, posterior pairs affixed to corolla tube, anterior pair or staminodes arising in throat; frequently anterior filaments each with distinct spur arising at or near base; anthers free or contiguous, cells divaricate. Stigma bilamellate; style slender, erect. Capsule globose, ovoid, obovoid, ellipsoid, or cylindric to narrowly so, bivalved. Seeds numerous, smooth to alveolate, variously shaped.

A genus of about 120 species from the tropics and sub-tropics and warm temperate regions.

1. Stamens 4, antheriferous, all fertile   -   -   -   -   -   -   -   -   -   -   2
– Stamens 2, fertile, staminodes 2   -   -   -   -   -   -   -   -   -   -   9
2. Anterior filaments distinctly clavate at or near base   -   -   -   -   -   3
– Anterior filaments bent or geniculate, not clavate   -   -   -   -   -   -   7
3. Plant decumbent; stems up to 60 cm. long, glabrous; capsule c. 6.5 mm. long, narrowly cylindric-ellipsoid   -   -   -   -   -   -   -   -   -   -   -   -   1. *whytei*
– Plant erect to spreading; stems up to 18 cm. tall, glandular-hispid to pubescent or glabrous; capsule 3–4 mm. long, not narrowly cylindric-ellipsoid   -   -   -   -   -   4
4. Stems much branched   -   -   -   -   -   -   -   -   -   -   -   -   5
– Stems rarely branched more than once, more usually simple   -   -   -   3. *bifolia*
5. Leaves 8–18 × 3.5–14 mm.; calyx divided below midway, tube not strongly ridged or shallowly winged   -   -   -   -   -   -   -   -   -   -   -   -   -   -   -   6
– Cauline leaves 5–12 × 0.3–1.75 mm.; calyx not divided to midway or below, tube strongly 5-nerved with nerves pronounced into shallow wings   -   -   -   -   -   8. *damblonii*
6. Plant procumbent, straggling; leaves subreniform, circular or broadly ovate; pedicels 7–15 mm. long   -   -   -   -   -   -   -   -   -   -   -   -   7. *subreniformis*
– Plant erect to spreading; leaves elliptic-oblong to ovate; pedicels 0.5–3 mm. long   2. *insularis*
7. Plant generally large-leaved; leaves spreading, 5–24 mm. wide, ovate   -   -   -   8
– Plant narrow-leaved; leaves 1–2.5 mm. wide, linear to linear-lanceolate   4. *oliveriana*
8. Stems decumbent or prostrate, up to 40 cm. long; leaves 23–32 mm. long, broadly ovate; calyx 8–9 mm. long; capsule c. 15 mm. long   -   -   -   -   -   -   -   5. *stictantha*
– Stems erect, up to 14 cm. tall; leaves smaller than above, ovate to subcircular; calyx 2.5–3.5 mm. long; capsule 4.5–6 mm. long   -   -   -   -   -   -   6. *nummularifolia*
9. Calyx lobed almost to base   -   -   -   -   -   -   -   -   -   -   -   10
– Calyx shortly lobed, never more than midway   -   -   -   -   -   -   -   11
10. Plant aquatic; stems naked or nearly so below apex   -   -   -   -   -   9. *conferta*
– Plant terrestrial; stems leafy throughout   -   -   -   -   -   -   10. *parviflora*
11. Flowers subsessile or very shortly pedicellate with pedicels not exceeding 1.5 mm.   -   -   -   -   -   -   -   -   -   -   -   11. *nana*
– Flowers markedly pedicellate with pedicels up to 20 mm. long   -   -   -   -   12
12. Plant much branched   -   -   -   -   -   -   -   -   -   -   -   -   13
– Plant simple-stemmed or branched occasionally from or towards base   -   -   14
13. Plant coarse; stems minutely short white hispid, densely branched from or towards base; valves of capsule erect after dehiscence   -   -   -   -   -   -   -   12. *pulchella*
– Plant more slender; stems minutely glandular-pubescent; valves of capsule markedly divergent after dehiscence   -   -   -   -   -   -   -   -   -   -   13. *exilis*
14. Basal leaves many, clustered or rosette-like; pedicels erect to spreading in fruit   14. *wilmsii*
– Basal leaves in one or two pairs, not clustered or rosetted; pedicels deflexed in fruit   -   -   -   -   -   -   -   -   -   -   -   15. *schweinfurthii*

1. **Lindernia whytei** Skan in F.T.A. **4**, 2: 340 (1906). Type from Uganda.
    Lindernia gossweileri S. Moore in Journ. Bot. **45**: 87 (1907). Type from Angola.
    Lindernia flava S. Moore in Journ. Linn. Soc., Bot. **40**: 153 (1911). Types: Mozambique, Mt. Maruma, 1065 m., *Swynnerton* 1922 (BM; K, syntypes); Zimbabwe, Chirinda, 1125 m., *Swynnerton* 1966 (BM; K, syntypes).

Decumbent or ascending annual; stems up to c. 60 cm. long, simple or branched, quadrangular, glabrous. Leaves sessile, 8–22 × 8–18 mm., opposite, sessile, broadly ovate, subcircular to ovate-lanceolate, obtuse to shortly apiculate at apex, rounded to cordate at base, dentate, glabrous or with few hairs on margin or major nerves beneath. Flowers terminally racemose, lax, pedicellate. Bracts smaller than leaves, ovate below becoming linear to subulate above. Pedicels 1.75–3 mm. long. Calyx 6–7 mm. long, 5-ribbed, 5-lobed, glabrous, or ciliate on ribs; lobes 3.5–5 mm. long, linear, acuminate, spreading, keeled, glabrous or ciliate. Corolla blue, lilac or purple, or yellow, 8–12 mm. long; tube 5–6 mm.

long, infundibuliform, minutely glandular-pubescent without; upper lip ovate, slightly emarginate, ciliate, pubescent without; lower lip trilobed. Anterior filaments 4–5(7) mm. long, arched with broadly clavate, glandular-papillose spur c. 1 mm. or more long produced above base; base of filament extended into pronounced antero-lateral ridge. Capsule c. 6.5 × 2 mm., narrowly oblong-ellipsoid.

Zimbabwe. E: Chimanimani (Melsetter) Distr., Tarka Forest Res., banks of Chisengu R., 1100 m., fl. & fr. v.1968, *Goldsmith 72/68* (K; P). **Mozambique**. Z: serra do Gúruè, E. of Picos Namuli, near source of R. Malema, c. 1800 m., fl. & fr. 8.xi.1906, *Swynnerton 1922* (BM; K, syntypes).

Also known from Kenya, Uganda, Tanzania, Rwanda and Angola. Streamsides and wet places in evergreen forest; up to 1150 m. in the Flora Zambesiaca area.

In his treatment of the genus *Lindernia* for Flora of Tropical Africa, Skan gave the corolla colour of *L. whytei* as blue, but mentioned in his discussion a Gossweiler collection from Angola which resembles this species and having yellow flowers. Unfortunately I have seen only two collections from the Flora Zambesiaca area and one, *Walters 2788*, lacks any descriptive notes while the other, *Goldsmith 72/68*, cites the flowers as being yellow. Although all of the East African material I have consulted has flowers of various shades of blue, I do not hesitate to widen the range of colour in my description to include both extremes.

This widening of the colour range is made even more valid by my inclusion of *L. flava* S. Moore under this name. This plant was described as being close to *L. gossweileri* and differing only in the size of the corolla, and the calyx lobes. As these characters are not constant in the large amount of material I have studied from East Africa, I am including both species under the name, *L. whytei*.

2. **Lindernia insularis** Skan in F.T.A. **4**, 2: 342 (1906). Type from Uganda.

Erect to spreading annual, (3)6–14(18) cm. tall, simple or branched; stems leafy below or towards base, quadrangular, minutely glandular to subglabrous; branches opposite or less frequently alternate. Leaves opposite, sessile, or lowermost with short petiole-like base, 8–18(25) × 3.5–13(18) mm., broadly ovate or elliptic-oblong, obtuse or rounded, rounded at base or lowermost subcuneate, shortly dentate or lower subentire, sparsely glandular-pubescent to subglabrous. Flowers in lax, terminal or lateral leafless racemes, pedicellate, bracteate. Bracts similar to leaves but smaller, becoming more so above, each member of pair of different size with smaller subtending flower. Pedicels 0.5–3 mm. long. Calyx 4–5 mm. long, minutely glandular-pubescent; lobes 3–4 mm. long, narrowly linear, acuminate widely spreading. Corolla yellow, orange-yellow or white, upper lip also recorded as purple [*King 300*]; upper lip oblong-ovate, lobes of lower lip with large, yellow clavate hairs c. 0.15 mm. long at throat, rounded. Anterior filaments 4–7 mm. long with papillose appendage c. 1 mm. long produced at base, antero-lateral ridge not evident. Capsule (2.5)3–4 × 3–4 mm., globose or globose-obovoid, shorter than calyx, glabrous.

Zambia. N: Mbala Distr., Lake Chila, 1525 m., fl. & fr. 9.iii.1952, *Richards 1040* (K). W: Solwezi Distr., Mulenga Protected Forest Area, c. 20 km. NW. of Kansanshi, fl. & fr. 19.iii.1961, *Drummond & Rutherford-Smith 7061* (K). C: 8.5 km. E. of Lusaka, 1280 m., fl. & fr. 6.ii.1956, *King 300* (K). S: Choma, Mapanza, 1065 m., fl. & fr. 8.iii.1958, *Robinson 2780* (K; SRGH). **Zimbabwe**. N: Mwami (Miami), fl. iv.1926, *Rand 61 & 62* (BM). **Malawi**. N: Mzimba Distr., Viphya Plateau, c. 58 km. SW. of Mzuzu, 1680 m., fl. & fr. 25.iii.1977, *Pawek 12521* (K). C: Lilongwe Distr., Dzalanyama For. Res., 1280 m., fl. & fr. 26.iii.1977, *Brummitt, Seyani & Patel 14938* (K). S: Kasupe Distr., between Kasupe and Liwonde, 685 m., fl. 20.iii.1977, *Brummitt, Seyani & Dudley 14897* (K). **Mozambique**. N: Lichinga (Vila Cabral), c. 80 km. from Missiao de Massangulo, c. 1300 m., fl. & fr. 26.i.1964, *Correia 149* (LISC).

Also known from Burundi, Kenya, Uganda and Tanzania. Open grassland and woodlands, on wet sandy or peaty soils; up to 1700 m.

3. **Lindernia bifolia** Skan in F.T.A. **4**, 2: 343 (1906). Type: Zambia, Kambole, c. 1525 m., *Nutt s.n.* in 1896 (K, holotype).

Erect, slender herb, (2.5)6–12(18) cm. tall, branched from base or at times above; stems quadrangular, very sparsely short spreading hispid, frequently mixed with short stipitate glands. Leaves sessile, basal in one or two pairs, 5–20 × 3–9 mm., broadly ovate to elliptic-ovate, thin, obtuse or acute at apex, rounded at base, short pubescent to subglabrous, obscurely few-toothed; one or two pairs of cauline leaves occasionally present, similar to basal but smaller. Flowers few, mostly alternate in terminal racemes, bracteate. Bracts 0.6–1.2(2.5) mm. long, subulate to linear-lanceolate. Pedicels 1–2.5 mm. long, slender. Calyx 3.25–5.25(6.25) mm. long, shortly hispid to glandular or subglabrous; lobes 2–4(5) mm. long, narrowly lanceolate. Corolla pale yellow to white, upper lip oblong-ovate, lateral lobes of lower lip elliptic, median lobe subcircular. Anterior filaments 4–5.5 mm. long with clavate appendage c. 0.8–1 mm. long produced at base;

antero-lateral ridge not evident. Capsule c. 3.5 × 1.5–2.5 mm., broadly ellipsoid to ellipsoid-globose, apiculate.

Zambia. N: Mbala Distr., top of Kalambo Falls, c. 1120 m., fl. & fr. 11.iv.1961, *Phipps & Vesey-FitzGerald* 3080 (K). W: 2 km. S. of Solwezi, fl. 20.iii.1961, *Drummond & Rutherford-Smith* 7107 (BR; K; P).
Known only from the Flora Zambesiaca area. Wet areas; up to 1500 m.
This plant is differentiated here from *L. insularis* only with difficulty. The specimens assigned by me under *L. bifolia* are very slender plants mostly with simple stems and, except for those subtending the lowest flower of the inflorescence, having equal-sized pairs of subulate bracts. In those plants under *L. insularis*, this character only becomes apparent when applied to the uppermost flowers of the inflorescence. In other respects including the shape and relative size of the filament appendages and the large clavate hairs at the throat of the corolla, both species are otherwise similar. At a later date it may be found better to combine them, but for the purpose of this work, I am keeping them as separate though doubtfully distinct.
Occasionally, as on some specimens from the collections of *Phipps & Vesey-FitzGerald* 3080 and *Richards* 4284, a pair of very small, long-petiolate leaves are present below the accepted basal leaves and at ground level. This is not constant in all collections and these leaves may either be early deciduous or lost during collecting.

4. **Lindernia oliveriana** Dandy in F.W. Andrews, Fl. Pl. Sudan **3**: 139 (1956). Type from Uganda.
   Vandellia lobelioides Oliv. in Trans. Linn. Soc. Lond. **29**: 120, t. 121B (1875) non F. Muell. (1858). Type from Uganda.
   Lindernia lobelioides (Oliv.) Wettst. in Engl., Pflanzenw. Ost-Afr. **C**: 357 (1895). —Skan in F.T.A. **4**, 2: 340 (1906). Type as above.

Erect or ascending herb, (6)15–35 cm. tall, annual, subglabrous, simple or branched usually at or towards base. Leaves sessile, (7)12–15(40) × (0.3)1–2.5(4) mm., opposite, linear to linear-lanceolate, acute, entire, glabrous except for few marginal hairs occasionally on young leaves. Flowers solitary, axillary, alternate. Pedicels 5–17 mm. long in flower, slender, suberect, up to 40 mm. long in fruit, stouter, spreading to reflexed. Calyx (3)5–7 mm. long, 5-lobed, slightly ridged; lobes (1)1.5–2.5 mm. long, lanceolate, acute, glabrous or shortly hispid on margins or central area. Corolla (6)10–13 mm. long, bright or violet-blue to mauve or brownish, with yellow or white tube; upper lip up to 8 × 6 mm., broadly oblong, lower lip trilobed, lobes rounded. Anterior filaments up to 6–7 mm. long, sharply geniculate above, gibbous, minutely densely papillose at base. Capsule 8.5–9(15) × (1) 3 mm., linear-oblong in outline, acute or acuminate.

Zambia. N: Mporokoso-Kawambwa road, 1200 m., fl. & fr. 17.iv.1957, *Richards* 9266 (K). W: between Kamakonde and Kamulende rivers, c. 6.5 km. SW. of Matonchi Farm, fl. 17.ii.1938, *Milne-Redhead* 4616 (K). C: Kabwe Distr., beyond Mulungushi R., 24 km. N. of Kabwe, fr. & fr. 23.ix.1947, *Brenan & Greenway* 7924 (K). E: Lundazi, c. 1100 m., fl. 1.vi.1954, *Robinson* 804 (K; SRGH). S: Namwala Distr., between Namwala and Baambwe, fl. & fr. 17.iv.1963, *van Rensburg* 2004 (K; SRGH). **Zimbabwe**. N: Mutoko Distr., Chitora R., 915 m., fl iv.1956, *Davies* 1921 (K; SRGH). C: Gweru (Gwelo) Distr., Mlezu School Farm, 1270 m., fl. & fr. 13.v.1976, *Biegel* 5301 (BR; K; SRGH). **Malawi**. N: Mzimba Distr., Katoto, c. 5 km. W. of Mzuzu, 1370 m., fl. 23.vii.1973, *Pawek* 7238 (BR; K; SRGH). C: Kasungu Distr., Lodjwa, c. 15 km. S. of Champira, 1370 m., fl. 5.vii.1976, *Pawek* 11470 (K; SRGH). S: Mulanje Distr., Likubula, 820 m., fl. 27.vi.1946, *Vernay* in Brass 16490 (K; NY; SRGH). **Mozambique**. N: Ribáuè, c. 60 km. from Nampula, c. 500 m., fl. 24.iii.1964, *Torre & Paiva* 11379 (LISC).
Also known from Nigeria, Ethiopia, Sudan, Angola and Zaire. Wet grasslands, marshes and streamsides; from 700–1500 m.
As I consider it here, this appears to be quite a diverse species varying from a slender, simple stemmed, erect herb to a coarser, decumbent to erect, much branched plant. In consequence, the size of all its vegetative and generative organs are variable in size, the range of which can be seen above. I have studied both material and drawings of type specimens of the names closely allied to *L. oliveriana*, but conclude that no other closely related taxon occurs in the Flora Zambesiaca area.
Unlike most other botanists who have studied this genus I am unable to link our material with *L. stuhlmannii* of Tanzania. The photograph of the type specimen, *Stuhlmann* 3550, once in the Berlin herbarium, shows a plant 14 cm. tall with very small leaves 2.5–3 mm. long. The flowers are borne on very short pedicels and appear to be erect at least in bud; in no instance is any flower, in bud or in fruit, shown as being deflexed. A fragment of the holotype deposited in the Kew herbarium shows leaves 2 mm. long, pedicels c. 1 mm. long carrying erect flower buds with calyces 2 mm. long, lobed for 0.8 mm. Wettstetter in his original description (Pflanzenw. Ost-Afr. **C**: 357) describes the pedicels as 2–3 times longer than the very short bracts (upper leaves) and these are shown in his subsequent illustration (Bot. Jahrb. **23**: t. 9, F–L) as no more than equalling them. He also considers them to be about equal to the length of the flowers which is not fact as seen from the Kew fragment. None of these characters is constantly evident in any of the material from the Flora Zambesiaca area

purportedly representing this name and I have chosen to exclude it from further consideration here. However, our material hitherto under the name *L. stuhlmannii* appears to be no more than a slender state of *L. oliveriana* which it matches in all other respects.

5. **Lindernia stictantha** (Hiern) Skan in F.T.A. **4**, 2: 339 (1906). Type from Angola.
   Ilysanthes stictantha Hiern, Cat. Afr. Pl. Welw. **1**: 765 (1898). Type as above.

Decumbent, prostrate or prostrate-ascending annual, sparingly branched from base, occasionally rooting at lower nodes; branches up to 40 cm. long, slender, subquadrangular, minutely pubescent to subglabrous. Leaves (14)23–32 × (7)11–22 mm., opposite, broadly ovate, acute to subobtuse, somewhat cuneate at base, serrulate, sparsely short pubescent especially on major nerves beneath, petiolate. Petiole 2–4 mm. long, somewhat flattened. Flowers solitary axillary, alternate, pedicellate. Pedicels 5–8 mm. long, slender. Calyx 5–9 mm. long, 5-ribbed, unequally 5-toothed; lobes or teeth (1.75)3.5–4 mm. long, lanceolate to lanceolate-subulate, glabrous or with few hairs on ribs. Corolla 6–10 mm. long, mauve or blue with white tube. Anterior filaments 3.5–4 mm. long, gibbous at base. Capsule c. 15 × 2–2.6 mm., linear-oblong in outline, acute.

Zambia. N: Chinsali Distr., Shiwa Ngandu, 1525 m., fl. & fr. 5.vi.1956, *Robinson* 1612 (K; SRGH). W: Mwinilunga Distr., c. 5–6.5 km. SE. of Angolan border and 1.5–6 km. SW. of Mujileshi R., 1290 m., fl. fr. 7.xi.1962, *Richards* 16943 (K). **Mozambique**. Z: serra do Gúruè, E. of Picos Namuli, near source of R. Malema, c. 1700 m., fl. & fr. 5.i.1968, *Torre & Correia* 16934 (LISC).
   Known also from Angola. Bogs, marshy ground and wet places; up to 1550 m.
   Plants from the Flora Zambesiaca region hitherto named as '*Lindernia diffusa* (L.) Wettst.' match neither Patrick Browne's type collection nor the original description. Instead they compare favourably with the *Welwitsch* collections from Angola which typify the name used here.

6. **Lindernia nummularifolia** (D. Don) Wettst. in Engl. & Prantl, Pflanzenfam. 4, **3B**: 79 (1891). —Engl., Pflanzenw. Ost-Afr. **C**: 357 (1895). —Hiern, Cat. Afr. Pl. Welw. **1**: 763 (1898). —Skan in F.T.A. **4**, 2: 341 (1906). Type from Nepal.
   Vandellia nummularifolia D. Don, Prodr. Fl. Nep.: 86 (1825). —Benth. in DC., Prodr. **10**: 416 (1846). —Hook. f., Fl. Brit. Ind. **4**: 282 (1884). Type as above.
   Pyxidaria nummularifolia (D. Don) Kuntze, Rev. Gen. **2**: 464 (1891). Type as above.

Erect annual herb, 3–14 cm. tall; stems simple or branched from base or above shortly hairy to subglabrous; branches 3–12 cm. long, ascending. Leaves subsessile, 9–15(122) × 5–13(24) mm., opposite, ovate to subcircular, acute or obtuse, serrate or crenate-serrate, short hairy on margins and major nerves beneath. Flowers solitary in axils of upper leaves, pedicellate. Pedicels (1)3–10(14) mm. long, very slender, often becoming stouter in fruit. Calyx 2.5–3.5 mm. long, sparsely pubescent to subglabrous; lobes 1–1.5 mm. long, unequal, broad, acute. Corolla blue, lilac or violet, c. 6 mm. long; upper lip entire, oblong; lobes of lower lip rounded. Anterior filaments c. 3 mm. long, swollen geniculate at base. Capsule 4.5–6 × 1.5–3 mm. broad, ovoid-cylindric, shortly beaked.

Zambia. N: Mbala Distr., Sansia Falls, Kalambo R., 1350 m., fl. & fr. 28.iii.1957, *Richards* 8916 (BR; K). W: Mwinilunga Distr., Kalenda Ridge, W. of Matonchi Farm, fl. & fr. 22.i.1938, *Milne-Redhead* 4276 (BM; BR; K). **Zimbabwe**. W: Matobo Distr., near Maleme Dam, fl. 3.ii.1962, *Wild* 5624 (K). E: Nyanga (Inyanga) Distr., Nyamkwarara R. (Numquarara) Valley, Stapleford, c. 700 m., fr. 3.iv.1962, *Wild* 5687 (K; SRGH). **Malawi**. N: Mzimba Distr., Mzuzu, Marymount, 1370 m., fl. & fr. 14.ii.1974. *Pawek* 8105 (P; SRGH). S: Zomba, Government Hostel grounds, 990 m., fl. & fr. 16.iii.1977, *Brummitt* 14866 (K; SRGH). **Mozambique**. N: Lichinga (Vila Cabral), serra de Massangulo, c. 1450 m., fl. & fr. 25.ii.1964, *Torre & Paiva* 10804 (LISC). Z: Gúruè, serra do Gúruè, c. 3 km. from waterfall on R. Lucungo, c. 1200 m., fl. 24.ii.1966, *Torre & Correia* 14854 (LISC).
   Known also from India, Sri Lanka, China and Thailand. In Africa from Sudan and Ethiopia, West, Central and East Africa and Angola. On wet, mostly acid soils of marshes and riversides, in crevices on granite outcrops; up to about 1600 m.

7. **Lindernia subreniformis** Philcox in Bol. Soc. Brot., sér. 2, **60**: 268 (1987). TAB. **23**. Type from Tanzania.

Creeping or straggling, procumbent annual to about 20 cm. tall, laxly branched throughout; stems up to 28 cm. long, leafy throughout, acutely quadrangular, minutely glandular-pubescent, sparsely above to subglabrous, denser below. Leaves sessile, (4)9–14 × (4)8–14 mm., opposite, subreniform, circular or broadly ovate, obtuse, base subcordate or truncate, dentate or broadly somewhat serrate, rarely subentire, glandular-pubescent especially on major nerves beneath. Flowers solitary in leaf axils throughout

66

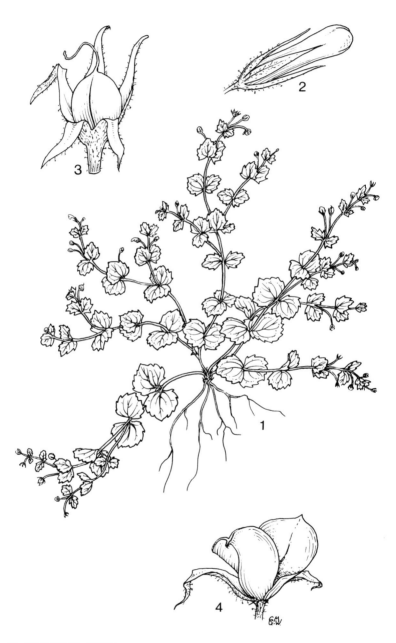

Tab. 23. LINDERNIA SUBRENIFORMIS. 1, habit (×⅔); 2, flower bud (×8); 3, fruit (×8); 4, dehisced fruit (×8), all from *Faulkner* 240.

most of stem, pedicellate. Pedicels 7–16 mm. long, slender, densely to subdensely glandular-pubescent with slender gland-tipped hairs. Calyx 2.4–3.25 mm. long, densely glandular-pubescent, 5-lobed almost to base, lobes 2–3 mm. long, narrowly lanceolate, acute, shortly ciliate, erect not spreading. Corolla white with pink, lilac to purple markings, or yellow, 4–5 mm. long; upper lip subcucculate, rounded to shallowly emarginate, lower lip trilobed with rounded lobes. Stamens four, fertile; anthers 0.5–0.6 mm. long; posterior filaments 0.5–1 mm. long, anterior filaments 1–2 mm. long, very slender with clavate appendage (0.1)0.5–0.6 mm. long at base, antero-lateral ridge prominent. Capsule subglobose, c. 2.5 mm. in diam., shorter than calyx.

Mozambique. Z: Mocuba Distr., Namagoa, 200 km. inland from Quelimane, c. 60–120 m., fl. & fr. vii.1945, *Faulkner* 240 (K).

Also known from Kenya, Tanzania and Zanzibar. Swamps, riversides or wet ground, or as a weed of ricefields (Zanzibar); sea-level to 380 m.

8. **Lindernia damblonii** Duvign. in Bull. Soc. Roy. Bot. Belg. **90**: 256 (1958). Type from Zaire.

Erect annual up to 14(28) cm. tall; stems very slender, quadrangular, branched towards base, less frequently simple, glabrous above, sparsely short-hispid below. Leaves 5–12(28) × 0.3–1.75(6) mm., opposite, linear-lanceolate to narrowly ovate-lanceolate, acute or obtuse, entire to sparsely sharply dentate. Flowers alternate in axils of bract-like, very reduced leaves, shortly pedicellate. Pedicels c. 1 mm. long. Calyx 3–5 mm. long, glabrous, strongly 5-nerved with nerves pronounced into narrow wings running full length of calyx tube; lobes 1–1.5 mm. long, subulate-lanceolate, flexuous. Corolla lilac or purple and white, 7–9 mm. long, upper lip shortly bifid, lobes rounded, lower lip trilobed, lobes rounded, palate with large, orange-yellow clavate hairs. Stamens four, fertile, anterior filaments c. 5 mm. long, slender, geniculate at base with short clavate spur c. 0.75–1 mm. long below geniculation. Capsule c. 4 × 1.5–2.25 mm. (? immature), shorter than calyx, broadly ellipsoid.

Zambia. W: Ndola Distr., Mkwera Falls, fl. & fr. 7.v.1986, *Philcox & Ngoma* 10231 (K; NDO). C: Serenje Distr., above Kundalila Falls, fl. & fr. 11.iv.1986, *Philcox, Pope & Chisumpa* 9874 (BR; K; NDO). **Mozambique**. N: Erati, Namapa, fl. 29.iii.1961, *Balsinhas & Marrime* 323 (K).

Known also from Zaire (type collection). Waterfalls and in wet pools on rocks; from 1200–1600 m.

Our material appears to differ from that of the type collection only in the calyx lobes being subulate-lanceolate, compared with those from Zaire being lanceolate. The extreme measurements of plant and leaf-size in the above description relate to a specimen from Mozambique, *Balsinhas & Marrime* 323, which in other characters matches my concept of this species, and which I would be unhappy to consider elsewhere. Further collections however may prove this to represent a distinct species.

9. **Lindernia conferta** (Hiern) Philcox in Bol. Soc. Brot., sér. 2, **60**: 268 (1987). TAB. **24**. Type from S. Africa.

Ilysanthes conferta Hiern in F.C. **4**, 2: 365 (1904). —Skan in F.T.A. **4**, 2: 347 (1906). Type as above.

Ilysanthes plantaginella S. Moore in Journ. Bot. **43**: 49 (1905). Type: Zimbabwe, Matopos (Matoppo) Hills, *Eyles* 47 (BM, holotype; SRGH, isotype).

Aquatic, glabrous, annual; stems 6–12(25) cm. long, simple or branched below, erect, naked below upper leaves except occasionally with remote pairs of much smaller cauline leaves present, rooting at or near base. Leaves: upper shortly petiolate or sessile, 6–18 × 3.5–8 mm., opposite, clustered at or towards end of stem, floating, obovate to obovate-oblong, obtuse, entire, slightly sheathing at base, somewhat fleshy, green to reddish-purple when living; lower usually sessile, c. 2–3(6) × 1 mm., ovate. Flowers axillary, subterminal, pedicellate. Pedicels 4–10(15) mm. long. Calyx 2–2.5 mm. long, deeply divided almost to base, glabrous; lobes 1.75–2.25 mm. long, 0.8–1 mm. wide, broadly oblong or ovate, obtuse. Corolla white with purplish markings to mauve or magenta, up to 14 mm. long, upper lip c. 4–5 mm. long, entire, broadly ovate, lower lip c. 8 mm. long, deeply trilobed. Fertile stamens two, filaments c. 2.5 mm. long; staminodes two, c. 1.25 mm. long, simple, apically incurved, minutely papillose, antero-lateral ridge up to 1.5 mm. long, obscure. Capsule up to 4 × 2.75–3 mm., broadly cylindric-ovoid, style persistent in fruit.

Botswana. N: Gidiba Is., fl. & fr. 14.ix.1976, *Smith* 1771 (K). **Zambia**. C: Kapiri Mposhi Distr., Lusaka to Kapiri Mposhi road, 1200 m., fl. & fr. 4.iv.1961, *Richards* 14922 (K). **Zimbabwe**. N: Mwami (Miami) Experimental Farm, 1370 m., fl. 7.iii.1947, *Wild* 1961 (K; SRGH 16597). W: Matobo Distr.,

68

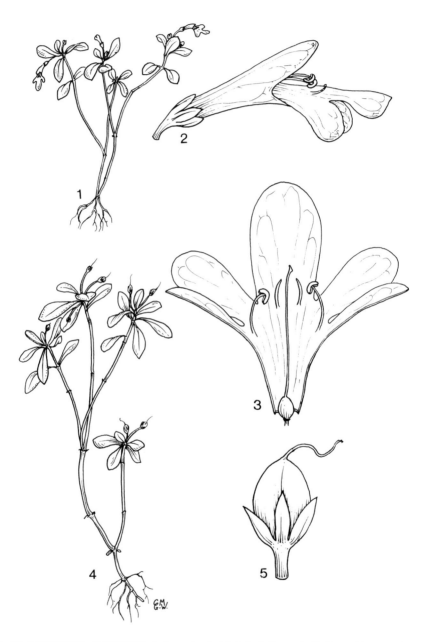

Tab. 24. LINDERNIA CONFERTA. 1, flowering habit (× ⅔); 2, flower (× 6); 3, corolla opened out showing androecium of 2 stamens and 2 staminodes and gynoecium (× 6), 1–3 from *Wild* 1961; 4, fruiting habit (× ⅔); 5, fruit (× 6), 4–5 from *Richards* 14922.

World's View, Matopos, fl. & fr. 19.iii.1966, *Mangena* 3939 (K; SRGH). C: Gweru (Gwelo) Distr., Mlezu, c. 18 km. SSE. of Kwekwe (Que Que), 1280 m., fl. & fr. 19.ii.1966, *Biegel* 919 (K; SRGH). E: Mutare Distr., Zimunya Res., 945 m., fl. & fr. 23.ii.1958, *Chase* 6831 (BR; K; P; SRGH). S: Mwenezi (Nuanetsi) Distr., Merrivale Ranch, fl. & fr. 4.iv.1967, *West* 7458 (K; SRGH).

Also known from S. Africa. Usually in ephemeral rain-filled pools on granite outcrops; up to 2100 m.

10. **Lindernia parviflora** (Roxb.) Haines, Bot. Bihar & Orissa **4**: 635 (1922). Type from India.
 Gratiola parviflora Roxb., Pl. Corom. **3**: 3, t. 203 (1811). Type as above.
 Bonnaya parviflora (Roxb.) Benth. in Wall. Cat. 3867 (1831). —Benth., Scroph. Ind.: 34 (1835). Type as above.
 Ilysanthes parviflora (Roxb.) Benth. in DC., Prodr. **10**: 419 (1846). —Skan in F.T.A. **4**, 2: 346 (1906). —Hepper in F.W.T.A., ed. 2, **2**: 365, fig. 289 (1963). Type as above.

Erect, decumbent or ascending annual, 5–15(25) cm. tall, glabrous; stems usually diffusely branched, glabrous to rarely slightly pubescent. Leaves sessile or shortly petiolate, 2–15(38) × 0.75–5(13) mm., opposite, ovate, ovate-lanceolate or elliptic, acute or obtuse, entire, or larger leaves occasionally minutely, shallowly dentate, 1–3(5)-nerved. Petiole 0–2 mm. long. Flowers axillary or in terminal racemes, slender pedicellate. Pedicels (5)7–12 mm. long in flower, extending in fruit 9–17(25) mm. long. Calyx (1.25)2–4 mm. long, glabrous, divided almost to base; lobes up to 0.5 mm. wide, lanceolate to linear-lanceolate, acute. Corolla white or pale blue, 4–7 mm. long, upper lip emarginate to shallowly bifid. Fertile stamens two, filaments 0.4–1 mm. long; staminodes two, 0.25–0.5(1) mm. long, slender, clavate at apex occasionally with short spur c. 0.75 mm. long, just below. Capsule 1.5–4.5(6) × 1.25–1.75(2.5) mm., obovate- or cylindric-ellipsoid to nearly subglobose, usually longer than calyx.

Botswana. N: Okavango, near Kinkogo Tsetse Camp, 900 m., fl. 14.iii.1961 *Richards* 14706 (K). **Zambia**. B: Sichili, fl. & fr. 5.vii.1963, *Fanshawe* 7902 (BR; K; SRGH). W: Mwinilunga Distr., by R. Matonchi, fl. & fr. 11.xi.1937, *Milne-Redhead* 3194 (BR; K). S: Choma, Mapanza, fl. & fr. 16.iii.1958, *Robinson* 2796 (K; SRGH). **Zimbabwe**. N: Binga Distr., valley S. of Binga Hill, c. 457 m., fl. & fr. 10.xi.1958, *Phipps* 1432 (BR; K; P; SRGH). W: Victoria Falls, Rain Forest, 915 m., fl. iv.1918, *Eyles* 1268 (K; SRGH). C: Harare Distr., fl. & fr. 3.iv.1952, *Wild* 3805 (K; SRGH). E: Chipinge Distr., Tanganda R. Valley, 730 m., fl. & fr. 17.v.1962, *Chase* 7736 (K). S: Mwenezi Distr., Malangwe R., SW. of Mateke Hills, 610 m., fl. & fr. 5.v.1958, *Drummond* 5579 (BR; K; SRGH). **Malawi**. S: Zomba Distr., W. of Lake Chilwa, 600 m., fl. & fr. 21.vi.1962, *Robinson* 5402 (K; SRGH). **Mozambique**. Z: above Morrumbala Marsh, fl. v. 1888, *Scott* s.n. (K). GI: Xai Xai (Vila João Belo), near Chongoene, fl. & fr. 16.vii.1944, *Torre* 6778 (LISC). M: Umbeluzi, fl. & fr. 7.iv.1949, *Myre* 500 (K).

Also known from West Africa, Sudan and Ethiopia, East Africa, Madagascar, Namibia and India. Sandy or muddy river- or streambanks in area of floodplain, wet areas around, and in ephemeral pools on, rocky outcrops, muddy margins of seasonal or permanent pans; 100–1450 m.

11. **Lindernia nana** (Engl.) Roessler in Mitt. Bot. Staatss. München **5**: 691 (1951). Type from Angola.
 Ilysanthes nana Engl., Bot. Jahrb. **23**: 505 (1897). —Hiern, Cat. Afr. Pl. Welw. **1**: 764 (1898); in F.C. **4**, 2: 365 (1904). —Skan in F.T.A. **4**, 2: 347 (1906). Type as above.

Erect annual or perennial, 2.25–3.5 cm. tall; stems quadrangular, minutely pubescent, leafy. Leaves sessile, glabrous, sparsely crenate-dentate, opposite, upper 4.5–11(16) × 1.75–5.5(9) mm., ovate, obtuse, lower 7–15 × 2–4 mm., petiolate, spathulate, obtuse. Petioles where present, up to 2–5 mm. long. Flowers solitary in axils of upper leaves, pedicellate. Pedicels 0.5–1.5 mm. long. Calyx 3–4 mm. long, glabrous, strongly keeled; lobes 1.25–1.75 mm. long, ovate-lanceolate, acute. Corolla white with pink or pale blue markings, c. 9 mm. long, upper lip bifid, lower lip trilobed, lobes rounded. Stamens two, filaments c. 0.75 mm. long, staminodes two, gibbous with minute filament c. 0.25 mm. long. Capsule 5–8 × 2–2.5 mm., narrowly ellipsoid, acuminate.

Zambia. E: Chipata Distr., Luangwa Valley, Chinzombo, c. 610 m., fl. & fr. 20.ii.1969, *Astle* 5478 (K). **Zimbabwe**. C: Harare Distr., fl. & fr. i.1952, *Greatrex* in GHS 35695 (K; SRGH). W: Matobo Distr., Besna Kobila Farm, c. 1460 m., fl. & fr. iii.1957, *Miller* 4142 (K; SRGH).

Also known from Uganda, Angola, Namibia and S. Africa. In shallow soil covering rocks; up to 1500 m.

12. **Lindernia pulchella** (Skan) Philcox in Bol. Soc. Brot., sér. 2, **60**: 268 (1987). Type: Malawi, Zomba Plateau, 1500–1800 m., *Whyte* s.n. (K, holotype).
 Ilysanthes pulchella Skan in F.T.A. **4**, 2: 348 (1906). Type as above.
 Ilysanthes purpurascens Hutch., Botanist in S. Afr.: 461 (1946). Type: Zimbabwe, near Runde (Lundi) R., 30.vi.1930, *Hutchinson & Gillett* 3269 (K, holotype; BM, isotype).

Ilysanthes saxatilis Norlindh in Bot. Not. **1951**: 115, figs. 4d–e (1951). Type: Zimbabwe, Nyanga (Inyanga) Distr., Mt. Inyangani, c. 2400 m., fl. & fr. 14.ii.1931, *Norlindh & Weimarck* 4969 (LD, holotype [n.v.]; K, isotype).

Ilysanthes pulchella subsp. *rhodesiana* Norlindh in Bot. Not. **1951**: 113, figs. 4 f & h (1951). Type: Zimbabwe, near Nyanga (Inyanga) village, c. 1700 m., xii.1930, *Fries, Norlindh & Weimarck* 3255 (LD, holotype; K, syntype (fruiting 3255b)).

Erect or procumbent annual or perennial, 4–8(18) cm. tall, branched from or towards base; stems quadrangular, variously covered with short, patent to somewhat retrorse, stiff white hairs. Leaves sessile or lower shortly petiolate, 5–10(14) × 1.5–2(4) mm. opposite, elliptic-oblong, obtuse, entire to sparsely crenate or crenate-dentate, glabrous to somewhat densely, short white hispid, especially beneath. Petioles, where present, up to 4 mm. long. Flowers solitary axillary, pedicellate. Pedicels 8–20 mm. long, slender. Calyx 4–5.5 mm. long, subglabrous to minutely pubescent; lobes 0.75–1.5 mm. long, ovate to ovate-elliptic, obtuse, glabrous on margin or shortly ciliate, unequal to subequal. Corolla blue to blue-mauve or violet, pink or white, 8–10(14) mm. long; upper lip deeply bifid, lower lip trilobed, lobes rounded, wavy-margined, ciliate. Stamens two, fertile, filaments 0.5–1(1.5) mm. long, anthers 1–1.5 mm. long; staminodes two, up to 1.75 mm. long, somewhat recurved, slightly clavate, geniculate above base but appearing gibbous due to tissue forming web in angle of geniculation, tissue produced laterally forming short, slender or blunt projection 0.1–0.35 mm. long, antero-lateral ridge barely evident. Capsule 5–6.5(9) × 1.5–2 mm., cylindric-ellipsoid, acute.

Zambia. C: Mkushi Distr., Fiwila, 1400 m., fl. & fr. 3.i.1958, *Robinson* 2582 (K). **Zimbabwe**. N: Mutoko Distr., Mudzi Dam, 1220 m., fl. & fr. 16.ii.1962, *Wild* 5670 (K). W: Matobo Distr., fl. 4.i.1948, *West* 2556 (K; SRGH). C: Harare Distr., 1525 m., fl. & fr. ii.1920, *Eyles* 2049 (K). E: Nyanga (Inyanga) Distr., Troutbeck, World's View, 2400 m., fl. & fr. 15.i.1959, *Lennon* 83 (K). S: Masvingo Distr., fl. 1909–1912, *Monro* 1518 (BM). **Malawi**. N: Zomba Plateau, near Chingwe's Hole, 1900 m., fl. & fr. 27.ii.1977, *Grosvenor & Renz* 982 (K; SRGH). **Mozambique**. Z: Alto Molócuè, c. 13 km. on road to Gile, c. 500 m., fl. & fr. 1.xii.1967, *Torre & Correia* 16306 (LISC). MS: Beira, *Corner* s.n. in 1961 or 1962 (K).

Also known from Ethiopia, East Africa, Angola and S. Africa (Transvaal). On damp rocks and wet, shallow sandy soils; up to 2400 m.

Norlindh (Bot. Not. **1951**: 113 (1951)) isolated his subsp. *rhodesiana* by reason of its densely hispid leaves and narrower and more acute calyx lobes. Here I dismiss these characters as they are encompassed by my general specific description, they being common throughout the range of material I have studied. He also described *Ilysanthes saxatilis* (ibid. p. 115) as distinct from *I. pulchella* by reason of the shorter and more obtuse calyx lobes and the glabrous, foveolate seeds. The character of the calyx differences occur within the range of the original material and description and in the case of both seed shape and indumentum, this appears to vary within the collections I have studied dependent on the maturity of the seeds. I have chosen to include both of these in synonymy under *L. pulchella*.

Hutchinson (1946) described *Ilysanthes purpurascens* as distinct by reason of the overall indumentum of short, retrorse hairs. The holotype of this name, *Hutchinson & Gillett* 3269 (K), shows this indumentum to be of short, white hairs which appear as a mixture of patent or somewhat retrorse individuals occurring densely except on the upper surface of the leaves. These hairs appear to match those on the lower parts of the stem of the holotype of *Ilysanthes pulchella* Skan (K) both in density and direction, and in all other characters and measurements both original descriptions and specimens compare. On dissection, flowers from both are found to have two fertile stamens and two staminodes, the latter matching in having similar swellings appearing gibbous, with a web of tissue in the geniculation being produced into a long or short proboscis-like appendage. I can see no reason to keep them separate and include Hutchinson's plant in synonymy here.

13. **Lindernia exilis** (Skan) Philcox in Bol. Soc. Brot., sér. 2, **60**: 269 (1987). TAB. **25**. Type from Nigeria.

Ilysanthes gracilis Skan in F.T.A. **4**, 2: 349 (1906). Type as above.

Erect annual or perennial, 5–7(14) cm. tall, minutely glandular-pubescent to subglabrous; stems quadrangular, simple or branched usually above, branches ascending. Leaves 2–10(14) × 0.5–3(5) mm., oblong-linear to lanceolate-oblong, obtuse, entire to shallowly crenate-dentate. Flowers solitary in axils of upper leaves, opposite, pedicellate. Pedicels 4–10 mm. long, slender. Calyx 2.25–5 mm. long, glabrous to minutely glandular-pubescent; lobes 0.5–1.2(2) mm. long, unequal, ovate-lanceolate, acute. Corolla blue, lilac-blue or violet to white, 7–11.5 mm. long; upper lip bifid with short acute lobes, lower lip with subcircular lobes. Stamens two, fertile, filaments up to 1.25 mm. long, anthers 0.8–1 mm. long; staminodes two, 1–1.5 mm. long, broadly gibbous with short

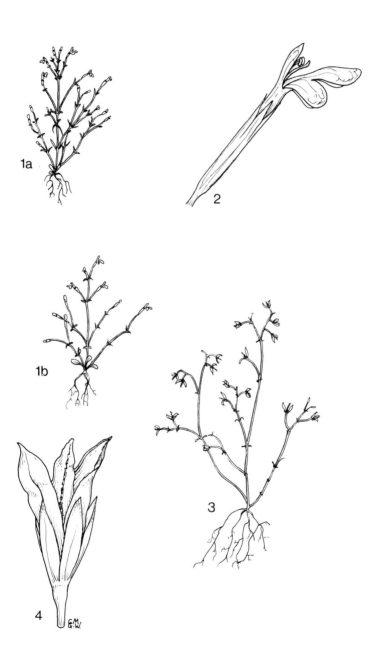

Tab. 25. LINDERNIA EXILIS. 1a & b, flowering **habit** (× ⅔); 2, flower (× 8), 1–2 from *Hooper & Townsend* 272; 3, fruiting habit (× ⅔); 4, dehisced **capsule** (× 8), 3–4 from *Robinson* 3627.

filaments, somewhat reflexed. Capsule (3.5)5.5–6(9) × 1–1.5 mm., oblong-linear in outline, valves markedly divergent when ripe.

Zambia. W: Mwinilunga Distr., Matonchi, Kalenda Plain, fr. 16.iv.1960, *Robinson* 3627 (K; SRGH). **Mozambique**. N: Mozambique Distr., Serra Merripa, Malema, 800 m., fl. & fr. 4.ii.1964, *Torre & Paiva* 10451 (LISC). Z: Ile, Errego, c. 900 m., fl. & fr. 4.iii.1966, *Torre & Correia* 15020 (LISC).

Also known from West Africa. Mostly in shallow, moist soil in crevices or depressions on rocky outcrops; up to 1200 m.

The plants from the Flora Zambesiaca area appear to be morphologically similar to the plant described by Skan (1906) as *Ilysanthes gracilis* from Northern Nigeria. As the combination *Lindernia gracilis* had already been made by Bonati (1927) based on *Vandellia gracilis* Bonati from Vietnam, I proposed the name *L. exilis* for our species. The collection by *Barter* from Nigeria and used by Skan, now typifies this species.

14. **Lindernia wilmsii** (Engl. ex Diels) Philcox in Bol. Soc. Brot., sér. 2, **60**: 268 (1987). Type from S. Africa.

    Ilysanthes wilmsii Engl. ex Diels in Engl., Bot. Jahrb. **26**: 123 (1898). —Hiern in F.C. **4**, 2: 366 (1904). Type as above.

    Ilysanthes muddii Hiern in F.C. **4**, 2: 366 (1904). Type from S. Africa.

Erect annual, 2.5–7 cm. tall, simple or branched towards base; stems quadrangular, glabrous to slightly minutely pubescent below, rigid. Leaves: basal where present, clustered, 4–7 × 1 mm., linear-spathulate narrowing into petiole-like base, minutely pubescent; cauline 1.5–5 × 0.8–1 mm., opposite, narrowly elliptic to almost linear, entire, obtuse, subglabrous. Flowers solitary, axillary, pedicellate. Pedicels up to 10 mm. long, slender, glabrous to minutely pubescent. Calyx (1.6)3–4 mm. long, glabrous to minutely pubescent, lobes 0.5–1.2 mm. long, deltate, acute or obtuse, glabrous on margin or minutely ciliate. Corolla various shades of blue and white, (once recorded as yellow), (3.5)5–7(8) mm. long, upper lip bifid, lower lip 3-lobed, lobes rounded. Stamens two, fertile, filaments up to c. 1 mm. long; staminodes two, 0.3–0.5 mm. long, clavate, erect or more frequently, slightly curved outwards, sparsely to densely clothed with minute clavate hairs which extend to palate between staminodes, short antero-lateral ridges barely evident. Capsule 3.5–6 × 1.75 mm., elliptic-ovate in outline, acute.

Zambia. N: near Mpika Airfield, fl. 13.i.1975, *Brummitt & Polhill* 13774 (K). C: Serenje, fl. 3.i.1963, *Fanshawe* 7403 (K). **Zimbabwe**. N: Shanawe R., fl. & fr. 14.i.1937, *Eyles* 8924 (K). W: Matobo Distr., Matopos, c. 1400 m., fl. & fr. iii.1918, *Eyles* 993 (BM; K; SRGH). C: Makoni Distr., Wick, c. 1800 m., fl. & fr. 9.ii.1931, *Norlindh & Weimarck* 4917 (K). E: Nyamkwarara (Nyumquarara) Valley, 915 m., fr. ii.1935, *Gilliland* 1558 (BM; K). **Mozambique**. MS: 16 km. N. of Marta, fl. & fr. 1.ii.1962, *Wild* 5631 (K; SRGH).

Also known from S. Africa. In shallow wet soil on or around granite rock outcrops; 900–1800 m.

15. **Lindernia schweinfurthii** (Engl.) Dandy in F.W. Andrews, Fl. Pl. Sudan **3**: 139 (1956). Type from Sudan.

    Ilysanthes schweinfurthii Engl., Bot. Jahrb. **23**: 504 (1897). —Skan in F.T.A. **4**, 2: 350 (1906). Type as above.

Erect, slender annual, (3)6–14 cm. tall; stems simple or branched at base, quadrangular, glabrous or slightly pubescent. Leaves 2–5 × 0.5–0.75(1) mm., oblong-ovate to linear-ovate, obtuse, entire, opposite, smaller above. Flowers few, distantly racemose, in axils of bract-like reduced leaves, pedicellate. Pedicels 4–6 mm. long, slender, somewhat deflexed in fruit. Calyx 1.5–3 mm. long, glabrous, prominently 5-nerved; lobes 0.4–0.5 mm. long, lanceolate, acute, ciliolate. Corolla blue or mauve and white, 4–4.5(9) mm. long, upper lip c. 2.5 × 0.5 mm., oblong, bifid at apex with lobes narrowly acute to acuminate, lower lip 3-lobed, lobes oblong, rounded. Stamens 2 fertile, posterior filaments c. 0.75 mm. long, staminodes two, 0.2(1) mm. long, 0.1(0.5) mm. wide, oblong, obtuse, papillose. Capsule c. 4 × 1.5 mm., linear-ellipsoid, acute.

Zambia. N: Mbala Distr., Katula Gorge, c. 1450 m., fl. & fr. 8.i.1952, *Richards* 410 (K). W: Mwinilunga Distr., E. of Matonchi Farm, fl. & fr. 12.ii.1938, *Milne-Redhead* 4554 (K).

Also known from West Africa, Cameroon, Congo, Zaire, Sudan, Ethiopia and Kenya. In wet sandy places and in waterfilled mossy hollows on granitic rocky outcrops; up to 1500 m.

Two very depauperate collections, *Greatrex* in GHS 18384 and *Whelan* 1222 are housed at Harare (SRGH) and may represent a new species. These plants, both from the Harare Distr., are small, low growing herbs of marshy ground and from the very poor material available appear to have four stamens, the posterior pair of which bear fertile anthers. The anterior pair have long, slender filaments terminated by sterile anthers densely covered in slender, minutely clavate, white hairs 0.5

mm. long. The paucity of the material precludes me from investigating further, but more material may prove that they represent a new species.

23. LIMOSELLA L.

Limosella L., Sp. Pl. **2**: 631 (1753). —Glück in Engl., Bot. Jahrb. **66**: 488–566, t. 6–8 (1934).

Small, glabrous, tufted, creeping or floating, marsh or aquatic herbs. Stemless or with slender stolon-like stems rooting at nodes. Leaves radical or fasciculate at nodes, rarely alternate on stems or branches, petiolate, erect or floating; lamina linear and petiole-like, spathulate or oblong-ovate. Peduncles axillary, ebracteate, often shorter than leaves, frequently deflexed in fruit. Flowers small. Calyx campanulate, thin, (4)5-lobed, persistent. Corolla tubular, campanulate, subrotate; limb 5-lobed, spreading, lobes subequal, rounded, ovate or oblong. Stamens four, didynamous, usually exserted; filaments filiform, inserted on corolla tube. Ovary shortly bilocular at base; ovules many; style short, incurved, apically thickened. Fruit capsular, ovoid to subglobose or spherical, bivalved, loculicidally dehiscent. Seeds numerous, striate, rugulose.

A genus of about 15 species mainly in temperate and cooler regions but also in tropical mountains.

1. Leaves cylindrical or subulate   -   -   -   -   -   -   -   -   1. *australis*
– Leaves distinctly laminate   -   -   -   -   -   -   -   -   -   -   - 2
2. Lamina cuneate or attenuate into petiole or narrow base; leaves rarely floating   2. *maior*
– Lamina rounded or subcordate at base; leaves usually floating   -   -   -   3. *capensis*

1. **Limosella australis** R.Br., Prodr.: 443 (1810). Type from Australia.
    Limosella subulata Ives in Trans. Med.-Phys. Soc. N.Y. **1**: 440 (1817). —Glück in Nat. Bot. Gart. Berl. **12**: 72 (1934); in Engl., Bot. Jahrb. **66**: 491, 546, tab.6 (1934). TAB. **26**, fig A. Type from North America.

Small creeping or submerged, amphibious herb, tufted. Rhizomes 1.5–2.5 cm. long in terrestrial plants, 1.5–3.8 cm. long when submerged. Leaves 1.5–8 cm. long, 0.4–1.5 mm. in diam., cylindrical or subulate. Flowers white, blue beneath, short to long pedicellate. Pedicels 5–35 mm. long. Calyx c. 1.5 mm. long, 5-lobed, lobes c. 0.5–0.75 mm. long, triangular, obtuse. Corolla tube 1.8–2.2 mm. long, lobes 4–5 subcircular, or somewhat oblong, c. 1.5 mm. long, minutely pilose towards throat. Style 2.5–2.75 mm. long in fruit, persistent. Capsule 2–2.5 mm. in diam., subglobose.

Zimbabwe. N: Mazowe (Mazoe), 1280 m., fl. & fr. ix.1906, *Eyles* 409 (SRGH). C: 9.5 km. S. of Gweru, 1340 m., fl. & fr. 26.ix.1965, *Biegel* 417 (SRGH). **Malawi**. N: Nyika Plateau, Lake Kaulime, 2150 m., fl. & fr. 24.x.1958, *Robson & Angus* 320 (K; LISC; SRGH).
Also in North America, Europe, Australia and areas of higher altitude elsewhere. In shallow lakes or on muddy lakesides; up to 2340 m. in the Flora Zambesiaca area.
This very complex taxon may eventually prove to be conspecific with *L. macrantha* R.E. Fries, but for the moment I prefer to keep it distinct by reason of its size and shape of flowers, morphology of the leaves and general habit as recorded by the collectors.

2. **Limosella maior** Diels in Engl., Bot. Jahrb. **26**: 122 (1898). —Hiern in F.C. **4**, 2: 357 (1904). —Skan in F.T.A. **4**, 2: 353 (1906). —Glück in Nat. Bot. Gart. Berl. **12**: 77 (1934); in Engl., Bot. Jahrb. **66**: 547 (1934). TAB. **26**, fig. B. Type from S. Africa.

Tufted marsh herb, minutely glandular, stoloniferous or not; stolons, where present, up to 6.5 mm. long. Leaves 25–45(140) mm. long, many, laminate, spathulate, oblong or oblong-spathulate; lamina 6–17(35) × 1.5–3.5(12) mm., usually ovate, obtuse or subacute, narrowed at base into erect, comparatively stout petiole. Pedicels (5)15–50 mm. long, erect in flower, deflexed in fruit. Calyx 2.5–3.5(5) mm. long; lobes 0.8–1.5 mm. long, triangular-ovate. Corolla c. 5 mm. long, lilac- or bluish-white; lobes c. 2.5 × 1.5 mm., ovate-circular to oblong, minutely pilose towards throat. Capsule c. 4 × 3 mm., broadly ovoid to subglobose.

Botswana. N: on Boteti R. below Samadupi Drift, fl. & fr. 26.ix.1973, *Smith* 698 (K; SRGH). SE: Boschwelatlon, fl. & fr. 7.x.1955, *McConnell* in GHS 68898 (K; SRGH). **Zimbabwe**. C: Harare Distr., Chinamora Res., 1370 m., fl. & fr. 23.ix.1955, *Wild* 4672 (BR; K; SRGH). E: Mutare Distr., Marange Res., S. of Mutare, 975 m., fl. & fr. 29.xi.1952, *Chase* 4725 (BM). **Malawi**. C: Dedza Distr., Ndebvu, Abona Dambo, Dzalanyama For. Res., fl. & fr. 29.vi.1967, *Salubeni*

74

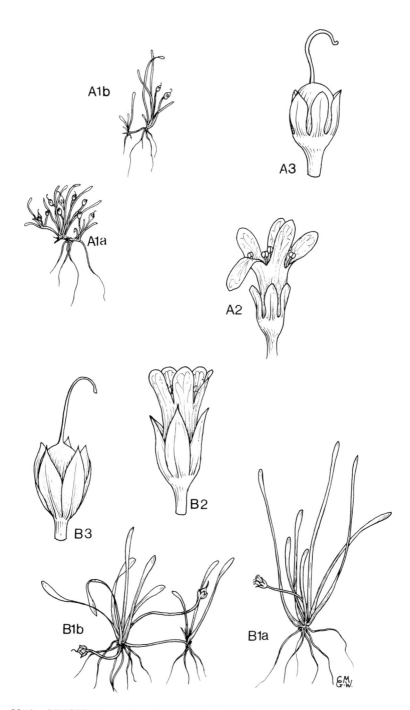

Tab. 26. A.—LIMOSELLA AUSTRALIS. A1a & 1b, habit (×⅔), from *Robson* 320; A2, flower (× 6); A3, fruit (× 6), A2–3 from *Richards* 22512. B.—LIMOSELLA MAIOR. B1a & 1b, habit (× ⅔); B2, flower (× 6); B3, fruit (× 6), B1–3 from *Fries* 2639.

760 (K; LISC; SRGH). **Mozambique**. T: 2 km. N. of Mlangeni, near Ntcheu-Dedza road, 1385 m., fl. & fr. 25.v.1970, *Brummitt* 11115 (BR; K; SRGH). MS: Manica, Serra Zuira, Tsetserra, 2100 m., fl. & fr. 6.xi.1965, *Torre & Paiva* 12767 (LISC).

Also known from Eritrea, Zaire, Kenya, Tanzania and S. Africa. Muddy margins of lakes and streams and in shallow running or standing water; 975–2350 m.

3. **Limosella capensis** Thunb., Prodr.: 104 (1800); in Fl. Cap., ed. Schultes: 480 (1823). —Hiern in F.C. **4**, 2: 358 (1904). —Skan in F.T.A. **4**, 2: 353 (1906). —Glück in Nat. Bot. Gart. Berl. **12**: 77 (1934); in Engl., Bot. Jahrb. **66**: 548, 559 (1934). Type from S. Africa.

Small, tufted, aquatic or marsh herb, often stoloniferous, rooting at nodes. Leaves 35–80(130) mm. long, crowded, erect, sublinear, elliptic-oblong, obtuse, usually attenuate at base into petiole, or floating blade 10–20(34) × 2.5–12(16) mm., ovate to broadly elliptic, obtuse to rounded at apex, broadly rounded, subcordate to broadly attenuate at base. Petiole 25–40(110) mm. long, filiform in floating specimens. Pedicels 1–10(20) mm. long, slender to filiform, erect in flower, becoming deflexed to spreading in fruit. Calyx 2–2.5 mm. long, 5-lobed, lobes 0.5–1 mm. long, broadly ovate. Corolla white with lobes pale blue or purplish beneath; lobes oblong, obtuse. Capsule 2.5–3 × 1.75–2 mm., ovoid-ellipsoid.

Botswana. N: Boteti R., Okavango Swamp, 914 m., fl. & fr. 15.ix.1970, *Wager* 35 (SRGH). SE: Thamaga, fl. 22.iii.1977, *Camerick* 114 (K).

Also known from Ethiopia, East and S. Africa. Muddy pools and slow running streams and also in temporary rock pools; from 1820–2740 m.

This controversial and difficult genus needs to be fully revised. Glück (1934) published a lengthy and apparently in-depth account of the genus but complicated matters further, by introducing infravarietal taxa covering the terrestrial, submerged or floating states within almost all the species. I have chosen here to discount all of these and to keep my specific descriptions broad enough to encompass his various forms.

24. SIBTHORPIA L.

Sibthorpia L., Sp. Pl.: 631 (1753); Gen. Pl., ed. 5: 279 (1754).

Perennial, small, creeping herbs; stems rooting at nodes. Leaves alternate, petiolate, suborbicular-reniform, crenate to incised. Flowers axillary, 4- to 8-merous, solitary or fasciculate, pedicellate. Pedicels ebracteate. Calyx 4–8-lobed, lobes somewhat unequal. Corolla rotate, tube short, lobes equalling calyx in number or one more, entire, subequal, spreading. Stamens 4–8, equalling corolla lobes or one or two less; filaments short, filiform, subequal; anthers sagittate; thecae contiguous at apex. Style short, stigma capitate. Capsule somewhat compressed, loculicidal. Seeds few, oblong-ovoid, reticulate or smooth.

A genus of 5 species occurring in Western Europe, the Azores, the Balearic Islands, and in tropical Africa and America.

Sibthorpia europaea L., Sp. Pl.: 631 (1753). —Hedberg in Bot. Not. 108, **2**: 168 (1955). —Hepper in F.W.T.A. ed. 2, **2**: 355 (1968). TAB. **27**. Type from Europe.

    Sibthorpia africana auct. non L. (1753).

    Sibthorpia europaea var. *africana* Hook. in Journ. Linn. Soc., Bot. **7**: 208 (1864). —Hemsl. & Skan in F.T.A. **4**, 2: 314 (1906). Type from West Africa.

    Sibthorpia australis Hutch., F.W.T.A. ed. 1, **2**: 221 (1931). Type from the Cameroons.

Prostrate, creeping herb rooting at nodes, somewhat pilose; stems slender up to 20 cm. long. Leaves petiolate, 5–15 mm. in diam., subcircular, crenate-dentate, finely to densely pilose above and below; petioles 5–30 mm. long, slender, pubescent. Flowers solitary, pedicels 2–5 mm. long. Calyx 1.5–2 mm. long, shortly 4–5-lobed, 4–5-nerved, pubescent, lobes short, obtuse to subacute. Corolla up to 2 mm. long, usually 5-lobed, yellow to purple, lobes short, obtuse. Stamens usually four, included. Capsule to about 1.5 mm. long, obovate, somewhat emarginate, pilose at apex.

Zambia. E: Nyika Plateau, below Rest House on path to N. Rukuru Waterfall, 2150 m., fl. 27.x.1958, *Robson* 402 (K). **Zimbabwe**. E: Nyanga (Inyanga) Distr., Pungwe R., c. 1850 m., fl. 21.ii.1946, *Wild* 851 (K; SRGH).

76

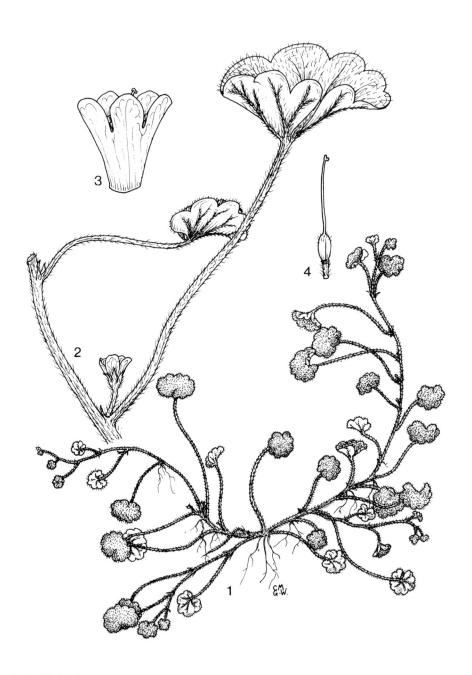

Tab. 27. SIBTHORPIA EUROPAEA. 1, habit (× ⅔); 2, portion of stem, with leaves (× 4); 3, corolla (× 12); 4, gynoecium (× 12), all from *Wild* 1390.

Also in Europe and the Azores and from Ethiopia to Tanzania. Recorded from Fernando Po. Streamsides, riverbanks and moist areas, usually in shade; up to 2150 m.

The plant described by Linnaeus (Sp. Pl.: 631 (1753) as *Sibthorpia africana* and referred to "*Shaw. afric.* 149, f. 149", has for a long while been considered by many botanists to be a strain or variant of *S. europaea* L. occurring in tropical Africa. Hedberg (Bot. Not. 108, fasc. 2, 164–167 (1955)) has made it abundantly clear that this yellow flowered plant is a well-defined species endemic to the Balearic Islands and not present on the African continent.

25. SCOPARIA L.

Scoparia L., Sp. Pl.: 116 (1753); Gen. Pl., ed. 5: 52 (1754).

Annual herb, or rarely small shrub, much branched, erect, glabrous or pubescent. Leaves opposite or verticillate, entire or dentate. Inflorescence raceme-like or flowers axillary, solitary or paired. Calyx 4–5-lobed, lobes ovate to lanceolate. Corolla regular, rotate, 4-lobed, throat bearded; lobes obtuse, subequal. Stamens 4, subequal, filaments filiform, anthers sagittate with divergent or parallel loculi. Style subclavate at apex, stigma truncate or emarginate. Capsule globose or ovoid. Seeds numerous, angular.

A genus of some 20 species native to tropical America, one species being widely spread throughout most of the Old and New World tropics.

Scoparia dulcis L., Sp. Pl.: 116 (1753). —Hemsl. & Skan in F.T.A. 4, 2: 354 (1906). TAB. 28. Type from Jamaica.

Woody perennial or subshrub to 60 cm. tall, erect, much-branched, glabrous or minutely hairy especially at nodes; stem and branches angular or ribbed. Leaves 1.2–4.5(5.5) × 0.2–0.9 cm., usually ternately whorled or occasionally opposite, narrowly oblanceolate to elliptic-lanceolate, narrowed towards base, apex acute or obtuse, crenate to clearly and sometimes deeply serrate in upper half, almost entire below, glabrous, densely punctate, nerves prominent below. Flowers solitary or more rarely paired in axils of leaf-like bracts, forming branched, many-flowered raceme-like inflorescence; pedicels up to 0.7 cm. long, filiform, glabrous. Calyx 1.5–2.5 mm. long, 4-lobed, lobes ovate to oblong, obtuse, shortly ciliate towards apex. Corolla 2.5–3.5 mm. long, white to pale mauve. Capsule 2–3.5 mm. long, ripening pale brown.

Zambia. B: 16 km. N. of Senanga, c. 1035 m., fl. 31.vii.1952, *Codd* 7287 (BM; K; SRGH). S: Namwala, c. 1000 m., fl. 9.i.1957, *Robinson* 2103 (K; SRGH). Malawi. S: Nsanje Distr., Makhanga, Chiponje Estate, fl. & fr. 22.vii.1975, *Seyani & Patel* 435 (K). Mozambique. N: c.35 km. from Montepuez to Namuno, c. 480 m., fl. & fr. 31.xii.1963, *Torre & Paiva* 9795 (LISC). Z: Namacurra, 30 m., fl. & fr. 31.i.1966, *Torre & Correia* 14275 (LISC).

Common throughout the tropics. Weed of wasteland, cultivation and grazed grasslands.

26. DIGITALIS L.

Digitalis L., Sp. Pl.: 621 (1753); Gen. Pl., ed. 5: 272 (1754).

Biennial or perennial herbs or rarely small shrubs. Leaves alternate or sometimes predominantly basal, simple, variously dentate to subentire. Flowers zygomorphic in terminal, bracteate, racemes. Calyx 5-lobed; lobes equal to subequal, shorter than corolla tube. Corolla tubular, cylindrical or inflated-globose, usually constricted towards base; limb usually bilabiate with lips spreading or erect, upper shorter. Stamens 4. Capsule conical to ovoid, septicidal. Seeds small, numerous.

A genus of 20 to 30 species from Europe, the Mediterranean and the Canary Islands. It has become naturalised elsewhere in the world, as an escape from cultivation.

Digitalis purpurea L., Sp. Pl.: 621 (1753). TAB. 29. Type from Southern Europe.

Biennial or perennial herb up to 75 (125) cm. tall, erect, shortly pubescent or lanate above, glabrescent below. Basal leaves long petiolate; lamina lanceolate-ovate to ovate, pubescent with indumentum of mixed short glandular and multicellular eglandular white

Tab. 28. SCOPARIA DULCIS. 1, fertile stem (×⅔), from *Robinson* 5412; 2, part of flowering branch (× 4); 3, flower, longitudinal section showing gynoecium (× 6); 4, corolla (× 6), 2–4 from *Fanshawe* 6827; 5, dehiscing capsule (× 8), from *van Rensburg* 2325.

Tab. 29. DIGITALIS PURPUREA. 1, flowering and fruiting branch (× ⅔); 2, habit (× 1/10); 3, flower (× 1); 4, flower, longitudinal section showing gynoecium (× 1); 5, young fruit (× 2), all from *Philcox, et al.* 8973.

hairs, 7.5–15 × 2–6.5 cm., finely serrate to crenate-dentate, obtuse at apex, cuneate at base, coarsely reticulate-veined; petiole 8.5–14 cm. long, indumentum similar to that of lamina; stem leaves few, similar to basal but much reduced in size, continuing above with much greater reduction as sessile, entire floral bracts. Inflorescence 15–50 or more-flowered, simple to rarely slightly branched, flowers drooping. Pedicels 6–18 mm. long. Calyx 5-lobed to base, 6–18 × 3–12 mm., ovate, strongly nerved, the uppermost smaller, more acute at apex. Corolla 4.0–4.5 cm. long, shortly 4-lobed, light purple usually crimson spotted, to rarely white. Stamens included. Capsule 14 × 8 mm., ovoid obtuse.

Zimbabwe. E: Mutare Distr., c. 1750 m., fl. 12.iii.1981, *Philcox & Leppard* 8973 (K; SRGH). **Malawi**. N: Nyika Plateau, Chelinda Area, 2250 m., fl. 22.xi.1967, *Richards* 22685 (K). S: Mulanje Mt., Luchenya Plateau, c. 1890 m., fl. 18.ix.1970, *Pawek* 3880 (K).
Distribution and ecology as for genus.

27. VERONICA L.

Veronica L., Sp. Pl.: 12 (1753).

Annual or perennial herbs, erect, prostrate or decumbent, glabrous or pubescent. Leaves opposite or occasionally verticillate or scattered, entire or variously dentate or crenate. Flowers in terminal or axillary racemes or rarely solitary-axillary, bracteate. Pedicels ebracteolate. Calyx 4- or 5-partite; where present, fifth posterior lobe usually smaller. Corolla 4–5-lobed; tube short, rarely exceeding calyx. Stamens 2, perfect, exserted; inserted on corolla tube at sides of upper lobe; anther-thecae divergent or parallel, obtuse. Style apically subcapitate. Capsule compressed or turgid, bisulcate, loculicidal. Seeds few to many.

A genus of about 300 species mostly from northern temperate regions with a few occurring in the southern hemisphere temperate zone. In the tropics they are usually only found as montane species.
Only three species are native to the Flora Zambesiaca area, but others are sometimes found as accidentally introduced weeds, or as garden escapes. To facilitate ease of identification of the native species, the commonest weed species, *V. persica* Poir. is included in the following key.

1. Stem erect, stout; leaves sessile, semi-amplexicaul; plant of riversides and
   wet places - - - - - - - - - - - - - - 1. *anagallis-aquatica*
 – Stems usually decumbent or procumbent, slender; leaves petiolate; plant not essentially of
   wet places - - - - - - - - - - - - - - - - 2
2. Leaves mostly alternate, but opposite only at base; flowers solitary axillary; pedicels (5)20–35
   mm. long; capsule 4–5 mm. long - - - - - - - - - - *persica*
 – Leaves opposite throughout; flowers in racemes; pedicels 0.5–10 mm. long; capsule 1.24–5 mm.
   long - - - - - - - - - - - - - - - - - - 3
3. Corolla 6–7 mm. long; pedicels 6–10 mm. long, slender; capsule 3.5–4 mm. long 2. *abyssinica*
 – Corolla barely 2 mm. long; pedicels 0.5–1 mm. long, stout; capsule 1.25–2.6 mm.
   long - - - - - - - - - - - - - - - - - 3. *javanica*

1. **Veronica anagallis-aquatica** L., Sp. Pl.: 12 (1753). —Benth. in DC., Prodr. **10**: 467 (1846) as "anagallis". —A. Rich., Tent. Fl. Abyss. **2**: 125 (1851) as "anagallis". —Skan in F.T.A. **4**, 2: 357 (1906) as "anagallis". —Walters & Webb, Fl. Europ. **3**: 248 (1972). Type from Europe.

Annual, somewhat succulent, glabrous to glandular-pubescent with stems creeping, rooting at lower nodes, then erect; erect stems 20–40(60) cm. tall, hollow, branched above. Leaves (18)35–70 mm. long, (3)10–20 mm. wide, opposite, ovate to lanceolate-oblong, shortly serrate to subentire; upper sessile, somewhat amplexicaul; lower long petiolate, obtuse. Racemes axillary, 3.5–10 cm. long, many-flowered; flowers blue or pale-blue, subopposite, bracteate. Bracts 2(5) × 0.25–0.4(1) mm., oblong-lanceolate to ovate-lanceolate. Pedicels 3.5–5(6) mm. long, slender, spreading, glabrous or occasionally minutely glandular-pubescent. Calyx 1.75–2 mm. long, lengthening to 3 mm. or more in fruit, lobes oblong-lanceolate, obtuse to subacute, connate in lower third. Corolla c. 4 mm. in diam.; lower lobe c. 1.5 mm. long, 1 mm. wide, upper and lateral lobes 2 mm. long, upper broadly ovate, lateral broadly elliptic. Capsule 3 mm. long, subspherical, minutely foveolate, strongly nerved.

Zambia. C: Lusaka Distr., Napombo R., SW. of Mwambula, 1200 m., fl. & fr. 5.xi.1972, *Strid* 2459 (K). **Zimbabwe**. C: Harare Distr., Avondale, fl. & fr. 10.iii.1983, *Philcox & Drummond* 9109 (K; SRGH). S: Gwanda Distr., Shashe R., fl. & fr. x.1954, *Davies* 813 (BR; SRGH).

81

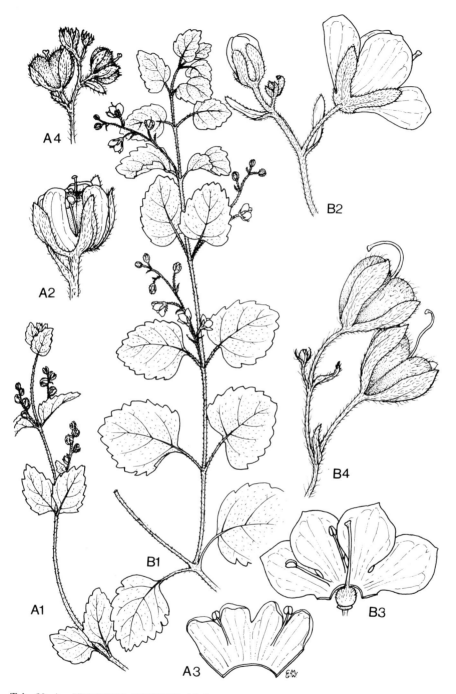

Tab. 30. A.—VERONICA JAVANICA. A1, flowering branch (×⅔), from *Drummond* 4962; A2, flower
(× 8); A3, corolla opened showing androecium (× 8); A4, capsules (× 4), A2–4 from *Fries et al*
3783. B.—VERONICA ABYSSINICA. B1, flowering branch (×⅔); B2, flowers (× 4); B3, corolla
opened showing androecium and gynoecium (× 4); B4, capsules (× 4), B1–4 from *Phillips* 1114A.

Known throughout most of the northern temperate areas of the world and widely spread in upland areas of Africa. Wet places, marshes, stream and riversides; in Flora Zambesiaca area from 600–1500 m.

2. **Veronica abyssinica** Fresen. in Bot. Zeit. **2**: 356 (1844). —Benth. in DC., Prodr. **10**: 490 (1846). —A. Rich., Tent. Fl. Abyss. **2**: 127 (1851). —Engl., Pflanzenw. Ost-Afr. **C**: 358 (1895). —Skan in F.T.A. **4**, 2: 358 (1906). —Rompp in Fedde Rep., Beih. **50**: 141 (1928). —Hedberg in Symb. Bot. Uppsala **15**: 169 (1957). —Hepper in F.W.T.A. ed. 2, **2**: 355 (1963). TAB. **30**, fig. B. Type from Ethiopia.
  Veronica petitiana A. Rich., Tent. Fl. Aybss. **2**: 127 (1851). Type from Ethiopia.
  Veronica africana Hook. f. in Journ. Linn. Soc., Bot. **7**: 208 (1864). Type from Cameroon.

Annual, prostrate herb, trailing or ascending, branched from or towards base; branches up to 7.5 cm. long or occasionally much more, often rooting at nodes, more or less densely pilose with long, white hairs; stem hollow, sometimes becoming somewhat woody. Leaves opposite, petiolate, lamina 15–45 × 10–36 mm., broadly ovate to ovate-elliptic, obtuse to subacute, shallowly cordate to subtruncate at base, at times subcuneate, shallowly to coarsely crenate-serrate; petiole 4–20 mm. long. Racemes axillary, 1–7.5 cm. long, 2–7-flowered; flowers blue, bracteate; bracts 3–4 mm. long, 0.5–1.3 mm. wide, linear-oblong, obtuse. Pedicels (2)6–10(15) mm. long, slender. Calyx 3.5–4.75 mm. long; lobes 1–2 mm. long, oblong to spatulate-oblong, obtuse, long-ciliate. Corolla 6–7 mm. long, upper lobe to 4 mm. wide, broadly ovate to subcircular. Capsule 3.5–4 × (3.5)4(5) mm., obcordate in outline, compressed, long pilose, or only ciliate on suture.

Zambia. N: Mbala Distr., Nkali dambo, c. 1500 m., fl. & fr. 5.i.1952, *Richards* 318 (K). **Zimbabwe**. E: Chimanimani (Melsetter), Glencoe For. Res., fl. & fr. 23.xi.1955, *Drummond* 4965 (K; SRGH). **Malawi**. N: Rumphi Distr., c. 29 km. S. of Chilindi, fl. & fr. 12.v.1977, *Phillips* 2296 (K; SRGH). S: Neno Distr., fl. & fr. 6.vi.1936, *Jackson* 1858 (BR; K).
Also known from Nigeria, Cameroon, Sudan, Ethiopia, Central and East Africa. In damp grasslands, forest fringes and clearings and roadsides; ascending to 2400 m.

3. **Veronica javanica** Bl., Bijdr.: 742 (1826). —Benth. in DC., Prodr. **10**: 489 (1846). —Skan in F.T.A. **4**, 2: 358 (1906). TAB. **30**, fig. A. Type from Java.
  Veronica wogorensis Hochst. ex A. Rich., Tent. Fl. Abyss. **2**: 126 (1851). Type from Ethiopia.
  Veronica chamydryoides Engl., Pflanzenw. Ost-Afr. **C**: 358 (1895). Type from Tanzania.

Annual, decumbent or ascending herb to about 15 cm. tall, much branched; branches up to 25 cm. long, usually densely greyish-white pilose. Leaves opposite; lamina 5.5–25 × 5–16 mm., broadly ovate, apex obtuse, truncate, subcuneate to subcordate, obtuse at base, coarsely creante-serrate, often less densely pilose above, subglabrescent; petiole 1.5–2(3.5) mm. long. Racemes axillary, 1.5–2.5 mm. long, few-flowered; flowers blue to whitish, bracteate. Bracts 2–2.3 × 0.4–0.5 mm., narrowly oblong, obtuse, shortly ciliate. Pedicels 0.5–1 mm. long, stout. Calyx 1.8–2.5 mm. long at flowering, up to 3.4 mm. long in fruit; lobes 0.5–1 mm. wide, oblong to obovate-oblong, obtuse, glabrous to sparsely pilose, ciliate. Corolla up to 2 mm. long. Capsule 1.25–2.6 × 2.25–2.8 mm., broadly obcordate in outline, compressed, glabrous, ciliate on suture.

Zimbabwe. C: Harare, c. 1480 m., fl. & fr. 23.viii.1978, *Biegel* 5682 (K; SRGH). E: Nyanga (Inyanga) Downs, 1800 m., fl. & fr. x.1961, *Wild* 5517 (K; LISC; SRGH). **Malawi**. S: Zomba Plateau, fr. 18.xii.1978, *Salubeni & Masiye* 2392 (SRGH).
In Africa from Eritrea, Ethiopia, Somalia, Kenya, Uganda and Tanzania; also known from India and the Far East. Forest clearings and borders, also recorded as a weed of cultivation; up to 1800 m.

28. MELASMA Berg.

Melasma Berg., Desc. Pl. Cap.: 162 (1767). —Melchior in Notizbl. Bot. Gart., Berlin **15**: 199–127 (1940).

Erect perennial herbs, erect to c. 100 cm. tall, coarse, scabrid or hispid. Leaves opposite, occasionally alternate within inflorescence, sessile or subsessile, entire or toothed. Flowers loosely arranged or clustered in terminal racemes, solitary in axils of leaves or bracts, pedicellate, bibracteolate. Calyx 5-lobed, campanulate, 10-ribbed, inflated in fruit; lobes erect, valvate, subequal. Corolla tubular-campanulate, 5-lobed, larger than calyx, not marcescent, lobes imbricate, widely spreading. Stamens 4, included, didynamous, subequal; anther thecae parallel, distinct, apiculate or acuminate. Style clavate, flattened,

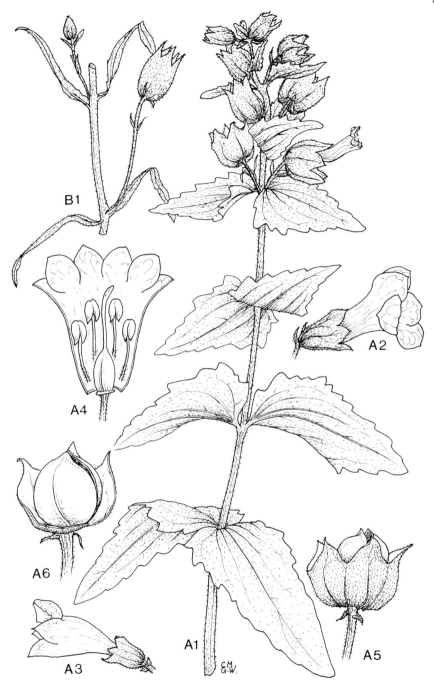

Tab. 31. A.—MELASMA CALYCINUM. A1, flowering branch (×⅔), from *Bramley* 2568; A2, flower
(× ⅓), from *Richards* 7002; A3, flower (× ⅓); A4, flower opened showing androecium and
gynoecium (× ⅓), A3–4 from *Lawton* 1326; A5, fruit (× 1); A6, fruit with part of calyx removed,
A5–6 from *Richards* 27445A. B.—MELASMA SCABRUM. B1, fruiting portion of branch (×⅔),
from *Corby* 708.

persistent, exserted from persistent calyx. Capsule loculicidal, included in calyx. Seeds numerous, 0.5–3 mm. long, straight or curved, outer testa transparent, extended at each end beyond seed nucleus, truncate.

A genus of about 20 species from tropical Africa and America.

Pedicels 2.5–10 cm. long; bracteoles arising well below calyx; leaves 2–6 mm. wide, narrowly-oblong
    or -lanceolate - - - - - - - - - - - - - - - 1. *scabrum*
Pedicels 0.4–1.2 cm. long; bracteoles arising just below calyx; leaves 7–28 mm. wide or more, ovate or
    oblong-ovate to ovate-lanceolate - - - - - - - - - 2. *calycinum*

1. **Melasma scabrum** Berg., Desc. Pl. Cap.: 162, t. 3, fig. 4 (1767). —Benth. in Comp. Bot. Mag. **1**: 202 (1836); in DC., Prodr. **10**: 338 (1846). TAB. **31**, fig. B. Type from S. Africa.
    Nigrina viscosa L., Mant.: 42 (1767). Type from S. Africa.
    Gerardia nigrina L.f., Suppl.: 278 (1781). —Thunb., Prodr. Pl. Cap.: 106 (1800). Type from S. Africa.

Erect or ascending perennial herb up to 100 cm. tall, almost simple or branched, short tuberculate hispid-scabrid almost throughout except flowers; stems ascending, leafy. Leaves 10–50 × 2–6 mm., opposite, narrowly-oblong or lanceolate, obtuse, slightly narrowed at base if at all, subentire to dentate, or occasionally lower shortly lobed, sessile or subsessile. Flowers in lax racemes 6–20 cm. long at ends of branches, long pedicellate. Pedicels (1)2.5–10 cm. long. Bracts subtending pedicels leaflike, becoming smaller above. Bracteoles two, opposite to subopposite in upper half or third of pedicel, but well below calyx, 4–7(13) × 0.5–0.75 mm., linear to narrowly linear-lanceolate, spreading. Calyx 15–22 mm. long in flower, campanulate-oblong, loose, somewhat angled, 10-nerved, densely reticulate-veined, somewhat concave at insertion of pedicel, 5-lobed, increasing to 30 mm. long in fruit, inflated; lobes 3–6 mm. long, broadly deltate, densely ciliate. Corolla white, yellow or purplish-brown, 20–38 mm. long, tube purple or red without, densely to subdensely long pilose without, extending to lobes; lobes 5, rounded, glabrous to subglabrous within. Capsule to 13 × 8 mm., broadly ellipsoid to subobovoid, slightly emarginate below persistent style.

Zimbabwe. C: Harare Distr., Chakoma Farm, fl. 26.iii.1977, *Grosvenor & Renz* 1299 (K; SRGH). E: Nyanga (Inyanga) Distr., Erin For. Res., c. 0.5 km. N. of Senyanga Estate, c. 1500 m., fl. & fr. 19.iii.1983, *Philcox* 9112 (BR; K; LISC; SRGH).
    Known also from S. Africa. Wet vleis, marshes and riversides; 1200–2100 m.

2. **Melasma calycinum** (Hiern) Hemsl. in F.T.A. **4**, 2: 362 (1906). TAB. **31**, fig. A. Type from Angola.
    Velvitsia calycina Hiern, Cat. Afr. Pl. Welw. **1**: 771 (1898). Type as above.

Erect perennial herb up to 75 cm. tall, scabrid almost throughout; stems leafy, simple or rarely branched above woody base. Leaves 25–65(87) × 7–28(40) mm. opposite, ovate or oblong-ovate to ovate-lanceolate, obtuse to rounded at apex, narrowed towards sessile or subsessile base, or occasionally subtruncate, crenate, shallowly, remotely denticulate or coarsely dentate to almost lobed, hispid-scabrid on both surfaces. Flowers alternate either in lax terminal racemes or clustered towards ends of stems, becoming laxer in fruit, pedicellate. Pedicels 4–8(12) mm. long in flower, extending to 45 mm. in fruit, bracteolate. Bracteoles 3–7 mm. long, up to 0.5 mm. wide, subopposite, narrowly linear, arising just below concave base of calyx. Calyx 14–20 mm. long in flower, campanulate, 10-nerved, subequally 5-lobed, densely reticulate-veined, becoming inflated in fruit; lobes 4–9 mm. long, broadly deltoid, not densely hairy. Corolla yellow to orange or white, reddish or purple veined, 20–30 mm. long, subglabrous to shortly glandular pubescent without; lobes 5, rounded to broadly ovate, up to 1 cm. in diam., glabrous or minutely pubescent above, shortly ciliate or not. Capsule 9–11 mm. long, c. 5 mm. broad, ovoid, somewhat compressed.

Zambia. N: Danger Hill, c. 16 km. N. of Mpika, 1370 m., fl. & fr. 26.x.1967, *Simon & Williamson* 1211 (SRGH). W: Mwinilunga Distr., R. Kasompa, fl. & fr. 31.x.1937, *Milne-Redhead* 3037 (K). C: Serenje Distr., 128 km. from Kapiri Mposhi, on Mpika road, 1200 m., fl. 2.xi.1962, *Richards* 16867 (K). **Zimbabwe**. C: Harare Distr., Cleveland Dam, fl. 16.xii.1945, *Wild* 564 (SRGH). **Malawi**. N: Mzimba Distr., Lunyangwa R., Mzuzu, 1370 m., fl. 29.xi.1972, *Pawek* 6040 (K; SRGH).
    Also known from Angola. Swamps, marshes, wet grassland bordering rivers; 1200–2250 m.

29. ALECTRA Thunb.

Alectra Thunb., Nov. Gen. Pl.: 81 (1784). —Benth. & Hook. f., Gen. Pl. **2**: 966 (1876).

Annual or perennial herbs, usually erect, simple or branched, usually hispid, root parasites, turning black when dry. Leaves opposite or alternate, sessile or shortly petiolate, often very small, scale-like, entire or toothed. Flowers solitary in axils of upper leaves or leaflike bracts, sessile or shortly pedicellate, yellow to orange with brown to reddish-purple veins, bibracteolate. Calyx campanulate, 5-lobed, 10-nerved, calyx not enlarging in fruit; lobes subequal, usually subequalling tube. Corolla thin, early marcescent, persistent in calyx, surrounding capsule, narrowly campanulate, slightly oblique, longer than calyx; lobes rounded, spreading to slightly recurved. Stamens 4, didynamous, at times subequal, attached towards base of corolla tube, included. Filaments all bearded, or only the two larger ones bearded or all glabrous, bearding may be overall or in lower half or upper half or just below anthers. Anthers coherent or connivent in pairs, thecae equal or unequal, apiculate or not. Ovary glabrous; style clavate above middle, recurved, included with stamens in persistent corolla. Capsule globose to broadly ovoid, included, loculicidal. Seeds very numerous, linear or clavate, straight or curved, slender, truncate.

A genus of some 30 species from tropical America, Asia and Africa and subtropical southern Africa.

1. Plants leafy; leaves well developed, not scale-like - - - - - - - - 2
 - Plants without well developed leaves; leaves scale-like, appressed to stem or spreading 9
2. Anther thecae apiculate - - - - - - - - - - - - - 3
 - Anther thecae not apiculate - - - - - - - - - - - - 8
3. Filaments clearly bearded - - - - - - - - - - - - 4
 - Filaments glabrous or not clearly bearded, very rarely with an occasional hair present 7
4. Stems hispid with clearly retrorse fine hairs, vegetative indumentum similar; flowers in spike-like compact racemes at ends of branches - - - - - - - - - 1. *rigida*
 - Stems pilose to hispid with patent or antrorse hairs or subglabrous, if hairs retrorse then coarse and thickened at base; flowers not compacted at ends of branches - - - - 5
5. Calyx stipitate glandular; corolla short glandular-pilose without - - - 2. *glandulosa*
 - Calyx and corolla eglandular - - - - - - - - - - - 6
6. Calyx 11.5–18 mm. long, lobes 5–10 mm. long - - - - - 3. *dolichocalyx*
 - Calyx much less than 11.5 mm. long - - - - - - - - 4. *sessiliflora*
7. Plants hispid-scabrid; hairs with large white tubercles at base; flowers subsessile in ± compact spikes at ends of branches - - - - - - - - - 5. *asperrima*
 - Plants pubescent to shortly hispid-pubescent; flowers pedicellate, pedicels 3–6 mm. long, pubescent - - - - - - - - - - - - - 6. *pubescens*
8. Filaments glabrous; branches floriferous almost throughout - - - - 7. *vogelii*
 - Filaments bearded; branches floriferous in upper part, rarely throughout 8. *picta*
9. Filaments glabrous - - - - - - - - - - - 9. *parasitica*
 - Filaments bearded - - - - - - - - - - - - 10
10. Plant to 10 cm. tall; stems glabrous; leaves 2–3.5 mm. long and wide, reniform, closely appressed; flowers crowded at end of stem - - - - - - - - 10. *bainesii*
 - Plant 10–45 cm. tall; stems shortly hispid-pubescent; leaves not as above nor closely appressed; flowers loosely arranged on stem - - - - - - 11. *orobanchoides*

1. **Alectra rigida** (Hiern) Hemsl. in F.T.A. **4**, 2: 369 (1906). Type from Angola.
  Melasma rigidum Hiern, Cat. Afr. Pl. Welw. **1**: 767 (1898). Type as above.

Erect annual herb, 10–50(250) cm. tall; stems slender, erect, simple or branched from about midway or above, hispid with retrorse hairs. Leaves sessile, all opposite, erect, subappressed, 10–20(30) × 1.5–3 mm., linear-lanceolate or oblong, obtuse, broadly caudate to rounded at base, fine appressed retrorse hirsute especially beneath; margin thickened, bristly with short tubercle-based coarse hairs, 2–3(4) thickened teeth on each side, subprominently 1-nerved. Flowers in subcompact spike-like racemes at ends of branches, shortly pedicellate, solitary in axils of leaf-like bracts, inflorescence becoming lax at fruiting. Pedicels 1–1.5 mm. long. Bracts 5–18 × 2–3.75 mm., ovate, acute or obtuse, 1-nerved, leaf-like, entire or with 1 or 2 teeth towards apex. Bracteoles 3–5 × 0.5–1.75 mm., linear. Calyx 6.5–7 mm. long, clearly 10-nerved, subglabrous to finely retrorse pubescent especially on nerves; lobes 2.5–3 mm. long, shorter than tube, triangular, obtuse, coarsely hairy on margin. Corolla yellow with purplish venation, to c. 12 mm. long, narrow at base widening into limb of short, rounded, subequal lobes. Filaments all glabrous, anther thecae apiculate. Capsule 5–6 × c. 3 mm., oblong-ovoid.

Zambia. N: Kawambwa, c. 1350 m., fl. & fr. 21.vi.1957, *Robinson* 2336 (K). W: Mwinilunga Distr., Zambezi Rapids, fl. 18.v.1969, *Mutimushi* 3282 (K). C: Chakwenga Headwaters, 100–129 km. E. of Lusaka, fl. & fr. 27.iii.1965, *Robinson* 6468 (BR; K; SRGH). **Zimbabwe**. C: Harare Distr., Chinamora Res., Ngomakurira, 1675 m., fl. 25.iii.1952, *Wild* 3787 (K; SRGH). **Malawi**. N: Mzimba Distr., Mzuzu, Katoto, fl. 10.vi.1972, *Pawek* 5444 (K).

Also known from Angola and Tanzania. Wet, boggy, acid-soil grasslands; 1350–1700 m.

2. **Alectra glandulosa** Philcox in Bol. Soc. Brot., sér. 2, **60**: 269 (1987). TAB. **32**. Type: Zambia, 42 km. from Mwinilunga on road to Solwezi, Mundwizi Dambo, 1700 m., fl. & fr. 17.v.1986, *Philcox, Pope, Chisumpa & Ngoma* 10350 (K, holotype; BR; LISC; MO; NDO; SRGH; isotypes).

Erect or decumbent annual herb, 6–14 cm. tall, simple or with branches up to 20 cm. long; branches mostly opposite, widely spreading, leafy. Leaves opposite, (10)15–30(50) × 1–4 mm., lanceolate-oblong to lanceolate-elliptic, retrorse hispid-pilose, short hispid-scabrid on margins and midrib beneath, glabrous above, apex acute or obtuse, narrowly or broadly cuneate at base, margin with 1–3(5) prominent teeth on each side. Flowers solitary in axils of leaf-like bracts, pedicellate. Pedicels 1–3(5) mm. long, slender. Bracts similar to leaves, longer or shorter than flowers. Bracteoles c. 6 × 0.5 mm., narrowly linear, minutely hispid on margin. Calyx 7–8.5(13) mm. long, prominently 10-nerved, stipitate-glandular mixed with short white hairs on nerves; lobes 1–1.75 mm. long, subequal, broadly triangular, obtuse, shortly hispid on margin. Corolla yellow, (10)15–19 mm. long, subdensely short glandular-pilose without; lobes 4–5 mm. long, rounded. Stamens subequal; filaments 3–6 mm. long, somewhat heavily bearded; anthers c. 1.5 mm. long, ellipsoid, apiculate. Capsule c. 11 × 7 mm., broadly ovoid, obtuse.

Zambia. N: Mbala Distr., Kambole Escarpment, 1828 m., fl. 22.iv.1969, *Richards* 24520 (K). W: Kafue R., 11 km. N. of Chingola, fl. & fr. 4.v.1960, *Robinson* 3703 (K).

Not known from outside Zambia. Moist grassland; from 1500–1830 m.

3. **Alectra dolichocalyx** Philcox in Bol. Soc. Brot., sér. 2, **60**: 269 (1987). TAB. **33**. Type: Zambia, Kabulamwanda, 120 km. N. of Choma, c. 1000 m., fl. & fr. 21.iv.1955, *Robinson* 1238 (K, holotype; SRGH, isotype).

Erect, prostrate or scrambling, short-rooted annual to c. 20 cm. tall; branches up to 18(40) cm. long, alternate, weakly medium- to long-pilose. Leaves 30–60 × 2–6 mm., linear-lanceolate, acute, subcordate at base, dentate with 3–7 teeth on each margin, glabrous above, pilose-scabrid on somewhat thickened margin and major nerves beneath. Flowers solitary in axils of leaves or leaf-like bracts, pedicellate. Pedicels 1.5–2.5 mm. long, to 4.5 mm. at maturity, stout. Bracts similar to leaves, equal to or longer than calyx. Bracteoles 10–12 mm. or more long, c. 1 mm. wide, pilose on margins. Calyx 11.5–18(21) mm. long, thin, clearly 10-nerved, long-pilose only on nerves and margins of lobes; lobes 5–10 mm. long, narrowly triangular. Corolla pale yellow or cream, 14–22 mm. long, glabrous without, eglandular. Stamens subequal, 4.5–5.5 mm. long; filaments bearded, anther thecae apiculate. Capsule c. 8.5 × 8.5 mm., compressed spherical to broadly ovoid.

Zambia. S: Kabulamwanda, 120 km. N. of Choma, c. 1000 m., fl. & fr. 21.iv.1955. *Robinson* 1238 (K; SRGH).

Also known from Kenya, Uganda and Tanzania. The above collection along with the East African material, has hitherto been referred to the Angolan species *A. aurantiaca* Hemsl. from which it differs in the linear-lanceolate sessile to subsessile leaves, larger calyx, very short root and prostrate habit.

4. **Alectra sessiliflora** (Vahl) Kuntze, Rev. Gen. Pl. **2**: 458 (1891). Type from S. Africa.
   Gerardia sessiliflora Vahl, Symb. Bot. **3**: 79 (1794). Type as above.

Erect annual, (5)15–50(60) cm. tall; stems straight, slender to stout, simple or branched, pilose to variously hispid. Leaves opposite, alternate within inflorescence, 14–25(55) × 8–18(28) mm., shape very varied from circular to ovate to broadly or narrowly lanceolate, subentire or crenate to coarsely toothed, acute or obtuse, sessile or subsessile to shortly petiolate, acute, cuneate, rounded to cordate at base, appressed or subappressed to spreading, hispid to subglabrous. Petiole 0–2(3) mm. long. Flowers solitary in axils of leaves or leaf-like bracts; pedicels absent or about 0.5 mm. long, rarely up to 1.5 mm. long.

Tab. 32. ALECTRA GLANDULOSA. 1, habit (× ⅔); 2, flower (× 2); 3, corolla opened showing androecium (× 2); 4, section through corolla showing gynoecium (× 3); 5, fruiting calyx (× 2); 6, capsule with calyx removed (× 2), all from *Robinson* 3546.

Bracts leaf-like, longer or shorter than flowers, usually coarsely toothed, coarsely hairy to subglabrous, ciliate. Bracteoles linear, subulate to filiform, equalling or slightly shorter than calyx, hairy ciliate to glabrous. Calyx 6–8 mm. long, 10-nerved, glabrous to ciliate on nerves and margins of lobes, very rarely overall; lobes 3–5 mm. long, subequal, triangular, acute. Corolla pale yellow to dull orange, at times with reddish-purple venation, campanulate, early marcescent, slightly or up to one third longer than calyx. Stamens unequal, longer filaments bearded, very rarely all glabrous; anther thecae apiculate. Capsule about 5.5 × 5.5 mm., spherical, glabrous.

This is a very complicated species which has, in the main, been clarified by Hepper (in Kew Bull. **14**: 405, 1960). In this work he acknowledges three varieties encompassing most of the names which have represented elements of this complex over the years. For the purpose of this work I have chosen to follow his treatment and apply his criteria here.

1. Leaves erect, sessile; lamina broadly cuneate at base, upper leaves never
   petiolate - - - - - - - - - - - - - - - var. *sessiliflora*
 – Leaves spreading, petiolate at least in the upper ones - - - - - - - 2
2. Leaves all petiolate, cuneate to rounded at base, oblong-lanceolate to lanceolate, rarely ovate;
   stem appressed-pubescent to subglabrous - - - - - - - var. *monticola*
 – Leaves sessile or petiolate, lower leaves truncate, rounded or cordate, sometimes with a petiole
   c. 1 mm. long, upper leaves often cuneate and petiolate; lamina subcircular or broadly-ovate to
   ovate; stem densely pubescent with erect to patent, or retrorse, coarse white
   hairs - - - - - - - - - - - - - var. *senegalensis*

Var. **sessiliflora** —Hepper in Kew Bull. **14**: 405 (1960).
   Alectra melampyroides Benth. in DC., Prodr. **10**: 339 (1846). Type from S. Africa.
   Melasma barbatum Hiern in F.C. **4**, 2: 374 (1904). Type from S. Africa.
   Alectra barbata (Hiern) Melch. in Notizbl. Bot. Gart. Berl. **15**: 25 (1940). Type as above.
   Alectra senegalensis var. *pallescens* Bonati in Bull. Soc. Bot. Fr. **74**: 96 (1927) nom. nud. —Melch. in Notizbl. Bot. Gart. Berl. **15**: 439 (1941).

Calyx ciliate on upper margin and rarely on nerves; filaments glabrous or the two longer bearded.

Zambia. B: 10 km. E. of Mongu, fl. & fr. 24.x.1965, *Robinson* 6694 (K). N: Mbala Distr., Kali (Nkali) Dambo, 1525 m., fl. & fr. 16.v.1955, *Richards* 5737 (K). W: Chingola, Mushishima, fl. & fr. 18.v.1968, *Mutimushi* 2611 (K). C: Great East Road, between Undaunda and Rufunsa, 1100 km., fl. 6.iv.1972, *Kornaś* 1531 (K). S: Livingstone, north bank of Zambezi R., 915 m., fl. ix.1912, *Rogers* 5780 (BM; K). **Zimbabwe**. N: Guruve (Sipolilo) Distr., Nyamunyeche Estate, fl. & fr. 5.iii.1979, *Nyariri* 740 (SRGH). W: Matobo Distr., Besna Kobila Farm, c. 1430 m., fl. ii.1954, *Miller* 2147 (K). C: Marondera Distr., Lendy Estate, c. 19 km. SSW. of Marondera, fl. & fr. 3.iii.1959, *Drummond* 5855 (K). E: Chimanimani (Melsetter) Distr., fl. ix.1953, *Williams* 140 (K). S: Masvingo Distr., Great Zimbabwe Nat. Park, fl. 29.iii.1973, *Chiparawasha* 643 (K). **Malawi**. N: Nkhata Bay Distr., Viphya, Luwawa Dam, 120 km. S. of Mzuzu, c. 1610 m., fl. 8.ii.1971, *Pawek* 4399 (K). S: Chikwawa Distr., Lower Mwanza R., 180 m., fl. 6.x.1946, *Brass* 18004 (K).
   Also known from S. Africa and Madagascar. Lake and riverside beaches, marshes, swamps and wet grasslands; up to 1610 m.

Var. **senegalensis** (Benth.) Hepper in Kew Bull. **14**: 405 (1960). Type from West Africa.
   Alectra senegalensis Benth. in DC., Prodr. **10**: 339 (1846). Type as above.
   Alectra cordata Benth. in DC., Prodr. **10**: 339 (1846). Type from Ethiopia.
   Alectra senegalensis var. *minima* A. Chev., Expl. Bot.: 475 (1920) nomen nudum.

Calyx ciliate on nerves and margins; 2 longer filaments always bearded; stem densely pubescent with erect, or retrorse, coarse white hairs.

Caprivi Strip: Singalamwe, c. 1010 m., fl. & fr. 1.i.1959, *Killick & Leistner* 3241 (SRGH). **Botswana**. N: Old Ngolchal Borokha (Sandadibide) junction, fl. 10.xii.1972, *Smith* 303 (SRGH). **Zambia**. B: c. 24 km. ENE. of Mongu, fl. & fr. 10.i.1959, *Drummond & Cookson* 6281 (K; LISC). N: Mbala Distr., near Welusi R., 86 km. N. of Kasama, 1750 m., fl. 29.iv.1986, *Philcox, Pope & Chisumpa* 10184 (K). W: Mwinilunga Distr., c. 1 km. from Kabompo Gorge, 1200 m., fl. 24.xi.1962, *Richards* 17311 (K). C: Kabwe Distr., Mwomboshi R., 12 km. N. of Chipembi, 1000 m., fl. & fr. 30.ix.1972, *Kornaś* 2195 (K). S: c. 20 km. N. of Choma, fl. 24.iii.1957, *Robinson* 2189 (K). **Zimbabwe**. N: Gokwe Distr., Sengwa Gorge, fl. 17.x.1968, *Jacobson* 248 (K). W: Matobo Distr., Besna Kobila Farm, 1430 m., fl. iii.1959, *Miller* 5180 (K). E: Mutare Distr., Vumba Mts., Castle Beacon, 1930 m., fl. & fr. 18.iii.1956, *Chase* 6035 (BM; K). S: Beitbridge Distr., near Tshiturapadsi (Chiturpadzi), 40 km. NNW. of Bubi (Bubye)-Limpopo R. confluence, fl. 12.v.1958, *Drummond* 5771 (K). **Malawi**. N: Rumphi Distr., Nyika Plateau, Sangule kopje, 8 km. SW. of Chelinda, 2280 m., fl. 15.v.1970, *Brummitt* 10734 (K). C: Dedza

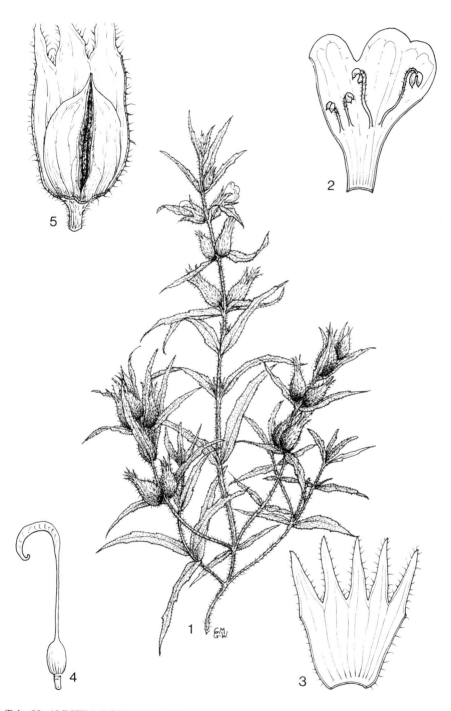

Tab. 33. ALECTRA DOLICHOCALYX. 1, flowering and fruiting stem (× ⅔); 2, corolla opened showing androecium (× 3); 3, calyx opened out (× 3); 4, gynoecium (× 3); 5, dehiscing capsule with part of calyx removed (× 3), all from *Robinson* 1238.

Distr., Chongoni Mt., E. slopes, 1750 m., fl. & fr. 3.viii.1978, *Iwarsson & Ryding* 878 (K). S: Mulanje Mt., below Thuchila (Tuchila) Hut, 1940 m., fl. & fr. 5.iv.1970, *Brummitt* 9625 (K). **Mozambique.** N: Malema, foot of Serra Mancuni, c. 20 km. from Mutuali, c. 900 m., fl. 17.iii.1964, *Torre & Paiva* 11238 (LISC). Z: Lugela Distr., Momba, fl. sine die, *Faulkner* 18A (K). T: c. 9 km. S. of Zobue, Kirk Range, 700 m., fl. 1.viii.1962, *Leach & Schelpe* 10504 (SRGH). MS: Gorongoza Mt., Gogozo summit, **fl. & fr.** iv.1972, *Tinley* 2501 (SRGH). GI: Ponta da Barra Falsa, Pomene R., 1–80 m., fl. 21.xi.1958, *Mogg* 28944 (K).

Also known from West and Central Africa, Sudan, Ethiopia and East Africa. Wet grassland, *Brachystegia* woodlands, stream- and lake-sides; from sea level up to 2280 m.

Var. **monticola** (Engl.) Melch. in Notizbl. Bot. Gart. Berl. **15**: 126 (1940); op. cit.:438 (1941). Type from Tanzania.

> *Glossostylis avensis* Benth., Scroph. Ind.: 49 (1835). Type from Asia.
> *Alectra indica* Benth. in DC., Prodr. **10**: 339 (1846). Type from Asia.
> *Melasma indicum* var. *monticolum* Engl., Bot. Jahrb. **30**: 402 (1901). Type from Tanzania.
> *Alectra communis* Hemsl. in F.T.A. **4**, 2: 372 (1906). Type: Malawi, without locality, 1891, *Buchanan* 520 (K, lectotype (chosen here); BM).

Calyx ciliate on nerves and margins; 2 longer filaments always bearded; stem more or less appressed pubescent to subglabrous.

Zambia. B: Masese, fl. & fr. 10.v.1961, *Fanshawe* 6548 (K). N: Chishimba Falls, 40 km. W. of Kasama, 1610 m., fl. 30.iv.1986, *Philcox, Pope & Chisumpa* 10214 (K). W: near Luakela Bridge, between Mwinilunga and Ikelenge, 1600 m., fl. & fr. 15.v.1986, *Philcox, Pope, Chisumpa & Ngoma* 10289 (K). **Zimbabwe.** N: Gokwe Distr., fl. & fr. 12.iii.1964, *Bingham* 1335 (SRGH). E: Chipinge Distr., Gungunyana For. Res., fl. xi.1967, *Goldsmith* 73/67 (BR; K; LISC; SRGH). **Malawi.** N: Nyika Plateau, 8 km. E. of Nganda by Wovwe R., 1920 m., fl. 3.viii.1972, *Brummitt, Munthali & Synge* WC145 (K). S: Mt. Chiradzulu, to 2130 m., fl. & fr. sine die, *Whyte* s.n. (K). **Mozambique.** Without further locality, fl. 8.xii.1910, *Dawe* 346 (K).

Widespread from West, Central and East Africa and Mauritius to India, Burma, Thailand and China; Philippines and Formosa. Wet or moist grassland and open woodland; up to 2130 m.

5. **Alectra asperrima** Benth. in DC., Prodr. **10**: 340 (1846). —Hemsl. in F.T.A. **4**, 2: 369 (1906). Type from Ethiopia.

> *Glossostylis asperrima* Hochst. in schaed. Schimperi Iter Abyss. Sectio secunda No. 1094 (1842–43) nom. nud.

Erect annual (10)20–55 cm. tall; stems erect, branched, scabrid. Leaves subopposite or alternate, 10–25 × 2.5–6 mm., lanceolate to oblong-lanceolate or narrowly ovate, sessile, roughly hispid-scabrid on both surfaces with hairs having large white tubercles at base, irregularly dentate, subprominently 3-nerved. Flowers almost sessile in subcompact spikes at ends of branches, individual flowers in axils of leaf-like bracts. Bracts longer than flowers, lanceolate, usually 1–3-toothed. Bracteoles up to 6 mm. long, linear or lanceolate, almost equalling calyx. Calyx 6–8 mm. long, subprominently 10-nerved, hispid especially on nerves; lobes 3.5–6 mm. long, deltoid, densely hispid-scabrid. Corolla yellow to orange-yellow with purplish striations. Filaments all glabrous; anther thecae shortly apiculate. Capsule 3–3.5 mm. long, 3–3.5 mm. in diam., globose.

Zambia. W: Mwinilunga Distr., beside Luakela Bridge on road from Mwinilunga to Ikelenge, 1600 m., fl. & fr. 15.v.1986, *Philcox, Pope, Chisumpa & Ngoma* 10291 (K). E: Chipata Distr., Luangwa R. Valley, fr. 13.vi.1970, *Sayer* 549 (SRGH). **Zimbabwe.** N: Bindura Distr., 970 m., fl. & fr. 16.vii.1969, *Mogg* 34342 (SRGH). C: Marondera Distr., c. 20 km. SSW. of Marondera, fl. & fr. 3.iii.1959, *Drummond* 5858 (K; SRGH). W: Matobo Distr., Besna Kobila Farm, c. 1450 m., fl. & fr. iv.1957, *Miller* 4296 (K; SRGH). **Malawi.** N: c. 6 km. N. of Karonga turn-off on Chisenga to Chitipa road, fl. & fr. 21.iv.1986, *Philcox, Pope & Chisumpa* 10061 (K).

Also known from Ethiopia, Somaliland, Sudan, Kenya, Uganda and Tanzania. In wet grasslands; up to 1700 m.

6. **Alectra pubescens** Philcox in Bol. Soc. Brot., sér. 2, **60**: 269 (1987). Type: Zambia, Mbala Distr., Chilongowelo Escarpment, 1500 m., fl. & fr. 6.iv.1962, *Richards* 16254 (K, holotype).

Erect or decumbent short-rooted annual to 20 cm. tall; branches 5–12 cm. long, opposite, widely spreading to 90° with main stem, shortly white hispid-pubescent to pubescent. Leaves 15–30 × 1.5–4 mm., opposite or subopposite, narrowly lanceolate, acute, broadly cuneate to subcordate at base, sessile, coarsely remotely dentate with 3–5 prominent teeth on each margin, densely short, fine hispid-pubescent above, more coarsely so on margin and major nerves beneath. Flowers solitary in axils of leaves or leaf-like bracts, pedicellate. Pedicels 3–6 mm. long, pubescent. Bracts similar to leaves, exceeding flowers or not. Bracteoles 4–5 × 0.5–0.8 mm., subopposite, arising on pedicel up to 1.5 mm. below calyx, linear, pubescent. Calyx 5–9 mm. long, generally short fine hispid-pubescent, eglandular or with few stipitate glands present towards base, prominently 10-ribbed; lobes 1.5–2.5 mm. long, triangular, acute. Corolla pale yellow to pale orange, c.11 mm. long, sparsely short glandular-pilose without. Filaments glabrous or very rarely subglabrous with occasional hair present, not clearly bearded; anther thecae apiculate. Capsule 7.5 × 4.5 mm., ovoid-ellipsoid.

Zambia. N: Kasanshi Dambo, 55 km. ESE. of Mporokoso, fl. & fr. 13.v.1962, *Robinson* 5173 (K). Known only from northern Zambia.

7. **Alectra vogelii** Benth. in DC., Prodr. **10**: 339 (1846). —Hemsl. in F.T.A. **4**, 2: 368 (1906). Type from Nigeria.

Erect annual, 15–30(50) cm. tall; stems branched or rarely simple, stout, straight, densely to somewhat densely patent-hispid, floriferous almost throughout. Leaves or leaf-like bracts in floriferous part of stem, spreading alternate, 15–35 × 3–9 mm., lanceolate or linear, subacute to obtuse, entire or with 2–5 large widely spaced teeth on each margin, sessile or narrowed into short petiole-like base, hispid-scabrid on both surfaces or subglabrous beneath, hispid only on margins and midrib; lower stem leaves upto 20(55) × 7–12(18) mm., ovate to ovate-lanceolate, obtuse, crenate to irregularly large toothed, otherwise similar to upper leaves. Flowers solitary in axils of upper leaves or leaf-like bracts, pedicellate. Pedicels (1)3–4 mm. long. Bracteoles 5–8 mm. long, up to 1 mm. wide, linear to narrowly spathulate, hispid-ciliate. Calyx (4.5)7–8.5 mm. long, somewhat densely hispid; lobes (1.5)2.75–5 mm. long, subequal, triangular, obtuse, densely hispid on margin. Corolla yellow to pale orange, purplish-veined, longer than calyx. Stamens subequal; filaments glabrous; anthers not apiculate. Capsule 5–6 × 5–6 mm., broadly ovoid to globose.

Botswana. N: Tsodilo Hill, fl. & fr. 8.ii.1964, *Guy* 180/64 (SRGH). SE: Sebele, 970 m., fl. 14.iii.1978, *Hansen* 3079 (K). **Zambia**. N: Samfya, fl. & fr. 9.v.1958, *Fanshawe* 4417 (BR; K). E: Chipata (Fort Jameson), fl. & fr. 9.v.1963, *van Rensburg* 2110 (K; SRGH). S: Kalomo, fl. 10.ii.1962, *Fanshawe* 9172 (SRGH). **Zimbabwe**. N: Trelawney, fl. & fr. 16.vi.1943, *Jack* 212 (K; SRGH). W: Lukosi R., Hwange, fl. 23.vi.1934, *Eyles* 8089 (K). C: Kadoma (Gatooma) Distr., fl. 22.iii.1950, *McKinstry* in GHS 27285 (K; SRGH). **Malawi**. C: Dedza Distr., Chongoni Forest School, fl. & fr. 16.iv.1968, *Jeke* 174 (K). S: Blantyre Distr., Matenje Road, 1–2 km. N. of Limbe, 1185 m., fl. 10.iii.1970, *Brummitt* 9005 (K). **Mozambique**. N: Cabo Delgado, c. 9 km. from Montepuez towards Nantulu, c. 410 m., fl & fr. 8.iv.1964, *Torre & Paiva* 11737 (LISC). Z: Mocuba Region, Namagoa, 200 km. inland from Quelimane, 60–120 m., fl. & fr. ii.1945, *Faulkner* 18 (K). MS: mouth of Zambezi R., Milambe, fl. 8.ii.1861, *Kirk* s.n. (K). GI: between Manjacaze and Coolela, 1 km. from Manjakaze, fl. & fr. 16.vi.1960, *Lemos & Balsinhas* 111 (K).
Also known from the Guinea Republic, Ghana, Nigeria, Portuguese Congo and East Africa. Mostly as a weed of cultivation parasitising various leguminous crops; otherwise recorded from grassy riverbanks and lake margins; up to about 1190 m.

8. **Alectra picta** (Hiern) Hemsl. in F.T.A. **4**, 2: 368 (1906). Type from Angola.
   Melasma pictum Hiern, Cat. Afr. Pl. Welw. **1**: 770 (1898). Type as above.

Erect annual, 15–45 cm. tall, branched; stems straight, stout to flexuous, patent-hispid, floriferous in upper part of stem, not generally throughout. Leaves or leaf-like bracts spreading, opposite or opposite or nearly so, 8–20(35) × 2–9 mm., elliptic to linear or linear-lanceolate, obtuse to rounded, entire to crenate or with two small teeth, lower shortly petiolate, hispid. Flowers solitary in axils of leaf-like bracts, pedicellate. Pedicels 1–5(7) mm. long. Bracteoles linear, shorter than bracts, subglabrous to shortly hispid. Calyx 5–7(8) mm. long, campanulate, shortly hispid; lobes 1.5–3.5(5) mm. long, subequal, broadly triangular, obtuse, shortly hisute on margin, often densely so. Corolla yellow with purplish veins, much longer than calyx. Stamens subequal; filaments all bearded, at times densely so; anthers not apiculate. Capsule about 4.5(9) × 4.5(9) mm., globose to broadly ovoid.

Botswana. N: Nata Village, fl. 17.iv.1976, *Ngoni* 506 (K). **Zambia**. B: Masese, fl. 6.ix.1969, *Mutimushi* 3608 (K). N: Mbala Distr., 70 km. from Nakonde on Mbala road, fl. & fr. 23.iv.1986, 1420 m., *Philcox, Pope & Chisumpa* 10090 (K; LISC; NDO; SRGH). S: Mumbwa, fl. & fr. 24.vi.1964, *van Rensburg* 2937 (K; SRGH). **Zimbabwe**. N: Darwin Distr., near Musengezi (Usengesi) Camp, fl. 11.v.1955, *Whellan* 906 (K; SRGH). W: Victoria Falls, fl. & fr. vii.1906, *Allen* 404 (SRGH). C: Chegutu

(Hartley) Distr., Poole Farm, fl. 21.iii.1946, *Hornby* 2899 (K; SRGH). E: Chimanimani (Melsetter) Distr., Mt. Peni summit, c. 1700 m., fl. 14.iv.1957, *Chase* 1407 (K). **Malawi**. C: Lilongwe, fl. 10.iv.1953, *Jackson* 1194 (BM). **Mozambique**. N: Ribáuè, Mt. Namatapun, c. 600 m., fl. & fr. 30.i.1964, *Torre & Paiva* 10329 (LISC). MS: Chimanimani Mts., between Skeleton Pass and The Plateau, fl. 8.iv.1967, *Grosvenor* 350 (K). GI: Gaza, Xai Xai (Vila João Belo), margin of R. Limpopo, fl. & fr. 10.ii.1942, *Torre* 3918 (LISC). M: Inhaca Is., c. 34 km. E. of Maputo (Lorenco Marques), 0–152 m., fl. 23.ix.1957, *Mogg* 27506 (K).

Known also from Angola and Tanzania. Streambanks, moist open woodland or scrubland, often on well drained soil; sea level up to 2280 m.

9. **Alectra parasitica** A. Rich., Tent. Fl. Abyss. **2**: 117 (1850). —Hemsl. in F.T.A. **4**, 2: 366 (1906). Type from Ethiopia.

Erect, (7)18–32 cm. tall; stems simple or branched, shortly pubescent to subglabrous. Leaves 5.5–7.5(17) × 1.5–2.5(4) mm., thick, spreading, subopposite to alternate, lanceolate-oblong, acute, remotely short-toothed, sessile, hispid to hispid-scabrid. Flowers solitary in axils of leaf-like bracts, shortly pedicellate; pedicels 1–1.5 mm. long. Bracts 2.5–6 × 0.5–1 mm., linear-lanceolate, hispid. Bracteoles 1.5–3 × 0.2-0.75 mm., linear to filiform, shortly ciliate. Calyx 3.5–6.5 mm., shortly hispid-pubescent to glabrous; lobes 1.5–3 mm. long, broadly triangular, shortly hispid-ciliate. Corolla yellow to orange with purplish venation, lobes subequal, rounded. Stamens subequal; filaments short, glabrous; anthers unequal, not apiculate. Capsule 4.5–6.5 × 5–6.5 mm., subspherical.

Zambia. N: Mbala Distr., Chilongowelo, 1440 m., fl. 27.iii.1957, *Richards* 8901 (K). S: Mumbwa, fl., *Macaulay* 1171 (K). **Zimbabwe**. E: Mutare (Umtali) Distr., Menine R., fl. & fr. 13.iv.1957, *Chase* 4486 p.p. (K). **Malawi**. N: Karonga Distr., Karonga, 610 m., fl. & fr. 27.iv.1977, *Pawek* 12733 (K; SRGH).

Also known from Ethiopia and Somalia. Streamsides and wet bush; up to about 1450 m.

10. **Alectra bainesii** Hemsl. in F.T.A. **4**, 2: 365 (1906). Type: Botswana, Kobie to North Shaw Valley, i–iii.1863, *Baines* s.n. (K, holotype).

Erect herb to 10 cm. tall, almost leafless; stem thick, glabrous, simple or rarely once-branched. Leaves 2–4.5 × 2.5–4 mm., scale-like, reniform, sessile, closely appressed. Flowers crowded at ends of stems, subsessile. Pedicels up to 0.5 mm. long. Bracts 5.5–6(14) mm. wide, broadly ovate, obtuse, entire or with one to two teeth on each side, 3–5-nerved, shortly hispid-pilose. Bracteoles 3.5–6 × 0.75–1 mm., oblong-spathulate, obtuse, shortly hispid-pilose. Calyx 5.5–7.5 mm. long, subglabrous to hispid; lobes equal to subequal, 1.3–2 mm. long, broadly triangular, obtuse. Corolla yellow-orange, 8–11 mm. long, deeply 5-lobed; lobes longer than tube, spathulate to rounded, undulate, strongly veined. Stamens appearing subequal; filaments 1.5–2 mm. long, heavily bearded below anthers, otherwise glabrous; anthers unequal c. 0.75–1 mm. long, cylindric-ovoid, obtuse, not apiculate. Capsule 5 mm. in diam., globose.

Botswana. N: between Lake Xau (Dow) and Tsokotse (Chukutsu) Pan, 74 km. W. of Letlhakane (Lothlekane), fl. & fr. 23.iii.1965, *Wild & Drummond* 7243 (K; SRGH). **Zimbabwe**. N: Makonde Distr., Great Dyke, fl. 8.iv.1981, *Philcox & Müller* 9077 (K).

Known from only one other collection (*Leach & Brunton* 9870 (SRGH)), from Makonde Distr., Zimbabwe.

11. **Alectra orobanchoides** Benth. in DC., Prodr. **10**: 340 (1846). Type from S. Africa.
   Celsia parvifolia Engl., Bot. Jahrb. **10**: 252 (1889). Type from Namibia.
   Alectra parvifolia (Engl.) Schinz in Verh. Bot. Ver. Brand. **31**: 195 (1890). —Hemsl. in F.T.A. **4**, 2: 366 (1906). Type as above.
   Melasma orobanchoides (Benth.) Engl., Pflanzenw. Ost-Afr. **C**: 359 (1895). Type as for *Alectra orobanchoides*.
   Alectra kilimandjarica Hemsl. in F.T.A. **4**, 2: 365 (1906). Types from Tanzania.
   Alectra kirkii Hemsl. in F.T.A. **4**, 2: 366 (1906). Type: Mozambique, Kongone, mouth of Zambezi R., 1.vii.1859, *Kirk* s.n. (K, holotype).

Erect herb, 10–45 cm. tall; stems branched from or near to base, or simple, shortly hispid-pubescent to subglabrous, rarely not drying black. Leaves 3.5–9(13) × (1)1.5–2 mm., appressed to subappressed or spreading, opposite or alternate, lanceolate, lanceolate-oblong to ovate-oblong, acute or obtuse, entire to crenate or remotely shortly toothed, sessile throughout, hispid or hispid-scabrid. Flowers solitary in axils of leaves or leaf-like bracts, pedicellate; pedicels 1–1.5 mm. long. Bracts leaf-like 3–6.5 × 1–2 mm., linear to oblong, hispid-pubescent. Bracteoles 2–4(5) × 0.25–1 mm., linear to filiform, shortly ciliate to glabrous. Calyx 4–6(7) mm. long, shortly hispid; lobes 1.5–2(3) mm. long, broadly triangular, obtuse, margin frequently thickened and scabrid. Corolla yellow to orange, dark purple striated, longer than calyx, lobes rounded. Stamens unequal; filaments slightly to heavily bearded; anthers obtuse, not apiculate. Capsule 4–6.5(9) × 3.25–6 mm., broadly ovoid to subglobose, glabrous.

Botswana. N: Okavango, near Nokaneng, 930 m., fl. & fr. 22.iii.1961, *Richards* 14837 (K). SE: Thalamabele-Mosu area, near Sowa pan, fl. 9.i.1974, *Ngoni* 281 (K). **Zambia.** N: Chilongowelo, 1460 m., fl. & fr. 21.iv.1952, *Richards* 1523 (K). W: Mwinilunga Distr., 96 km. S. of Mwinilunga on Kabompo road, fl. & fr. 3.vi.1963, *Loveridge* 757 (BR; K; LISC; SRGH). C: Lusaka Distr., 85 km. from Lusaka to Petauke, above Chinyunyu Hot Springs, fl. 12.ii.1975, *Brummitt & Lewis* 14325 (K). E: Chipata District, near Mfuwe, Luangwa Valley, c. 610 m., fl. & fr. 17.vii.1969, *Astle* 5713 (K). S: 20 km. N. of Choma, 1280 m., fl. 7.iv.1958, *Robinson* 2836 (K; SRGH). **Zimbabwe.** N: Mutoko Distr., Nyamahere Hill, fl. 14.iii.1978, *Pope* 1658 (K). W: Hwange Distr., Lupane-Gwayi (Lutope-Gwai) R. junction, c. 550 m., fl. 26.ii.1963, *Wild* 6023 (K). C: Gweru Distr., Gweru Kopje, c. 1460 m., fl. 5.iii.1967, *Biegel* 1977 (K). E: Chipata (Fort Jameson), fl. 8.iii.1949, *Shepheard* s.n. (K). S: Beitbridge Distr., Nulli (Nule) Hills, c. 40 km. ENE. of Beitbridge on Tshitura-padsi road, c. 550 m., fl. & fr. 20.iii.1967, *Rushworth* 456 (K). **Malawi.** N: Chitipa Distr., 16 km. down Songa stream from Crossroads, 1066 m., fl. 22.iv.1975, *Pawek* 9484 (K). C: Dowa Distr., Chimwere, 13 km. N. of Lombadzi towards Ntchisi, 1370 m., fl. 24.iii.1970, *Brummitt* 9356 (K). S: Machinga Distr., 1 km. W. of Machinga (Kasupe), 720 m., fl. 14.vi.1970, *Brummitt* 11391 (K). **Mozambique.** Z: Mocuba, 200 km. inland from Quelimane, Namagoa, 60–120 m., fl. & fr. vi–vii.1944, *Faulkner* 175 (BR; K; SRGH). MS: Chimoio, between Chimoio (Vila Pery) and Revuè, fl. 27.iv.1948, *Barbosa* 1588 (LISC). GI: Gaza, Chibuta, Changane, fl. 15.vii.1944, *Torre* 6756 (LISC). M: Namaacha, Changalane, Estatuene, fl. 10.v.1969, *Balsinhas* 1486 (LISC).

Also known from Kenya, Uganda, Tanzania, Angola, Namibia and S. Africa. In open woods and scrublands, rocky hillsides to flood plain grassland (also on cultivated land); up to 1800 m.

This complicated taxon has long been the source of great confusion. Prior to Hemsley's treatment of the genus in F.T.A. (1906) when he discribed several new species by reason of their habit and density of branching and inflorescence, many botanists had kept much of the material seen by him under the name *A. orobanchoides* Benth. Since the turn of the century much material has been added to the collections and Melchior (1941) followed Hemsley in most of his views upholding his names. I have seen most of the material related to these works and much more, and have decided for the purpose of this treatment to return them under the earliest name, *A. orobanchoides*. I consider that because of the diversity of all the material I have studied over the whole range of the complex, the broader pre-Hemsley concept is more accurate and workable, although this complex clearly needs much more examination.

30. HARVEYA Hook.

Harveya Hook., Ic. Pl. t.118 (1837). —Benth. & Hook. f., Gen. Pl. **2**: 967 (1876).

Parasitic herbs, erect or ascending, usually glandular-hairy. Leaves opposite or alternate usually at least the lower reduced to scales. Flowers usually large, solitary-axillary or in terminal spikes or racemes, sessile or pedicellate, bibracteolate or ebracteolate. Calyx tubular or tubular-campanulate, 5-dentate or 5-lobed to below middle. Corolla tube usually narrow at base, often much enlarged above and towards throat, incurved or slightly incurved; lobes 5, erect-spreading, flat, entire or undulate-crisped on margin. Stamens 4, didynamous, usually included or barely exserted; anthers bithecal with thecae parallel or transverse, one perfect, acuminate or mucronate at base, the other empty, longer, subulate-acuminate. Style incurved. Stigma clavate or oblong, tongue-shaped. Capsule ovoid, ellipsoid or subspherical, somewhat compressed, loculicidal. Seeds numerous, oblong in outline, truncate.

A genus of about 40 species mostly from S. Africa with one in the Mascarenes.

1. Plant acaulescent or with stems to 2.5 cm. tall; flowers bibracteolate   -   -   - 1. *randii*
 – Plant caulescent with stems 6–50 cm. tall; flowers ebracteolate   -   -   -   -   - 2
2. Stems stout, 20–50 cm. tall, to 6 mm. in diam.; upper leaves 30–55 mm. long; capsule ovoid to ellipsoid   -   -   -   -   -   -   -   -   -   -   -   -   -   -   - 2. *obtusifolia*
 – Stems slender, 6–16 cm. tall, 1–2.5 mm. in diam.; upper leaves 6–20 mm. long; capsule subspherical   -   -   -   -   -   -   -   -   -   -   -   -   -   - 3. *huillensis*

1. **Harveya randii** Hiern in Journ. Bot. **41**: 197 (1903); in F.C. **4**, 2: 411 (1904). TAB. **34**, fig. A. Type from S. Africa.

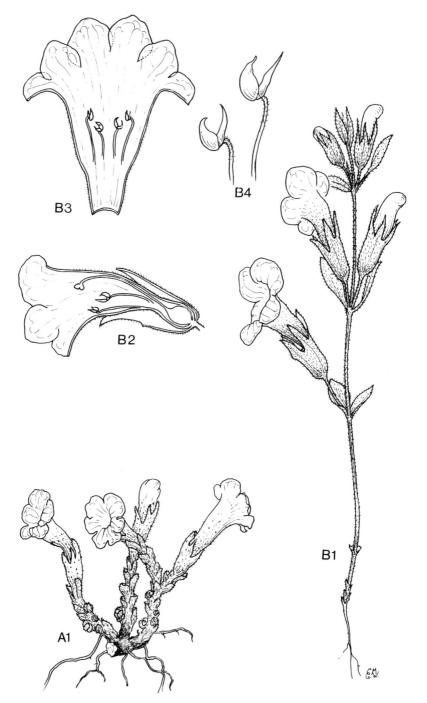

Tab. 34. A.—HARVEYA RANDII. A1, habit (×⅔), from *Crook* M194. B.—HARVEYA HUILLENSIS.
B1, habit (×⅔); B2, flower, longitudinal section showing gynoecium (×1); B3, corolla opened
out showing androecium (× 1); B4, stamens (×4), B1–4 from *Pawek* 10781A.

Dwarf, subherbaceous annual, 4–7.5 cm. tall when flowering; stems none or to c. 2.5 cm. long, erect, fleshy, scaly. Leaves all reduced to scales. Scales many, crowded, appressed, 3–6 mm. in diam., subcircular to broadly ovate, sessile, entire, viscid-puberulous, fleshy. Flowers several, crowded in axils of scale-like bracts, subsessile, bibracteolate. Bracts c. 10 × 6.5 mm., broadly ovate-oblong, minutely apiculate, viscid-puberulous. Calyx 16.5–24 mm. long, campanulate-oblong, glandular-pubescent without; lobes (5.5)7–10 mm. long, lanceolate-oblong, obtuse, two shorter and more connate than the others. Corolla pink with whitish to cream throat; tube c. 3 cm. long, cylindrical-funiculate, gently curved, c. 3.5 mm. wide at base, widening to 12 mm. at throat, glandular-pubescent without; limb 25–30 mm. in diam.; lobes c. 9 × 7 mm., obovate-circular, undulate-crenulate. Mature capsule not seen.

Zimbabwe. E: Chimanimani (Melsetter) Distr., Mt. Musapa, 2135 m., fl. 10.x.1950, *Wild* 3555 (K; SRGH). **Mozambique**. MS: Manica, Serra Zuira, 2000 m., fl. 10.xi.1965, *Torre & Paiva* 12842 (LISC). Also known from S. Africa. Montane grasslands; up to 2135 m.

2. **Harveya obtusifolia** (Benth.) Vatke in Bremen Abh. **9**: 130 (1885). —Skan in F.T.A. **4**, 2: 437 (1906). Type from Madagascar.
　　Gerardia obtusifolia Benth. in Hook., Comp. Bot. Mag. **1**: 211 (1836). Type as above.
　　Sopubia obtusifolia (Benth.) G. Don, Gen. Syst. **4**: 560 (1837). Type as above.
　　Aulaya obtusifolia (Benth.) Benth. in DC., Prodr. **10**: 523 (1846). Type as above.

Annual, more or less densely glandular-pilose herb; stems erect, simple 20–50 cm. tall, to c. 6 mm. in diam. Leaves opposite to subopposite; upper 30–55 × 5–14 mm., lanceolate to oblong-lanceolate, entire, obtuse or acute, cuneate at base, scabrid on margin and midrib beneath; lower small, scale-like. Flowers solitary in axils of upper leaves, opposite, curved, pedicellate. pedicels 7–11 mm. long, erect, lengthening to 20 mm. or more at maturity. Calyx 15–20 mm. long, sparsely to densely glandular-pilose; lobes 5–10 mm. long, unequal, lanceolate-triangular, acute. Corolla with yellowish-green tube flushed with pink, lobes similar but occasionally with more intense pink; tube 20–30 mm. long, c. 3 mm. wide at base, widening to c. 8–12 mm. at mouth, glandular-pilose to subglabrous; lobes 5–8(10) mm. in diam., subcircular, spreading; limb to 25 mm. in diam. Capsule 10 × 7 mm., ovoid or broadly ellipsoid, glabrous, shortly beaked.

Malawi. N: Karonga Distr., fl. & fr. vi.1958, *Jackson* 2241 (K; SRGH). S: Likhubula to Chambe, 1525 m., fl. & fr. 21.iii.1958, *Chapman* 549 (SRGH).
Also known from the Yemen, Ethiopia, Sudan, East Africa and Madagascar. Dry grasslands and open woodlands or low bush; up to 1600 m. in the Flora Zambesiaca area.

3. **Harveya huillensis** Hiern, Cat. Afr. Pl. Welw. **1**: 780 (1898). —Hemsl. & Skan in F.T.A. **4**, 2: 437 (1906). TAB. **34**, fig. B. Type from Angola.

Annual herb, 6–16 cm. tall when flowering, densely to more or less densely glandular-pilose. Stems erect, simple (1)1.5–2.5 mm. in diam. Leaves opposite, 6–20 × 1.5–7 mm., oblong-ovate to narrowly obovate, entire, acute or obtuse, cuneate at base, sessile; lower smaller to scale-like. Flowers solitary in axils of upper leaves, pedicellate, ebracteolate. Pedicels 3–7(10–17) mm. long, spreading, densely glandular-pilose. Calyx 18–24(27) mm. long, somewhat inflated, more or less densely glandular-pilose; lobes 4–8(14) mm. long, unequal, lanceolate-triangular, acute or obtuse. Corolla rose-pink to magenta or white with whitish tube, yellow or cream throat; tube (16)20–30(33) mm. long, glandular-pilose especially adaxially above calyx, glabrescent elsewhere; lobes (4–6)8–10 mm. long, obovate, entire; limb c. 15 mm. in diam. Capsule (? immature) c. 7 × 7.5 mm., compressed subspherical, glabrous.

Zambia. N: Mbala Distr., Lumi R. marsh 1680 m., fl. 30.iii.1957, *Richards* 8944 (K). S: Choma Distr., c. 18 km. N. of Choma, 1280 m., fl. 23.iii.1957, *Robinson* 2166 (SRGH). **Malawi**. N: Nkhata Bay Distr., Viphya, c. 57 km. SW. of Mzuzu, 1675 m., fl. 21.iii.1971, *Pawek* 4509 (K; SRGH). S: Zomba Distr., Mlomba, fl. 2.ii.1955, *Jackson* 1443 (K). **Mozambique**. N: Lichinga (Vila Cabral), Massangulo, c. 1300 m., fl. 25.ii.1964, *Torre & Paiva* 10774 (LISC).
Also known from Angola. Sandy to moist areas in open woodland; 1250–1680 m.
There are four collections represented by material at Kew from the northern region of Zambia, *Richards* 8310A, 12607, 18889 and *Fanshawe* 1958, which, except for their smaller overall size and white flowers, appear to be correctly placed under this name. They are duly included here.

31. BUCHNERA L.

Buchnera L., Sp. Pl.: 630 (1753); Gen. Pl., ed. 5: 278 (1754).

Annual or perennial, often hemi-parasitic, herbs, usually scabrid and rigid, usually becoming black upon drying; stems erect, ascending or prostrate, simple or branched from the base or below the inflorescence, glabrous to hispid-scabrid often with callus-based hairs. Leaves alternate, subopposite to opposite on stem, rosulate below, simple, sessile to shortly petiolate, filiform to lanceolate, elliptic or ovate-spathulate, apex acute or obtuse, entire to irregularly and coarsely dentate, 1–5-nerved; lower usually broader with upper narrower and passing into bracts. Inflorescence terminal, spicate or with flowers in terminal or axillary globose or subglobose clusters, or flowers solitary-axillary; spikes compact to lax; flowers subtended by one bract and usually two bracteoles, blue to violet or purple, rose to scarlet, yellow to white. Calyx tubular, 10 (rarely 7 or 8)-nerved or ribbed, sometimes obscurely so, 4–5-dentate, glabrous, pilose, hispid-scabrid or glandular; teeth short-deltoid to lanceolate-acuminate, erect, spreading or reflexed. Corolla tubular; tube cylindric, slightly longer to 2–3 times longer than calyx, straight, slightly curved to arcuate, externally glabrous to variously softly pilose-pubescent; limb subequally 5-lobed, spreading; lobes obovate to subcircular to oblong, entire or emarginate; throat densely pilose or not. Stamens 4, included, slightly didynamous; anthers monothecous, dorsifixed. Stigma narrowly cylindric. Style apex clavate, entire. Ovules numerous in each locule. Capsule ovoid, oblate to cylindric, shorter or longer than calyx, loculicidal. Seeds numerous.

A genus of some 100 species from the tropics and subtropics, mostly from the Old World.

1. Flowers solitary in leaf axils	1. prorepens
– Flowers in spikes, heads, etc.	2
2. Inflorescences globose or subglobose	3
– Inflorescences spicate	6
3. Calyx 4-lobed	2. foliosa
– Calyx 5-lobed	4
4. Inflorescence clusters terminal and axillary	5. nuttii
– Inflorescence clusters terminal at ends of branches only	5
5. Corolla tube glabrous without	3. cryptocephala
– Corolla tube hairy without	4. capitata
6. Calyx 4-lobed (at times appearing so in B. arenicola and B. randii)	7
– Calyx 5-lobed	24
7. Inflorescences quadrangular	8
– Inflorescences cylindrical, not clearly quadrangular	13
8. Corolla tube hairy without	9
– Corolla tube glabrous or minutely glandular without	10
9. Stems patent-hispid throughout; bracts (at least the upper) apically trilobed; corolla tube c. 7 mm. long	6. trilobata
– Stems glabrous except at extreme base where sparsely patent hirsute; bracts not trilobed	7. peduncularis
10. Corolla tube minutely glandular	8. quadrifaria
– Corolla tube glabrous	11
11. Stems simple; leaves tightly appressed to stem, glabrous; plant drying pale yellow-brown; bracts not bicoloured	9. descampsii
– Stems simple or branched; leaves not tightly appressed to stem, scabrid; bracts usually bicoloured	12
12. Bracts subreniform, mucronate, pubescent, hispid-ciliate	10. bangweolensis
– Bracts ovate to elliptic-rhombic or 3-lobed, long-acuminate, glabrous or glandular	8. quadrifaria
13. Spikes densely long white silky-pilose	11. splendens
– Spikes not as above	14
14. Spikes densely compacted, not becoming lax at fruiting, except occasionally for one or two basal flowers	15
– Spikes elongated and more slender, lower flowers becoming somewhat lax and remote at fruiting	17
15. Stems almost simple, sparsely hirsute to glabrous	14. arenicola
– Stems usually branched, minutely pale brown pubescent	16
16. Stems generally retrorsely pubescent throughout; corolla tube pilose without	12. crassifolia
– Stems bifariously pubescent; corolla tube glabrous without	13. nitida

17. Corolla tube hairy without - - - - - - - - - - - - - 18
– Corolla tube glabrous without - - - - - - - - - - - 21
18. Stems simple, rarely branched, patent-hirsute or with large multicellular hairs 19
– Stems branched, indumentum not as above - - - - - - - - 20
19. Stems glabrous except at base where sparsely patent-hirsute; corolla tube 4.5–5.5 mm. long, limb 2.25–4 mm. in diam. - - - - - - - - - - - - 7. *peduncularis*
– Stems minutely retrorse-pubescent with large multicellular hairs present; corolla tube 6.5–10.5 mm. long, limb 8–13 mm. in diam. - - - - - - 15. *rungwensis*
20. Stems short hispid-pubescent; throat of corolla subglabrous, not villous or pilose - - - - - - - - - - - - - - - 16. *nigricans*
– Stems hispid-pilose; throat of corolla densely pilose - - - - - 17. *randii*
21. Calyx 6–9 mm. long - - - - - - - - - - - 16. *nigricans*
– Calyx 2.25–4 mm. long - - - - - - - - - - - - 22
22. Calyx 2.25–3.25 mm. long - - - - - - - - - - - - 23
– Calyx 4 mm. long; corolla tube 6–7 mm. long, limb 6–7 mm. in diam. 20 *ciliolata*
23. Calyx lobes 0.6–0.8 mm. long, broadly deltate to lanceolate; corolla tube 5–7 mm. long; limb 1.25–4 mm. in diam.; capsule oblong-ellipsoid, slightly emarginate 18. *strictissima*
– Calyx lobes 1–1.5 mm. long, narrowly lanceolate; corolla tube 4.5 mm. long; limb 1.5–2.5 mm. in diam.; capsule ovoid to subglobose - - - - - - - 19. *buchneroides*
24. Bracteoles absent - - - - - - - - - - - - - 25
– Bracteoles present - - - - - - - - - - - - - 28
25. Corolla tube glabrous - - - - - - - - - - 23. *humpatensis*
– Corolla tube pubescent - - - - - - - - - - - - 26
26. Plant up to 12 cm. tall - - - - - - - - - 21. *androsacea*
– Plant 30–70 cm. tall - - - - - - - - - - - - 27
27. Corolla tube c. 5 mm. long; limb to 2.25 mm. in diam.; bracts bi- or trifid at apex - - - - - - - - - - - - 22. *ruwenzoriensis*
– Corolla tube 9–10.5 mm. long; limb 8–13 mm. in diam.; bracts entire 24. *ebracteolata*
28. Inflorescence quadrangular or eventually so - - - - - - - 29
– Inflorescence cylindrical, not quadrangular - - - - - - - 31
29. Plant densely leafy throughout, most leaves imbricate; bracts c. 8 × 5 mm. 25. *lippioides*
– Plant with stems sparsely leafy, especially above; bracts 3–7.5 × 1–2 mm. - - 30
30. Lower stem leaves 15–55 × 2–9 mm., imbricate, ovate-oblong to obovate, apiculate to mucronate - - - - - - - - - - - - 26. *wildii*
– Lower stem leaves 12–40 × 1–2 mm., not imbricate, linear to oblong-linear 27. *welwitschii*
31. Corolla tube glabrous without - - - - - - - - - - 32
– Corolla tube pubescent without - - - - - - - - - - 44
32. Plant to c. 7.5 cm. tall; basal leaves (where present) drying pale brown - - 33
– Plant much larger than 8 cm. tall - - - - - - - - - 34
33. Plant c. 7.5 cm. tall; stem prostrate or ascending; cauline leaves 11–35 mm. long, oblong-spathulate; basal leaves not always evident; corolla tube occasionally short-pilose with minute stipitate glands - - - - - - - - - - 28. *multicaulis*
– Plant 2.5–4.5 cm. tall, erect; cauline leaves 4–6.5 mm. long, elliptic or ovate-elliptic; basal leaves usually persistent, evident, up to 6 × 3.5 cm.; corolla tube glabrous - - - 29. *hockii*
34. Plant drying black - - - - - - - - - - - - - 35
– Plant not drying black - - - - - - - - - - - - 43
35. Corolla tube markedly arcuate; leaves mostly 0.5–1 mm. wide, subfilamentous - - - - - - - - - - - - 30. *chisumpae*
– Corolla tube straight or rarely slightly curved, never arcuate; leaves wider, never filamentous - - - - - - - - - - - - - - 36
36. Stems leafy throughout; leaves opposite, most in upper two thirds, equal to or longer than internodes; spikes with flowers imbricate until flowering - - - - 31. *lastii*
– Not as above - - - - - - - - - - - - - - 37
37. Stems simple; inflorescence few-flowered - - - - - - - - 38
– Stems branched; inflorescence many-flowered - - - - - - - 39
38. Flowers scarlet to orange-red, in one to three remote pairs - - - 33. *geminiflora*
– Flowers blue to purple, in few-flowered spikes - - - - - 34. *subglabra*
39. Inflorescence erect, slender-spicate; flowers more or less appressed, somewhat imbricate; branches not widely spreading; corolla tube 5.5–6 mm. long, limb c. 5.5 mm. in diam. - - - - - - - - - - - 32. *leptostachya*
– Inflorescence of widely spreading branches, laxly flowered; corolla tube 6–18 mm. long; limb 3-30 mm. in diam. - - - - - - - - - - - - 40
40. Calyx tube hispid to pilose - - - - - - - - - - - 41
– Calyx tube glabrous - - - - - - - - - - - - 42
41. Calyx tube hispid to hispid-pilose; corolla tube 6–7.5 mm. long, limb 3–5 mm. in diam.; plant to 100 cm. tall - - - - - - - - - - - - 35. *hispida*
– Calyx tube appressed short, white pilose; corolla tube 9–10 mm. long, limb c. 11.5 mm. in diam.; plant to 55 cm. tall - - - - - - - - - - 36. *laxiflora*

42. Plant to 75 cm. tall; corolla tube 7–10 mm. long, limb 6–10 mm. in diam.   37. *eylesii*
 – Plant to 200 cm. or more tall; corolla tube 12–18 mm. long, limb 18–30 mm.
    in diam. - - - - - - - - - - - - - - - 38. *speciosa*
43. Stem simple or rarely branched, drying pale brown; corolla tube 5.5–6 mm. long; limb
    c. 2 mm. in diam. - - - - - - - - - - - - 39. *candida*
 – Stem branched, drying greyish-green; corolla tube 9.5–10.5 mm. long, limb 9–14 mm.
    in diam. - - - - - - - - - - - - - - 40. *pulcherrima*
44. Leaves with smaller leaves or branches in axils; flowers closely clustered in spike; calyx lobes
    variously divergent - - - - - - - - - - - - - - 45
 – Leaves without smaller axillary leaves or branches; calyx lobes not spreading - - 48
45. Leaves narrowly linear, flexuous; inflorescence rarely exceeding 4.5 cm. in length at fruiting;
    calyx lobes widely spreading with reflexed apices at fruiting   41. *chimanimaniensis*
 – Leaves not flexuous; inflorescence up to 26 cm. long - - - - - - - 46
46. Inflorescence less than 5 cm. long at fruiting; bracts 5.5–6 mm. wide, rhombic to reniform-
    rhombic, strongly parallel-nerved - - - - - - - - - 42. *nervosa*
 – Inflorescence 5–26 cm. long at fruiting; bracts up to 1.75 mm. wide, lanceolate to
    ovate-lanceolate - - - - - - - - - - - - - - 47
47. Inflorescence not more than 10 cm. long at fruiting; plant 6–25 cm. tall; calyx with spreading or
    upwardly directed hairs - - - - - - - - - - 43. *henriquesii*
 – Inflorescence 20–26 cm. long at fruiting; plant 30–45 cm. tall; calyx with downwardly directed
    hairs - - - - - - - - - - - - - - 44. *longispicata*
48. Plant c. 7.5 cm. tall; stem prostrate or ascending; corolla tube short-pilose, hairs intermixed with
    minute stipitate glands - - - - - - - - - - 28. *multicaulis*
 – Plant taller; stem erect; corolla tube not glandular - - - - - - 49
49. Plant extremely slender; stem leaves one to two pairs in upper half or two thirds with internodes
    up to 20 cm. long - - - - - - - - - - - 45. *attenuata*
 – Plant less slender; upper stem leaves many more than two pairs, internodes much
    shorter - - - - - - - - - - - - - - - 50
50. Calyx irregularly and unevenly lobed, 5-lobed sometimes apparently 4-lobed with fifth lobe
    much reduced or absent - - - - - - - - - - - - 51
 – Calyx regularly and equally 5-lobed - - - - - - - - - - 53
51. Calyx glabrous - - - - - - - - - - - - 14. *arenicola*
 – Calyx variously pubescent - - - - - - - - - - - - 52
52. Calyx 4–6 mm. long, obscurely 10-nerved; lobes not broadly divergent at
    fruiting - - - - - - - - - - - - - - 17. *randii*
 – Calyx 6.5–9 mm. long, ± prominently 10-nerved; lobes broadly divergent at
    fruiting - - - - - - - - - - - - - - 26. *wildii*
53. Stems simple, 12–36 cm. tall; bracts 2.5–6 mm. long - - - - - 54
 – Stems branched or only occasionally simple; bracts up to 6.5 mm. wide; bracteoles up to
    33 mm. wide - - - - - - - - - - - - - - 56
54. Stems patent-hispid; calyx and bracts hispid; corolla tube 5.5–6.5 mm.
    long - - - - - - - - - - - - - - 46. *pusilliflora*
 – Stems glabrous to sparsely appressed hispid; corolla tube 7–10.5 mm. long - - 55
55. Bracts 5–6 × 5.5–6.5 mm., rhombic to reniform-rhombic; corolla tube 10–10.5 mm.
    long - - - - - - - - - - - - - - 42. *nervosa*
 – Bracts 2.8–4 × 2 mm., ovate; corolla tube 7–7.5 mm. long - - - - 34. *subglabra*
56. Bracts 5.5–6.5 mm. wide; bracteoles 2–3 mm. wide - - - - - 42. *nervosa*
 – Bracts and bracteoles much narrower - - - - - - - - - - 57
57. Inflorescence small, not exceeding 1.5 cm. in length; flowers remaining compact with the lower
    ones apparently not becoming remote after flowering - - - - 47. *granitica*
 – Inflorescence larger with spikes up to 25 cm. long or more - - - - - 58
58. Leaves all entire; plant rarely more than 40 cm. tall - - - - - - 59
 – Leaves at least some coarsely, irregularly dentate; plant usually exceeding 50 cm. in height,
    frequently up to 100 cm. or more - - - - - - - - - - 60
59. Corolla tube c. 5 mm. long, lobes 1–1.5 × 0.8 mm. - - - - - 48. *albiflora*
 – Corolla tube c. 10 mm. long, lobes 4 × 3 mm. - - - - - 49. *namuliensis*
60. Inflorescence spike not (or rarely) exceeding 4 cm. in length; flowers remaining compact, never
    becoming generally lax; leaves and calyx long white hispid with coarse, patent or antrorse,
    callus-based hairs - - - - - - - - - - - - 17. *randii*
 – Inflorescence spike usually longer than 10 cm., up to 40 cm. long; flowers initially compact, soon
    becoming lax throughout; leaves and calyx appressed hispid-scabrid - - 35. *hispida*

1. **Buchnera prorepens** Engl. & Gilg in Warb., Kunene-Samb.-Exped., Baum: 368 (1903). —Skan in
   F.T.A. **4**, 2: 398 (1906). Type from Angola.

   Annual or perennial, prostrate herb; stems many, arising from woody tap-root usually
much branched, terete, subglabrous to short or long pilose. Leaves: lower 14–42 × 5–12
mm., opposite, obovate, obtuse, narrowing into petiole-like base, entire to remotely,
minutely dentate, obtuse to rounded at apex, 1–3-nerved with midrib occasionally

pinnately-nerved, glabrous to more or less densely hispid to long pilose; upper leaves considered here as bracts. Flowers pedicellate, solitary in axils of leaf-like bracts throughout greater part of stem. Pedicels 1–1.5 mm. long. Bracts 6–10 × 1.75–3.5 mm., oblong-lanceolate, obovate or elliptic, 1-nerved, entire to laxly minutely dentate, glabrous, ciliate to hispid-pilose. Bracteoles 4.5–5.5 × c. 0.6 mm., linear to linear-subulate, subglabrous. Calyx (4.5)6–7.5(11) mm. long, obscurely 10-ribbed, 5-lobed, pilose below with long, white hairs; lobes (1.5)3.5–5 × (0.5)0.8–1 mm., lanceolate, glabrous. Corolla blue, mauve to purple; tube 7–10 mm. long, straight except base where markedly curved, subglabrous to densely pilose without, densely pilose at throat; limb 5.5–8.5 mm. in diam.; lobes 2.5–4 × 1.5–3.8 mm., circular to broadly obovate, unequal. Capsule c. 4 × 2.8–3.8 mm., obovoid-cylindrical to subcircular.

Zambia. B: Mongu Distr., c. 20 km. NE. of Mongu, fl. 10.xi.1959, *Drummond & Cookson* 6275 (K; LISC; M; SRGH). N: Chinsali Distr., Machipara Hill, Shiwa Ngandu, 1500 m., fl. 16.i.1959, *Richards* 10694 (BM; K; SRGH). W: Mwinilunga, fl. & fr. 30.ix.1937, *Milne-Redhead* 3447 (BM; BR; K). C: Mkushi, Fiwila, 1370 m., fl. 3.i.1958, *Robinson* 2580 (K). S: Machili, fl. 6.x.1960, *Fanshawe* 5824 (K; SRGH).

Also known from Angola. On well drained soils in grassland, open woodland and rocky hillsides; from c. 950–1590 m.

2. **Buchnera foliosa** Skan in F.T.A. **4**, 2: 389 (1906). Type: Malawi, Tanganyika Plateau, Chitipa (Fort Hill), *Whyte* s.n. (K, holotype).
    Buchnera quadrangularis S. Moore in Journ. Bot. **57**: 217 (1919). Type from Angola.

Annual herb, 40–60 cm. or more tall, stout, erect; stems branched with branches opposite, mostly above, erect-spreading, obscurely quadrangular, markedly leafy, white long subappressed-hispid, to subglabrous in parts. Leaves 20–65(90) × (3)6–25(42) mm., opposite, oblong-obtuse below, lanceolate-acute above, narrowing at base, 3–5-nerved, entire or remotely toothed, sparsely appressed pubescent above, hispid with short hairs beneath (especially on major nerves). Inflorescence of compact heads of spikes, 10–30 × 10–22 mm., corymbosely arranged, quadrangular, markedly so in fruit. Bracts varied, lowest leaf-like to c. 25 mm. long, lanceolate, upper 5–8 × 3–4 mm., broadly ovate to circular, apex acute to acuminate, occasionally keeled at apex, hispid. Bracteoles 4–5 × 0.6–1 mm., distinctly keeled, hispid. Calyx 5.5–7 mm. long, subquadrangular, 4-lobed, 4-nerved, hispid mainly on nerves, split on upper and lower side to c. 2.5 mm.; lobes c. 1.5 mm. long, narrowly deltate acute. Corolla blue to occasionally purple, or white; tube 6(8) mm. long, slender, straight, cylindric, very sparsely pilose or pubescent without, sometimes appearing glabrous; limb 2.5–4 mm. in diam.; lobes 1.5–2(3) × 1.5–2 mm., obovate or obovate-cuneate. Capsule c. 4.5 mm. long, ovoid, shortly rostrate.

Zambia. N: Mbala Distr., top of Katanga Falls, Kambole, 1500 m., fl. 5.vi.1957, *Richards* 10031 (K). W: Mwinilunga Distr., Kabompo road 27 km. from Mwinilunga, fl. & fr. 6.vi.1963, *Edwards* 677 (BR; K; SRGH). C: Chakwenga Headwaters, 100–129 km. E. of Lusaka, fl. & fr. 25.viii.1963, *Robinson* 5614 (K). **Malawi**. N: Chitipa (Fort Hill), fl. & fr. vii.1896, *Whyte* s.n. (K).
Also known from Tanzania. *Brachystegia* woodland and grassland; 1000–1600 m.

3. **Buchnera cryptocephala** (Baker) Philcox in Kew Bull. **40**: 606 (1985). Type: Zambia, Fwambo, Lake Tanganyika, *Carson* 59 (K, holotype).
    Lobostemon cryptocephalum Baker in Bull. Misc. Inf., Kew **1894**: 30 (1894). Type as above.
    Buchnera pulchra Skan ex S. Moore in Journ. Linn. Soc., Bot. **37**: 190 (1905); in F.T.A. **4**, 2: 383 (1906). Type from Uganda.

Annual herb, (7)20–60(180) cm. tall, erect, drying green or brown, not black; stems simple or less frequently branched above, scabrid, moderately to densely leafy throughout. Leaves 12.5–65 × 1.5–13 mm., opposite or alternate, ovate-elliptic to oblong, obtuse, shortly mucronate or not, entire, 3(5)-nerved, hispid-scabrid. Inflorescence spicate (8)15–38 × 8–35 mm., terminal, globose, hemispheric to cylindric, densely compactly flowered, not quadrangular, usually subtended by numerous upper leaves. Bracts (5.5)7.5–9 × (1.5)2–3 mm., lanceolate, ovate or elliptic-lanceolate, keeled, acuminate, pilose-hispid to subglabrous. Bracteoles (4.5)6–8 × 0.5 mm., linear to linear-lanceolate, hispid to pilose. Calyx (5)7–8.5 mm. long, 5-lobed, prominently 10-nerved, hispid-scabrid especially on nerves; lobes 1.5–2.75 mm. long, ovate or ovate-lanceolate to linear-triangular, hispid-ciliate. Corolla blue to purple; tube to about 10 mm. long, glabrous or minutely glandular without, throat pilose; lobes to 4.5 × 4 mm., broadly obovate, subequal. Capsule 4.25–5 × 2–2.5 mm., ovoid-ellipsoid, glabrous.

Var. **cryptocephala** — TAB. **35**, fig. B.

Up to 90 cm. tall, branched below; stems leafy throughout; bracts and bracteoles generally pilose-hispid.

Zambia. N: Mbala Distr., Zombe Plain, 1500 m., fl. 14.ii.1969, *Sanane* 40 (K; P). W: Solwezi Distr., Chifubwa, fl. & fr. 5.i.1962, *Holmes* 1448 (BR; K). C: c. 9 km. E. of Lusaka, fl. & fr. 10.ii.1956, *King* 306 (K). E: Nyika Plateau, 2000 m., fl. 8.v.1952, *White* 2783 (K). **Zimbabwe**. N: Mwami (Miami) Experimental Farm, 1370 m., fl. & fr. 7.iii.1947, *Wild* 1798 (K; SRGH). **Malawi**. N: Nyika Plateau, between Kasaramba and Chelinda, fl. & fr. 10.i.1967, *Hilliard & Burtt* 4413 (E; K). C: Lilongwe Distr., Bunda, 23 km. S. of Lilongwe, 1175 m., fl. & fr. 31.iii.1970, *Brummitt* 9562 (K).

Also known from Uganda, Tanzania and Angola. Wet grasslands, woodlands and swamps; 600–2300 m.

Var. **mwinilungensis** Philcox in Bol. Soc. Brot., sér. 2, **60**: 269 (1987). TAB. **35**, fig. A. Type: Zambia, 28 km. E. of Mwinilunga, fl. & fr. 17.iv.1960, *Robinson* 3667 (K, holotype; SRGH, isotype).

Up to 1.8 m. tall, erect; stems branched above, leafy below, sparsely so above; inflorescence not surrounded at base by bract-like leaves; calyx, bracts and bracteoles glabrous except on major nerves and margins; calyx lobes regularly ovate-triangular, cuspidate.

Zambia. W: Mwinilunga Distr., between Mukimina and Kanku, S. of Mwinilunga, fl. & fr. 24.viii.1930, *Milne-Redhead* 954 (K).

Known in *Brachystegia* woodlands from the Mwinilunga Distr. of Zambia with one collection made in Zaire.

This plant differs from the typical variety in having an almost glabrous inflorescence with the indumentum usually confined to the margins and midribs of the calyx lobes, bracts and bracteoles. The calyx lobes are almost regularly broad triangular-ovate. The stem is densely leafy below; above and especially on the inflorescence branches the leaves are very much reduced both in size and number, internodes are up to 55 mm. or more long. The inflorescence is not subtended by the large concentration of bract-like leaves seen in the typical variety.

Having seen a number of collections of this plant I do not feel that these characters justify the recognition of a separate species. However, in view of its restricted occurrence (with one exception) to the Mwinilunga region of Zambia and its distinct appearance, I have decided to give it varietal rank.

After further study it may be found necessary to include the closely related *B. quangensis* Engl. as a variety of this species.

4. **Buchnera capitata** Benth. in DC., Prodr. **10**: 495 (1846). —Engl., Pflanzenw. Ost-Afr. **C**: 359 (1859). —Skan in F.T.A. **4**, 2: 381 (1906). —Hepper in F.W.T.A. ed. 2, **2**: 369 (1963). Type from Madagascar.

Annual herb, 30–50(95) cm. tall, erect; stems simple or occasionally branched above, somewhat furrowed, glabrous to shortly pubescent above. Leaves 10–60(80) × 2–14(18) mm., opposite, or upper alternate, oblong-lanceolate to linear obovate-oblong, obtuse, subamplexicaul, entire to shallowly crenate, occasionally shortly and remotely toothed, sparsely to more or less densely scabrid, minutely hispid on margins and nerves beneath particularly in upper leaves, (1)3–5-nerved, sometimes obscurely so. Inflorescence 6–45 × 10–18 mm., terminal, globose, hemispheric to cylindric, densely compactly flowered, not quadrangular. Bracts 3–5 × c. 1 mm., oblong-lanceolate or lanceolate, acuminate, pilose. Bracteoles 3–4.5(6) mm. long, linear to filiform, pilose. Calyx c. 4.5 mm. long, 5-lobed, pilose, obscurely 10-nerved; lobes 1.5–2 mm. long, subulate. Corolla usually white, occasionally blue to violet; tube 3.5 mm. long, pilose without; lobes 2–2.5 mm. long, up to 2 mm. wide, obovate. Capsule c. 2.75 mm. long, up to 2 mm. wide, ovoid-cylindric, obtuse.

Zambia. N: Mbala Distr., Lumi R., Kawimbe, 1740 m., fl. & fr. 27.iii.1959, *Richards* 12273 (BR; K). W: 8 km. S. of Mufulira, 1300 m., fl. 27.iii.1960, *Robinson* 3432 (K; SRGH). C: Walamba, fl. & fr. 22.v.1954, *Fanshawe* 1234 (BR; K). **Malawi**. N: Rumphi Distr., near Mwenembe Forest, fl. 9.iv.1981, *Salubeni* 3067 (SRGH). S: Machinga (Kasupe) Distr., Kawinga, 8 km. NE. of Ntaja near Mikoko Bridge, fl. 16.ii.1979, *Blackmore, Brummitt & Banda* 446 (BM; K). **Mozambique**. Z: Munguluni (M'Guluni), fl. Sept., 915 m., *Faulkner* 49 (K).

Also widely spread in West, Central and East Africa; known also from Sudan, Ethiopia, Angola and Madagascar. Swamps, bogs and marshy grasslands; from 900–1800 m.

5. **Buchnera nuttii** Skan in F.T.A. **4**, 2: 388 (1906). Type: Zambia, Urungu, Fwambo, 1600 m., *Nutt* s.n. (K, holotype).

Perennial, up to 150 cm. tall, erect, rigid; stems branched usually in upper part or

Tab. 35. A.—BUCHNERA CRYPTOCEPHALA var. MWINILUNGENSIS. A1, habit (× ⅔); A2,
bract (× 4), A1–2 from *Hooper & Townsend* 337. B.—BUCHNERA CRYPTOCEPHALA var.
CRYPTOCEPHALA. B1, bract (× 4); B2, flower (× 4); B3, corolla opened out showing
gynoecium (× 4), B1–3 from *Chapman* 1951.

occasionally simple, branches spreading, hispid-scabrid-pilose with white ascending hairs. Leaves 18–50(90) × 2–10(30) mm., opposite or subopposite, lanceolate, acute, narrowed at base to broadly ovate-elliptic, obtuse, entire with one to two obscure shallow lobes or toothed, 3–5-nerved, more or less appressed pilose-scabrid above, hispid-scabrid beneath, especially on nerves. Inflorescence of densely flowered, irregular, sessile or shortly pedunculate, axillary compound heads up to 40 × 25 mm., individual flowers not becoming lax when fruiting. Bracts 3–4 × 2–2.5 mm. or more wide, ovate-lanceolate, acuminate, keeled, subglabrous, ciliate. Bracteoles 1.75–3.75 × 0.5–1.5 mm., narrowly lanceolate, keeled, ciliate. Calyx 3.75–4.5 mm. long, sparsely pubescent, 5-lobed; lobes 0.8–1.5 mm. long, ovate-lanceolate, acuminate, ciliate. Corolla white or occasionally lavender, with a mauve or bluish tube, frequently with lobes bordered with mauve or blue, fragrant; tube 4–5(6) mm. long, rather wide, funnel-shaped, straight, glabrous without except for few hairs immediately below sinuses, throat densely pilose with purplish hairs spreading well onto limb; limb 3–4 mm. in diam.; lobes c. 1.5(3) × 1–1.8 mm., obovate. Capsule 2.5–3 × c. 2 mm., broadly ovoid-globular.

Zambia. N: Mbala Distr., Sunzu Mt., Chisau Gorge, c. 1500 m., fl. 30.v.1969, *Sanane* 762 (K; P). C: Chakwenga Headwaters, 100–129 km. E. of Lusaka, fl. 25.viii.1963, *Robinson* 5614 (SRGH). S: Zimba, fl. & fr. 1930, *Hutchinson & Gillett* 3508 (K). **Malawi**. N: Mzimba Distr., Lwanjati Pass, c. 4.5 km. NE. of Katete, fl. 5.vii.1976, *Pawek* 11440 (K; MO).
Also known from Tanzania and Kenya. Grasslands and open *Brachystegia* woodlands; 1300–1980 m.
All material from the Flora Zambesiaca area shows the corolla tube to be externally glabrous except for the occasional hair below the sinuses, whereas the Kenyan and Tanzanian material has many more hairs in evidence spreading well down the tube.

6. **Buchnera trilobata** Skan in F.T.A. **4**, 2: 378 (1906). Type: Malawi, Nyika Plateau, 2130 m., fl. Sept. 1902, *McClounie* 55 (K, lectotype, chosen here); without locality, *Whyte* s.n. (K, lectoparatype).

Annual herb, 13.5–52 cm. tall, erect; stems simple below, occasionally sparingly branched above, patent-hispid. Leaves: cauline, 10–30 × 2–5 mm., lanceolate to elliptic-lanceolate, obtuse, scabrid-hispid above, patent hispid only on median nerve and margin beneath, entire: basal leaves where present (*Iwarsson et al.* 766), 2.7 × 2.0 cm., subcircular, with a short petiole-like base c. 1.5 mm. long. Spikes 5–25 × 5–8 mm., terminal or lateral, densely compacted, clearly or obscurely quadrangular. Bracts 4–6 × 3–3.5 mm., ovate-lanceolate, acuminate, at least upper apically trilobed, long-ciliate especially on the middle lobe. Bracteoles 3–3.5 mm. long, lanceolate, acuminate, long-ciliate. Calyx 5.5–6 mm. long, clearly 4-lobed; lobes 3–4 mm. long, narrowly lanceolate, long-ciliate only on margin. Corolla pink, pinkish-purple to mauve or white; tube c. 7 mm. long, pubescent without; lobes 3–4 × 2.5 mm., throat pilose. Capsule 3–3.5 × c. 2 mm., cylindric-ovoid, glabrous.

Zambia. N: Mafinga, fl. & fr. 24.v.1973, *Chisumpa* 51A (K). W: Solwezi Distr., Mutanda, fl. 1930, *Milne-Redhead* 549 (K). **Malawi**. N: Chitipa Distr., 4 km. WNW. of Muzengapakweru, 2285 m., Nyika Plateau, fl. & fr. 25.vii.1972, *Synge* WC307 (BR; K; SRGH). S: Mangochi Distr., Fort Mangochi, 1500 m., fl. & fr. 11.vi.1978, *Iwarsson & Ryding* 766 (K).
Known only from Zambia and Malawi. Montane grasslands and *Brachystegia* woodlands; up to 2500 m.

7. **Buchnera peduncularis** Brenan in Mem. N.Y. Bot. Gard. **9**: 12 (1954). Type: Malawi, Rumphi Distr., Nyika Plateau, Nchenachena, 1340 m., *Brass* 17371 (K, holotype; BR; NY; SRGH, isotypes).

Annual herb, 20–45(65) cm. tall, erect; stems slender, simple or rarely branched, glabrous throughout except for small area near base where sparsely patent hirsute. Leaves 6–20(25) × 0.5–3(4.5) mm., opposite, oblong or oblong-lanceolate, entire, acute, appressed hirsute above, short hispid scabrid on midrib and margins of lower surface, 1-nerved. Spikes 5–15(25) mm. long, cylindrical to subcapitate, compact, occasionally somewhat quadrangular at fruiting stage, not becoming especially lax when fruiting. Bracts 2.5–3.5 × 2–3 mm., obtriangular, abruptly cuspidate (cusp 0.8–1.4 mm. long, acute), glabrous to subglabrous, ciliate, somewhat concave. Bracteoles 2–3 × 0.25–0.7 mm., lanceolate or linear-lanceolate, ciliate. Calyx 2.25–3 mm. long, 4-lobed with sinus 1.5–2.5 mm. deep; lobes 1.25–1.5 mm. long, narrowly lanceolate, acute, ciliate. Corolla blue to mauve or purple; tube 4.5–5 mm. long, pubescent without; limb 2.25–4 mm. in diam.; lobes c. 1.5 × 1–1.2 mm., circular-elliptic, entire or slightly emarginate. Capsule 2.3–2.8 × 1.5–2 mm., subspherical-ovoid, beak c. 0.2 mm. long.

Zambia. N: Mansa Distr., fl. 3.v.1964, *Fanshawe* 8521 (SRGH). W: 4.8 km. E. of Mufulira, 1220 m., fl. & fr. 13.vi.1948, *Cruse* 371 (K). C: Lusaka Distr., 17 km. E. of Lusaka, 1220 m., fl. & fr. 29.vi.1958, *Best* 140 (BR; K). E: Lundazi Distr., fl. 22.vi.1967, *Hilundu* 27 (K; SRGH). **Malawi**. N: Nyika Plateau, Nchenachena, 1340 m., fl. & fr. 21.viii.1946, *Brass* 17371 (BR; K; NY; SRGH).

Known also from Zaire. In *Brachystegia* woodland, open grasslands and sandy areas; 1200–1350 m.

8. **Buchnera quadrifaria** Baker in Bull. Misc. Inf., Kew **1895**: 71 (1895). —Skan in F.T.A. **4**, 2: 378 (1906). Type: Zambia, Fwambo, 1894, *Carson* 100 (K, lectotype, chosen here); N. of L. Nyasa, *Thomson* s.n. (lectoparatype).

Annual herb, 20–68 cm. tall, erect; stems simple or more rarely branched, quadrangular, glabrous or occasionally shortly pubescent immediately below insertion of leaves, glabrous or scabrid on prominent angles. Leaves (2.5)5–14 × 0.75–1 mm., sparse, alternate or opposite, subulate or linear, sparsely, minutely scabrid to glabrous, entire, 1-nerved. Spikes (4)10–27 × 4–5 mm., broad, densely compacted, quadrangular. Bracts 3.5–5 × 2.5–3 mm., broadly ovate to elliptic-rhombic, acuminate, at times somewhat trilobed towards base with lobes or margins dark coloured as with midrib, glabrous or minutely granular-glandular. Bracteoles 3–4 mm. long, linear-lanceolate, shallowly carinate, glabrous to finely glandular. Calyx 4–5 mm. long, 4-lobed, deeply divided on lower side, glabrous to subglabrous; lobes 3–4 mm. long, subulate-lanceolate, glabrous except for a few small hairs on margins. Corolla blue to mauve or purple or white; tube (4)7–8 mm. long, glabrous or glandular, throat pilose; limb 4–6 mm. in diam.; lobes 1.5–2.5 mm. wide, upper lobes elliptic-obovate, lower lobes oblong. Capsule 2.5 × 1.5 mm., broadly ellipsoid, glabrous.

Zambia. N: Mporokoso Distr., Kansanshi Dambo, 55 km. ESE. of Mporokoso, fl. & fr. 20.v.1961, *Robinson* 4667 (BR; K). E: Nyika, fl. 26.vi.1966, *Fanshawe* 9743 (K; SRGH). **Malawi**. N: Nkhata Bay Distr., N. Viphya, 8 km. NW. of Chikwina, 1270 m., fl. & fr. 22.v.1970, *Brummitt* 11051 (K). C: Dedza Distr., Chongoni Forestry School, fl. & fr. 10.v.1967, *Salubeni* 701 (K; SRGH).

Also known from Zaire, Burundi and Tanzania. Bogs, marshes, damp grasslands and open woodlands; between 1250–2300 m.

9. **Buchnera descampsii** De Wild. & Ledoux in De Wild., Contrib. Fl. Katanga, Suppl. 3: 131 (1930). Type from Katanga.

Annual herb, 14–58 cm. tall, erect; stems simple, glabrous, tetragonal, drying pale fawnish-green. Cauline leaves 2.5–3.5 × 0.5–1 mm., linear-lanceolate, glabrous, closely appressed, opposite, 1-nerved; basal leaves 2–4.5 × c. 2 mm., ovate-caudate, subsessile, 3-nerved, shortly hispid, glabrescent. Spikes 5–15(40) × c. 7–8 mm., pale yellowish-brown when dry, densely compacted, quadrangular. Bracts 2.5–3.5(4) × 1.25–1.75 mm., ovate, long-acuminate, 5-nerved, minutely stipitate-glandular. Bracteoles 3.5–4.5 × 0.5–1 mm., linear-subulate, scarious, minutely glandular. Calyx 4.5–5.5 mm. long, scarious, 7–8-nerved, 4-lobed, subglabrous to sparsely minutely glandular; lobes 2.5–3.5 mm. long, subulate, minutely short hispid on margin. Corolla violet; tube 7–9 mm. long, cylindric, subglabrous to minutely glandular without, throat pilose; limb up to c. 6 mm. in diam.; lobes 2–3.5 × 2–3.5 mm., obovate to subcircular. Capsule c. 2.5 × 1.5 mm., obovoid, short-beaked.

Zambia. W: Solwezi Distr., R. Chikundwi, fl. & fr. 17.vii.1930, *Milne-Redhead* 733 (K).

Also known from Zaire. From grasslands and open woodland.

There is clearly a very close relationship between *B. descampsii* and *B. quadrifaria*, but I feel that, from the material I have seen, they are best kept separate. As De Wildeman observed following his description of *B. descampsii*, the plants differ mainly in both the size of the leaves and the nervation of the bracts. In *B. descampsii* the cauline leaves are from 2–4.5 mm. long and tightly appressed to the stem while in *B. quadrifaria* they are somewhat larger from 5–8.5 mm. long and clearly spreading-ascending. The nervation of the bracts is somewhat less clear as in *B. descampsii*, which De Wildeman claims has five nerves, there is frequently a narrowing of the organ giving clarity to only three. In *B. quadrifaria*, the nerves are admittedly more clearly defined, but occasionally there appears a further pair where the shape is broadened almost into a wing-like lower margin.

In the course of my studies of this complex a third element has come to light, this in the form of a collection, *Milne-Redhead* 852 from Zambia. This plant has all the first appearances of *B. descampsii* but differs only in the shape of the bracts, which in this case are cuneate and apically truncate-cuspidate and prominently 3-nerved. The bracteoles are clearly 1-nerved with the nerve not impressed but rather excavated in the upper half. I choose to withhold any decision as to its novelty until more material is seen, and have only to record its existence.

10. **Buchnera bangweolensis** R.E. Fries, Wiss. Eregbn. Schwed. Rhod.-Kongo-Exped. 1911–1912, **1**: 291 (1916). Type: Zambia, L. Bangweulu, *Fries* 892 (K, lectotype).

Annual herb, 30–65(75) cm. tall, erect; stem slender, simple but occasionally branched within inflorescence, glabrous except for few patent hairs towards base. Leaves 4–24 × 0.2–1(1.5) mm., 1-nerved, subappressed to stem, linear, scabrid, hispid-ciliate. Spikes 8–25(35) mm. long, to 6.5 mm. wide, subquadrangular, compact, not becoming lax in fruit. Bracts 3.5–5 × 3–4 mm., broadly subreniform with a short broad apical mucro, shortly pubescent, hispid-ciliate, not keeled. Bracteoles 2.5–3.5 × 0.5–0.8 mm., linear-lanceolate, acute, hispid-ciliate. Calyx (2.5)3.5–4 mm. long, 4-lobed, 4-nerved; lobes lanceolate 1.5–2 mm. long, ciliate, erect, sinus c. 2–2.5 mm. deep. Corolla pink to mauve or pale purple; tube 5.5–6 mm. long, glabrous without, slightly short-pilose at throat; limb 1.5–2 mm. in diam.; lobes c. 1 × 0.7–0.8 mm., subobovate to oblong. Capsule 2.8–3 × 1.6–1.8 mm., broadly ovoid, slightly emarginate; beak to c. 0.4 mm. long.

Zambia. N: Mansa Distr., Samfya, Lake Bangweulu, c. 1200 m., fl. & fr. 8.x.1947, *Greenway & Brenan* 8192 (K). W: Mwinilunga Distr., Sinkabolo Dambo, fl. & fr. 21.xii.1937, *Milne-Redhead* 3761 (BR; K). C: Mkushi Distr., Fiwila, 1400 m., fl. & fr. 9.i.1958, *Robinson* 2706 (K; SRGH). **Zimbabwe**. N: Goromonzi Distr., Ruwa Valley, fl. & fr. 29.iii.1970, *Linley* 506 (SRGH). C: near Harare, fl. & fr. 18.x.1936, *Eyles* 8782 (K).
Known only from Zambia and Zimbabwe. In peaty bogs and wet grasslands; 1200–1600 m.

11. **Buchnera splendens** Engl., Bot. Jahrb. **18**: 71 (1894). —Skan in F.T.A. **4**, 2: 380 (1906). Type from Angola.

Annual herb, 35–95 cm. tall, erect, long white pilose; stems simple, leafy. Basal leaves 20–40 × 10–26 mm., broadly elliptic, entire, sparsely to slightly densely hispid-scabrid, 3–5(7)-nerved; lower cauline leaves opposite, 15–45 × 2.5–5 mm., upper leaves alternate, 15–53 × 2–5(8) mm., all linear-oblong to lanceolate, obtuse, entire, hispid-scabrid to long pilose-scabrid, 3-nerved, uppermost obscurely so. Spike 10–35 mm. long, up to 18 mm. broad with flowers densely compacted, not markedly quadrangular, densely white silky-pilose. Bracts c. 12–15 mm. long, lanceolate, densely pilose, upper smaller. Bracteoles 5–6 × 0.5 mm., linear-subulate, densely pilose. Calyx 7–8 mm. long, cylindrical, densely pilose, 4-lobed; lobes c. 2 mm. long, subulate-lanceolate. Corolla pale blue, through darker blue to pink; tube c. 10 mm. long, slender, pilose above and without, throat pilose; lobes (3)5–7 mm. long, up to 5 mm. wide, obovate. Capsule 4–4.8 × 2–2.5 mm., cylindric to subconical, apex cuspidate, glabrous.

Zambia. N: Mbala Distr., Kambole Escarpment, 1500 m., fl. 7.vi.1957, *Richards* 10064 (K).
Also known from Angola and Zaire. Open woodlands; up to c. 1500 m.

12. **Buchnera crassifolia** Engl., Bot. Jahrb. **30**: 403 (1901). —Skan in F.T.A. **4**, 2: 385 (1906). Type: Malawi, *Whyte* s.n. (K, neotype chosen here).

Annual herb, 20–48 cm. tall, erect; stems branched above or more rarely simple, minutely pale brown, retrorse pubescent throughout. Leaves 8–35 × 1–3.5(6.5) mm., opposite or subopposite, oblong-elliptic to narrowly oblong-ovate, obtuse, obscurely 1–3-nerved, glabrous, shortly ciliate, basal leaves where present similar but broader. Spikes 15–35(45) × (5)8–12 mm., compact, not generally becoming lax at fruiting, occasionally only lower flowers becoming remote. Bracts 4.75–8 × 1.75–2.5(3) mm., lanceolate, pilose, ciliate. Bracteoles (3)4–6 × 0.5–1 mm., linear-lanceolate, pilose. Calyx (6.5)8–9.5 mm. long, obscurely 10-nerved, 4-lobed, hirsute-pilose with pale whitish brown antrorse subappressed hairs or subglabrous; lobes 1.5–2 mm. long, acutely lanceolate, ciliate, sinus present on adaxial side. Corolla mauve to purple, or magenta; tube (8.5)9–1 mm. long, slender, pilose without, throat densely pilose; limb (7)8.5–10.5(12) mm. in diam.; lobes to 6.5 × 4.5 mm. Capsule c. 6.5 × 2 mm., oblong-subcylindric, apex rounded to shortly rostrate.

Zambia. N: Mafinga, fl. 26.viii.1958, *Lawton* 465 (K). **Malawi**. N: Rumphi Distr., Nyika Plateau, 18 km. W. of Nchenachena Mt., 2400 m., fl. & fr. 5.v.1952, *White* 2593 (BR; K).
Known only from submontane grasslands; up to 2400 m.
In the absence of the holotype, presumably destroyed in Berlin by war action, or of any existing isotype material, I have chosen the un-numbered collection of *Whyte* made in Malawi (Nyasaland) and cited by Skan, as the neotype housed at Kew.

13. **Buchnera nitida** Skan in F.T.A. **4**, 2: 389 (1906). Type: Malawi, without further locality, fl. 1903, *McClounie* 123 pro parte (K, holotype).

Annual herb, 16–30 cm. or more tall, erect; stem woody, much branched or more rarely simple, arising from thickened woody base, bifariously, minutely pubescent, branches leafy throughout. Leaves 12–20 × 2–4.5 mm., opposite, linear-lanceolate, acute, glabrous to somewhat glossy, 1-nerved with small, leafy branches arising in axils. Inflorescence of few-flowered spikes or spicate clusters, much shortly branched. Bracts 5.5–7.5(8) × 2–3 mm., ovate to oblong-lanceolate, glabrous, ciliate. Bracteoles 4–5 × 0.5–1 mm., linear-lanceolate, glabrous. Calyx (5)7–9.5 mm. long, 5-nerved, 4-lobed, glabrous; lobes 1.5–2 mm. long, narrowly ovate-acuminate, shortly ciliate. Corolla pinkish-mauve or white; tube (7)10–11 mm. long, glabrous without, throat sparsely long pilose; limb 8–10.5 mm. in diam.; lobes 4–6.5 × 3–4 mm., obovate to oblong, unequal. Capsule not seen.

Zambia. N: Mafinga Hills, fl. 21.v.1973, *Chisumpa* 30 (K). **Malawi**. Without further locality, fl. 1903, *McClounie* 123 (K).

From the account of his travels, *McClounie* collected his material from the Nyika Plateau in Malawi during his survey of the area in September 1903.

14. **Buchnera arenicola** R.E. Fries, Wiss. Ergebn. Schwed. Rhod.-Kongo-Exped. 1911–1912, **1**: 292 (1916). Types: Zambia, Mbabala (Mbawala) Island, Lake Bangweulu, *Fries* 779 et 779a (B†).

Perennial 10–20 cm. tall, woody; stems erect or ascending, almost simple, subterete, sparsely hirsute or almost glabrous. Leaves 15–30 × 1 mm., opposite or in axillary fascicles, linear-acute, somewhat curved, convolute when dry, glabrous or pilose towards base, rarely ciliate. Spikes up to 40 × 6–7 mm. at fruiting, ovoid or shortly cylindric, densely flowered, occasionally with one or two flowers becoming remote at maturity. Bracts 4.5–5 × 2–2.5 mm., rhombic-ovate, acuminate, minutely ciliate, otherwise glabrous. Bracteoles c. 4 mm. long, lanceolate, acute. Calyx 8–9.5 mm. long, glabrous, subequally 5-lobed or with 4 subequal lobes and one much reduced in size or even absent; lobes 1–1.5 mm. long. Corolla white; tube c. 8 mm. long, sparsely pilose without; limb 12–14 mm. in diam.; lobes up to 4 mm. wide, obovate. Capsule 3–3.5 mm. long, ovoid, slightly acute.

Zambia. N: Mbabala (Mbawala) Island, Lake Bangweulu, fl. & fr., *Fries* 779, (B†). Known only from Zambia.

15. **Buchnera rungwensis** Engl., Bot. Jahrb. **30**: 403 (1901). —Skan in F.T.A. **4**, 2: 384 (1906). Type from Tanzania.

Annual herb, 16–36 cm. tall, erect; stems one to many arising from a rootstock, usually simple but occasionally branched below inflorescence, retrorsely, minutely pubescent with large, patent, multicellular hairs more or less densely present. Leaves opposite, (9)12–32 × 1.5–4(6) mm., narrowly elliptic-ovate, obtuse, obscurely 1–3-nerved, glabrous to subglabrous above, densely pilose beneath, especially on major nerves, ciliate; basal leaves where present, similar but broader. Spikes 15–40 × 5–10(12) mm., compact or with lowest flowers lax or somewhat remote at anthesis. Bracts 8–12.5 × 2.5–4 mm., lanceolate, imbricate, glabrous or subglabrous above, fairly long pilose beneath. Bracteoles 4.5–6.5 × 0.75(1.25) mm., long ciliate, linear-lanceolate. Calyx (6.5)8–9.5(10.5) mm. long, 4-lobed but occasionally with rudimentary fifth lobe represented by bristly tuft, densely pilose with coarse, antrorse, spreading hairs; lobes ciliate, (0.75)1.5–2.5 mm. long, narrowly triangular, calycal sinus present. Corolla pale lilac or mauve to purple, tube (8)9–11(13) mm. long, short pilose without, throat pilose; limb 8–13 in diam.; lobes 3.25–4.5 mm. long, c. 4.5 mm. wide at most. Capsule c. 6 × 2 mm., oblanceolate, tapering towards base, short beak present.

Malawi. N: Nyika Plateau, between Kasaramba and Chelinda, fl. 10.i.1967, *Hilliard & Burtt* 4411 (K).

Known also from Tanzania. Montane grasslands; 1800–2750 m.

16. **Buchnera nigricans** (Benth.) Skan in F.T.A. **4**, 2: 390 (1906). Type from Angola.
  Stellularia nigricans Benth. in Hook. Ic. Pl. **1880**: t. 1318 (1880) ex descr. *"nigrescens"* sphalm.
  —Hiern, Cat. Afr. Pl. Welw. **1**, 3: 774 (1898), as *"nigrescens"*. Type as above.
  Benthamistella nigricans (Benth.) Kuntze, Rev. Gen. Pl. **2**: 458 (1891). Type as above.

Annual herb, 25–40 cm. tall, erect, branched, short, hispid-pubescent. Leaves 30–45 × 1.5–4 mm., opposite, linear to linear-lanceolate, acute, entire, shortly hispid-pubescent, scabrid, 1-nerved. Spikes terminal, 3–9(12) cm. long, slender, cylindric, densely flowered or less so. Bracts c. 4.5–5.5(7.5) × 1.5–2.5 mm., ovate to ovate-lanceolate, acute to acuminate, pilose especially on three raised dark nerves, densely ciliate. Bracteoles 3–4(5.5) × 0.6–1 mm., lanceolate, acute, pilose, somewhat keeled, ciliate. Calyx 6–6.5(9) mm. long, 7–8-nerved, deeply divided on upper side, glabrous to subglabrous below on tube, 4-lobed; lobes 2–2.5 mm. long, ovate-lanceolate to lanceolate-acuminate, shortly pilose, ciliate. Corolla yellow, through greenish to purple; tube 7.5–9 mm. long, glabrous below, minutely to shortly pubescent above, markedly curved from midway or above, slightly constricted at throat, throat subglabrous, not pilose or villous; lobes 2–2.5 mm. long, oblong, fleshy, margins involute, undulate. Style 2–2.5 mm. long. Capsule c. 6 × 1.25 mm., oblong-cylindric.

Zambia. N: Mbala Distr., Kawimbe Rocks, c. 1750 m., fl. & fr. 18.vi.1970, *Sanane* 1218 (K; SRGH).
Also known from Angola. In moist shrubby pastures and thickets, and woodlands on sandy, stony soils.
Some clarification is needed here of the correct epithet to be used for this taxon. Bentham, when he described the original species, used the name *Stellularia nigricans* as the heading for his article in Hooker's Icones. Following his generic description he precedes his specific description with the name *Stellularia nigrescens*. This would appear to be the acceptable name by reason of its application to the description, but the illustration on which the article is based bears the name *S. nigricans*. This gives more reason for using the latter name, and any further doubt as to his intentions are finally settled by his use of the name *Stellularia nigricans* on the holotype.
I do not follow Skan's distinction of *B. nigricans* from *B. benthamiana*, as the characters he used are easily encompassed by my broader species description above.

17. **Buchnera randii** S. Moore in Journ. Bot. **38**: 467 (1900). —Skan in F.T.A. **4**, 2: 387 (1906).
—Norlindh & Weimarck in Bot. Not. **98**: 122 (1951). Type: Zimbabwe, Harare, July 1898, *Rand* 573 (BM, holotype).

Annual herb, 10–75 cm. tall, erect; stems slender, simple or more usually branched especially above, pilose-hispid. Upper cauline leaves 9–50 × 3–7 mm., opposite, narrowly linear, obtuse, entire, densely long white-hispid with callous-based patent or antrorse hairs, 1-nerved; middle and lower cauline leaves up to 70 × 12 mm., narrowly linear-obovate, obtuse, entire or rarely coarsely dentate with one or two teeth on each margin, densely hispid, 3-nerved; basal leaves where present, up to 35 × 15 mm., ovate to obovate-oblong, obtuse, entire to coarsely dentate, hispid. Spikes 5–38 × 5–12 mm., terminal or lateral, densely compacted except occasionally lower flowers becoming more remote during fruiting. Bracts 4–6.5 × 1.5–2.5(3) mm., ovate-lanceolate, hispid-scabrid, long white hispid usually on margins. Bracteoles 2–4 × c. 0.5–0.7 mm., subulate, white hispid-scabrid. Calyx 4–6 mm. long, white hispid-strigose, obscurely 10-nerved, 5-lobed, frequently with 4 lobes subequal, c. 1.6 mm. long, with fifth lobe much smaller, c. 0.8–1 mm. long, narrower; all lobes narrowly lanceolate. Corolla light blue, through dark blue to mauve and purple, occasionally white; tube (5.5)7–8 mm. long, slender, pilose without, throat densely pilose, limb 2.25–3(6) mm. in diam.; lobes obovate-oblong, obtuse, subequal. Capsule 3.5–4.5 × 2–2.4 mm., ovoid, glabrous.

Botswana. N: Xharatshaa, Gomoti R., fl. 12.v.1976, *Smith* 1727 (BR; K; SRGH). SE: Morale Research St., 980 m., fl. 3.iii.1978, *Hansen* 3359 (K). **Zambia**. B: Masese, fl. 14.iii.1961, *Fanshawe* 6435 (SRGH). N: Kambole Escarpment, c. 1500 m., fl. & fr. 24.iv.1969, *Sanane* 659 (K). W: Kamwedji R., fl. & fr. 21.vi.1953, *Fanshawe* 110 (BR; K; SRGH). C: Mkushi Distr., Muchinga Escarpment, 850–1050 m., fl. & fr. 19.iv.1972, *Kornaś* 1593 (K). E: Lukusuzi Game Res., c. 1030 m., fl. 6.v.1970, *Sayer* 203 (SRGH). S: c. 17 km. NE. of Choma, c. 1280 m., fl. & fr. 11.vii.1930, *Hutchinson & Gillett* 3541 (K). **Zimbabwe**. N: Mwami (Miami), 1370 m., fl. 15.v.1946, *Wild* 1158 (K; SRGH 15096). W: Matobo Distr., Matopos Nat. Park, Hazelside, fl. & fr. 28.iii.1978, *Mahlangu* 280 (SRGH). C: Mt. Hampden, N. of Harare, fl. & fr. 9.v.1934, *Gilliland* 119 (BM; K; SRGH). E: Mutare Distr., Odzani Falls, c. 0.5 km. below falls, fl. 30.iv.1950, *Chase* 2177 (BM; K; SRGH). S: Masvingo Distr., c. 32 km. N. of Masvingo, fl. 4.v.1962, *Drummond* 7951 (SRGH). **Malawi**. N: Mzimba Distr., Mzuzu, Marymount, 1370 m., fl. & fr. 20.ix.1972, *Pawek* 5743 (K; MO; SRGH). C: Nkhota Kota Distr., Ntchisi Mt., 1400 m., fl. & fr. 24.vii.1946, *Brass* 16911 (BM; BR; K; MO; NY; OXF; SRGH). S: Machinga Distr., Kawinga, 8 km. NE. of Ntaja near Mkoko Bridge, fl. 16.ii.1979, *Blackmore et al.* 449 (K). **Mozambique**. N: border of Lake Malawi (Nyasa), 600 m., fl. & fr. vii.1955, *Gomes e Sousa* 1517 (K; COI).
Also recorded from Tanzania and, doubtfully Zaire. Grassland and open woodland; from 600–1650 m.

18. **Buchnera strictissima** Engl. & Gilg in Warb., Kunene-Samb.-Exped. Baum: 367 (1903). —Skan in F.T.A. **4**, 2: 384 (1906). Type from Angola.

Annual herb, 9–20(55) cm. tall, erect; stems usually simple but sometimes sparingly branched, very slender, glabrous or with very few minute stiff hairs, very scattered. Leaves 4.5–16 × 0.5–1.5(2) mm., opposite, linear, entire, acute or subobtuse, scabrid, occasionally with few minute stiff hairs on margins and midrib, 1-nerved. Spikes 5–35 mm. long, up to 6 mm. broad, compact becoming slightly laxer at maturity. Bracts 2–2.8 × 2–2.6 mm., broadly ovate to subcircular, shortly apiculate, margin whitish, shortly ciliate. Bracteoles 1.25–2 × 0.5–0.75 mm., oblong-lanceolate, acute, white margined, ciliate. Calyx 2.8–3.25 mm. long, 4-lobed with pronounced sinus c. 1.75 mm. deep, bearing a minute fifth lobe or denticle c. 0.1 mm. long at base; lobes 0.6–0.8 mm. long, broadly deltoid to lanceolate, minutely hispid-ciliate. Corolla magenta to reddish-purple; tube 5–5.75(7) mm. long, glabrous without, very shortly pilose at throat; limb 1.5–2.5(4) mm. in diam.; lobes 1–2 × 0.5–1.5 mm., obovate to oblong, entire or bifid, Capsule 2–3 × 1.5–2.5 mm., oblong-ellipsoid, ± emarginate, beak up to 0.6 mm. long.

Zimbabwe. W: Matobo Distr., Besna Kobila Farm, c. 1450 m., fl. & fr. v.1955, *Miller* 2871 (K). C: Marondera Pasture Research St., fl. & fr. 18.ix.1931, *Rattray* 420 (K).
Also known from Angola. Moist grasslands; from 1200–1450 m.

19. **Buchnera buchneroides** (S. Moore) Brenan in Mem. N.Y. Bot. Gard. **9**: 13 (1954). Type: Zimbabwe, Mazowe (Mazoe), *Eyles* 366 (K).
  Eylesia buchneroides S. Moore in Journ. Bot. **46**: 311 (1908). —Eyles in Trans. Roy. Soc. S. Afr. **5**: 475 (1916). —Norlindh & Weimarck in Bot. Not. **98**: 124 (1951). Type as above.

Annual herbs, 20–48 cm. tall, erect; stems slender, simple or occasionally branched, glabrous to subglabrous with occasional short white patent hairs. Leaves 10–35(45) × 0.5–2.5(3.5) mm., opposite or subopposite, spreading, linear to narrowly linear-lanceolate, acute, short white hispid-scabrid on midrib and within margins beneath, only within margins above, 1-nerved. Spikes 5–30(55) mm. long, up to 8 mm. wide, cylindrical, compact, becoming lax especially towards base when fruiting. Bracts 3.25–4 × 2–2.5(3.5) mm., broadly ovate, acuminate, concave, glabrous except for median nerve which is shortly white-hispid, hispid-ciliate. Bracteoles 1.5–2.5 × 0.5 mm., lanceolate, hispid-ciliate. Calyx 2.25–3 mm. long, 4-lobed with pronounced sinus 1.5 mm. deep; lobes 1–1.5 mm. long, ciliate, narrowly lanceolate. Corolla blue to purple; tube 4.5 mm. long, glabrous without, throat pubescent; limb 1.5–2.5 mm. in diam.; lobes 0.8–2 × 0.25–0.5 mm., oblong, entire or bifid. Capsule 2–3 × 1.8–2 mm., broadly ovoid to subglobose, beak 0.3–0.6 mm. long.

Zambia. N: Mbala Distr., from Kambole to Kalongola, 1650 m., fl & fr. 31.i.1959, *Richards* 10822 (K). C: Mkushi Distr., Fiwila, c. 1200 m., fl. 5.i.1958, *Robinson* 2616A (K; SRGH). W: Mwinilunga Distr., Sinkabolo Dambo, fl. & fr. 21.xii.1937, *Milne-Redhead* 3760 (BR; K). **Zimbabwe**. N: Mazowe (Mazoe), c. 1350 m., fl. 20.iv.1906, *Eyles* 366 (K). W: Matobo Distr., Besna Kobila Farm, c. 1430 m., *Miller* 1675 (SRGH). C: Harare Distr., R. Ruwa, 1370 m., fl. & fr. 21.i.1946, *Wild* 704 (K; SRGH). **Malawi**. C: Dedza Distr., Chongoni Forest, fl. & fr. 24.iii.1969, *Salubeni* 1282 (K; SRGH).
Known only from the above area. Swamps, marshes and wet grasslands frequently on sandy soil; 1200–1700 m.

20. **Buchnera ciliolata** Engl., Bot. Jahrb. **18**: 69 (1894). —Hiern, Cat. Afr. Pl. Welw. **1**: 775 (1898). —Engl. & Gilg in Warb., Kunene-Samb.-Exped. Baum: 366 (1903). —Skan in F.T.A. **4**, 2: 379 (1906). Type from Angola.

Annual herbs, 8–30 cm. tall, erect, slender; stems simple or branched, glabrous, terete, almost filiform. Leaves 8–30 mm. long, up to 1 mm. wide, opposite or subopposite, linear, entire, acute, glabrous to sparsely scabrid. Spikes 3–15(25) × 2–5 mm., densely compacted, lowest flowers becoming more remote at fruiting. Bracts c. 2.5 × 1.5 mm., ovate, acute or acuminate, occasionally minutely toothed towards apex, glabrous, margin very minutely hispid. Bracteoles 2 × 0.5 mm., linear, glabrous, margin minutely hispid. Calyx 4 mm. long, 4-lobed, slightly bipartite, glabrous, obscurely 8-nerved; lobes 1 × 0.5 mm., ovate, obtuse to acuminate. Corolla blue-violet; tube 6–7 mm. long, glabrous without, throat not evidently pilose; limb 6–7 mm. in diam.; lobes 3–5 mm. long, up to 2.25 mm. wide, entire or emarginate, or bifid at apex. Capsule c. 3 × 1.75 mm., slightly compressed ellipsoid-cylindrical, emarginate, glabrous.

Zambia. N: Mbala Distr., Mbala, road near pan, 1524 m., fl. 1.iii.1955, *Bock* 287 (SRGH). S: c. 35 km. N. of Choma, c. 1120 m., fl. 17.v.1954, *Robinson* 761 (K; SRGH).
Also known from Angola. Drying dambos and dryer grasslands; 1200–1980 m.

21. **Buchnera androsacea** Merxm. in Trans. Rhod. Sci. Ass. **43**: 117 (1951). Type: Zimbabwe, Marondera (Marandellas), Cave Vlei, 3.ix.1941, *Dehn* 431 (SRGH, isotype).

Annual or perennial herbs up to 8 cm. tall; stems several from rootstock, erect, loosely long-patent pilose. Leaves: cauline opposite or alternate, 5.5–6 × 2–2.5 mm., ovate-elliptic, obtuse, entire, pilose, 1-nerved; basal leaves 8–18 × 4–8(10) mm., entire, circular to broadly ovate, 3–5-nerved, margins somewhat thickened. Inflorescence of few-flowered short spikes about 1 cm. long; flowers sessile. Bracts (3.8)4.5–6 × (1.5)1.75–3.5 mm., broadly ovate-lanceolate, somewhat concave, finely pilose-pubescent. Bracteoles absent. Calyx 4.5–6 mm. long, 5-lobed, nervation obscured by indumentum being finely pilose-pubescent; lobes 0.8–1.5 mm. long, lanceolate, pilose, ciliate. Corolla blue or mauve; tube 6.5–7.5 mm. long, pubescent without, long pilose at throat; limb 3–5 mm. in diam.; lobes 2–3 × 1–1.5 mm., obovate, rounded. Capsule (immature) 2 × 0.8 mm., cylindric.

Zimbabwe. C: Enterprise Distr., fl. 1.ix.1946, *Greatrex* in GHS 15319 (K; SRGH).
Known only from the central area of Zimbabwe, where it occurs in grasslands.
Similar to *B. hockii* De Wild., but the basal leaves do not turn black on drying.

22. **Buchnera ruwenzoriensis** Skan in F.T.A. **4**, 2: 378 (1906). Types from Uganda.

Annual herbs, (10)30–65 cm. tall, erect; stems simple or occasionally branched above, patent to retrorse hispid-pilose. Leaves: cauline 18–40 × 2.5–5 mm., opposite or subopposite, lanceolate, acute or obtuse, entire, somewhat densely long pilose to hispid-pilose above, more sparsely so beneath; basal leaves 25–30 × 8–12 mm., elliptic to oblong, obtuse, entire to sparsely dentate, 3–5-nerved, scabrid above, pilose beneath mainly on nerves. Spikes 10–26(30) mm. long, quadrangular, terminal or lateral, with flowers dense, compact. Bracts 3–5.5 mm. long, up to 1.75 mm. wide, lanceolate to ovate-lanceolate, upper at least with the apices bi or trifid, hispid to hispid-scabrid. Bracteoles absent. Calyx 4–4.5 mm. long, 5-lobed, 10-nerved, densely long, soft pilose; lobes 0.6–1.25 mm. long, lanceolate, uneven, frequently with one lobe very much shorter. Corolla blue to purple, or white; tube c. 5 mm. long, pubescent without above; limb 1.6–2.25 mm. in diam.; lobes 1–1.3 × c. 1 mm., obovate to circular, margins slightly involute or crenate when dry. Capsule 4.5–5 × 2 mm., cylindric-ovoid.

Zambia. N: Mbala Distr., Chitembwa, fl. & fr. 30.v.1955, *Nash* 120 (BM). W: Mufulira, fl. & fr. 11.vi.1934, *Eyles* 8162 (BM; K; SRGH).
Also known from Angola and Uganda. Rocky or grassy slopes, often on clayey soils; up to 1550 m.

23. **Buchnera humpatensis** Hiern, Cat. Afr. Pl. Welw. **1**: 777 (1898). —Skan in F.T.A. **4**, 2: 393 (1906). Type from Angola.

Annual herbs, 8–38 cm. tall, erect, slender; stems branched with branches ascending, angular, glabrous. Leaves 10–40 × 1.5–6 mm., opposite or superior alternate, lanceolate or linear-lanceolate, narrowed at base, entire or remotely, minutely dentate, 1-nerved, variously covered with callous scales especially on margin. Inflorescence of long slender spikes, 5–12(16) cm. long, at maturity flowers very lax, opposite to almost opposite. Bracts 2.5–4.5 × 1.25–2 mm., broadly ovate to ovate-elliptic, concave, minutely toothed or hispid on margins. Bracteoles absent. Calyx (2.5)3–4 mm. long, 5-lobed, 5-nerved, glabrous; lobes 0.5–1.25 mm. long, ovate, acute. Corolla pink or blue, lilac to purple; tube 2.25–2.6 mm. long, glabrous; limb 1.25–1.5 mm. in diam.; lobes c. 0.6–1 mm. long, narrowly obovate. Capsule 1.9–2.25 × 1.8–2.4 mm., subglobular to ellipsoid, somewhat emarginate.

Zambia. B: Mongu Distr., Lealui, fl. & fr. 5.iii.1966, *Robinson* 6871 (K; MO; SRGH). N: Mbala Distr., Kawimbe, Lumi R. Marsh, 1740 m., fl. & fr. 27.iii.1959, *Richards* 12282 (K). W: Kasempa Distr., 11 km. E. of Chizela (Chizera), fl. & fr. 27.iii.1961, *Drummond & Rutherford-Smith* 7433 (K; SRGH). S: Mazabuka Distr., 1060 m., fl. & fr. 8.iii.1958, *Robinson* 2785 (K). **Malawi**. N: Rumphi Distr., c. 22 km. N. of Rumphi, fl. & fr. 26.ii.1978, 1100 m., *Pawek* 13918 (BR; K; SRGH).
Also known from Angola. Wet grasslands and pools; up to 1750 m.

24. **Buchnera ebracteolata** Philcox in Kew Bull. **42**: 384 (1987). TAB. **36**. Type: Zambia, Chilongowelo, Plain of Death, 1460 m., 5.v.1955, *Richards* 5518 (K, holotype; BR, isotype).

109

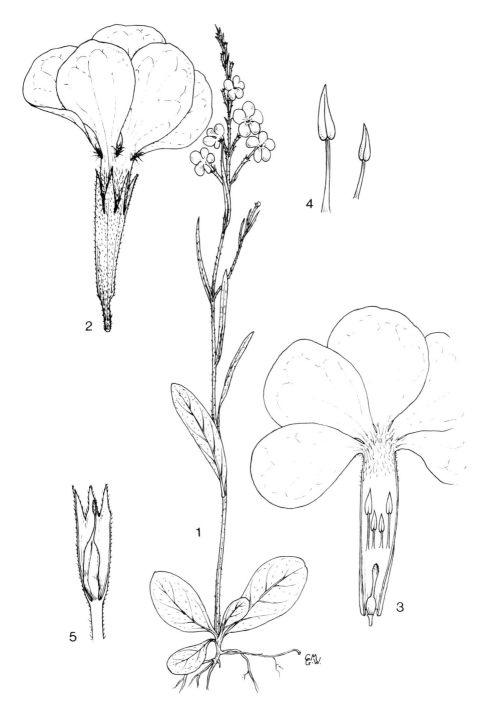

Tab. 36. BUCHNERA EBRACTEOLATA. 1, habit (× ⅔); 2, flower (× 4); 3, corolla opened out showing gynoecium (× 4); 4, anthers (× 8); 5, young fruit (× 4), all from *Richards* 4510.

Annual or possibly perennial herbs, 30–70 cm. or more tall, erect, branched; stems terete, appressed to subpatent white pilose or hispid throughout, some hairs being swollen at base; branches alternate, spreading-ascending. Leaves opposite, subopposite or alternate; upper leaves 30–60(80) × 3–7.5 mm., linear-oblong to narrowly ovate-elliptic, obtuse, shortly apiculate, narrowing into a petiole-like base, entire, appressed-hispid on both surfaces with hairs more patent beneath, margins short hispid-ciliate, 1-nerved; lower leaves up to c. 45 × 20 mm., broadly ovate-elliptic to obovate, obtuse, rounded, narrowed at base, with 1–3 major nerves, median nerve pinnately branched, entire or crenate or bearing 1 or 2 blunt teeth, indumentum as for upper leaves, but becoming scabrid. Inflorescence of long interrupted spikes 12–30 cm. long; flowers alternate or more rarely almost opposite, sessile or if subsessile, pedicels c. 0.5 mm. long, stout, pubescent. Bracts 2.4–4(5) × 1–1.5(2) mm., ovate-lanceolate, acuminate, concave, hispid, ciliate. Bracteoles absent. Calyx 7–8 mm. long, slightly curved, densely almost appressed hispid throughout with short white, frequently callous-based, antrorsely directed hairs, never retrorse, 5-lobed, 10-nerved; lobes 1.5–2(3) mm. long, narrowly lanceolate, subequal, always forwardly directed, never becoming reflexed. Corolla blue to violet or mauve; tube 9–10.5 mm. long, cylindrical, widening at mouth, somewhat curved, densely shaggy-pilose without, throat densely pilose within; limb 8–13 mm. in diam.; lobes 4.5–8 × 2–5 mm., obovate, unequal. Capsule 5–7 × 1–1.5(1.8) mm., oblong in outline, subacuminate at apex, remains of style persistent.

Zambia. N: Mbala Distr., Chilongowelo, 1440 m., fl. & fr. 9.v.1957, *Richards* 9627A (K). Known only from northern Zambia. In open grasslands and woodlands; 750–1500 m.

25. **Buchnera lippioides** Vatke ex Engl., Bot. Jahrb. **18**: 68 t. 3, figs. O, P (1893). —Engl. & Gilg in Warb., Kunene-Samb.-Exped. Baum: 368 (1903). —Skan in F.T.A. **4**, 2: 377 (1906). Type from Angola.
 Buchnera trinervia Engl., Bot. Jahrb. **23**: 512 (1897). —Hiern, Cat. Afr., Pl. Welw. **1**: 777 (1898). Type from Angola.

Annual or perennial herbs, erect to 1 m. or more tall; stem usually branched above, rarely simple, leafy, hispid. Leaves ± opposite, 40–70 × 8–12(20) mm. towards base of stem, decreasing in size above, obovate-oblong to linear-elliptic, obtuse at apex, narrowed at base, entire to occasionally dentate with 1–few obscure teeth, hispid-scabrid on both surfaces, less so above, 3-nerved, loosely imbricate. Spikes 10–20(45) mm. long, up to 20 mm. broad, terminal or axillary, eventually quadrangular. Bracts to about 8 × 5 mm., broadly ovate to ovate-circular, apex acute, hispid-pilose to scabrid (generally or frequently only on major nerves) ciliate. Bracteoles similar but narrower. Calyx 8–9 mm. long, 5-lobed, 10-ribbed, pilose on ribs; lobes c. 1–1.5 mm. long, broadly lanceolate-ovate, obtuse to broadly acute. Corolla pale mauve to dark blue or white; tube 10–11 mm. long, sparsely, minutely glandular-pubescent without, throat pilose; limb 14–18 mm. in diam.; lobes 6–7 mm. wide, subcircular-obovate. Capsule 3–5 × 1.5–2 mm., ovoid-cylindric, glabrous.

Zambia. N: Mbala Distr., c. 1500–1800 m., fl. iii.1934, *Gamwell* 193 (BM). S: near Mumbwa, fl. 1911, *Macaulay* 371 (K).
The name *B. lippioides* was first published and attributed to Vatke in a key to the African species of the genus by Engler (1893). It is considered to have been validly published by reason of the characters used in this key although no further description was included and no material cited by which to typify the name. Skan (1906) cites three specimens as representing *B. lippioides* only two of which, *Welwitsch* 5883 and 5885, Engler or Vatke could have seen before 1893. The Berlin material of these collections was destroyed by war action, but isosyntypes of these numbers exist at the BM and K, from which I have chosen the BM specimen of *Welwitsch* 5885 to be the lectotype.

26. **Buchnera wildii** Philcox in Kew Bull. **42**: 208 (1987). TAB. **37**. Type: Zimbabwe, Chimanimani Distr., Chimanimani Mts., near source of Bundi R., 1828 m., 8.v.1958, *Chase* 6907 (K, holotype; SRGH, isotype).
 Buchnera multicaulis var. *grandifolia* Norlindh in Bot. Not. **1951**, 2: 123 (1951). Type: Zimbabwe, Nyanga Distr., c. 2300 m., 4.ii.1931, *Norlindh & Weimarck* 4820 (K, isotype).

Annual or perennial herbs up to 65 cm. tall, erect or ascending; stems 1–several arising from woody rootstock, simple or branched near or towards base, or occasionally within inflorescence, long pilose to hispid-pilose, frequently intermixed with shorter appressed hairs, sparsely leafy above with long internodes up to 8(11) cm. long. Leaves: upper cauline opposite to subopposite, up to c. 10 × 1.5 mm., linear, patently hispid-pilose; lower

Tab. 37. BUCHNERA WILDII. 1, habit (× ⅔); 2, flower (× 3); 3, corolla opened out showing gynoecium (× 3); 4, capsule with calyx (× 6), all from *Wild* 4485.

leaves opposite to alternate, 15–40(55) × 2–5(9) mm. long, fine pilose to varying density on both surfaces, ovate-oblong, obovate to linear-oblong, 1–3-nerved, entire or occasionally shallowly crenate, obtuse, shortly apiculate to mucronulate. Spikes terminal, 10–50 mm. long at flowering, up to 90(180) mm. long in fruit, up to 15 mm. wide, cylindrical, compact with lower flowers occasionally becoming remote at fruiting, obscurely quadrangular. Bracts 4–7.5 × 1.2–1.5(2) mm., hispid-pilose, lanceolate. Bracteoles 2.5–5 × 0.3–0.5 mm., linear, hispid-ciliate. Calyx 6.5–9 mm. long, obscurely to prominently 10-nerved, densely to subdensely patent hispid-pilose, 5-lobed, frequently with adaxial lobe much reduced, even causing calyx to appear 4-lobed; lobes (0.5)2–4 mm. long, narrowly triangular to subulate, broadly divergent especially at fruiting, calycal sinus not present. Corolla blue, mauve, purple or white; tube 7–8.5 mm. long, slightly densely pilose without, throat pilose; limb (7)10–13(17.5) mm. in diam.; lobes 2.5–6 × 2–3.5 mm., oblong to obovate, unequal. Capsule 4.8–5.5 × 2.5–3 mm., cylindrical to broadly cylindric oblong-ovoid, apex truncate, shortly rostrate, c. 0.5 mm. long.

Zimbabwe. E: Chimanimani Distr., Chimanimani Mts., Upper Bundi Valley, fl. & fr. 3.ii.1957, *Goodier* 193 (K; SRGH). **Mozambique**. MS: Chimanimani Mts., c. 1525 m., fl. & fr. 6.vi.1949, *Wild* 2896 (K; SRGH).

Known only from the Eastern Highlands of Zimbabwe and the adjacent area in Mozambique. Grasslands; up to 2440 m.

A specimen collected by *Pawek* 2091, from the Nyika Plateau in Malawi appears to belong to this species, but it is far out of its geographical range.

27. **Buchnera welwitschii** Engl., Bot. Jahrb. **18**: 71 (1894). —Skan in F.T.A. **4**, 2: 386 (1906). Type from Angola.

Annual herbs, 12–44 cm. tall, erect; stems simple, rarely branched, slender, white to fawn sparingly patent to antrorse hirsute, denser at base. Leaves: cauline opposite to subopposite, 2–3 pairs only, 12–40 × 1–3(8) mm., oblong-linear to linear, obtuse, 1-nerved, white hirsute-scabrid; basal leaves 17–40 × 10–25 mm., elliptic-oblong to broadly ovate or subcircular, obtuse, entire to crenate, hirsute-scabrid above, pale brown hispid-scabrid beneath especially on major nerves, 3–5-nerved. Spikes 8–30 mm. long, to c. 8 mm. wide, flowers crowded or compact, somewhat quadrangular, occasionally base of spike lax with flowers of lowest internode c. 45 mm. distant from next above. Bracts 3–5.5 × 1–2 mm., lanceolate, concave, pilose. Bracteoles (2)3.5–4.5 × (0.25)0.6 mm., linear to subulate, pilose. Calyx 4.5–5.5 mm. long, pilose, 5-lobed; lobes 0.8–1.5(2.5) mm. long, unequal, subulate. Corolla blue; tube 4.5–5.5(8) mm. long, pilose without, densely so at throat; limb 3–3.5(8) mm. in diam.; lobes 1.5–2(4) × 1–2(3) mm., obovate. Capsule 4–5 × 1.5–2.5 mm., ovoid-cylindric.

Zambia. W: Solwezi Distr., Mutanda Bridge, fl. & fr. 26.vi.1930, *Milne-Redhead* 539 (K). C: Mumbwa Distr., between Landless Corner and Mumbwa, fl. 19.iii.1963, *van Rensburg* 1706 (K; SRGH). Also known from Angola and Tanzania. Grasslands and open woodland.

28. **Buchnera multicaulis** Engl., Bot. Jahrb. **18**: 69, t. 3, fig. A (1894). —Skan in F.T.A. **4**, 2: 396 (1906). —Norlindh & Weimarck in Bot. Not. **98**: 122 (1951). Type from Angola.

Annual herbs to c. 7.5 cm. tall, erect, ascending or prostrate; stems many, 3–7.5 cm. long arising from a short woody base, terete, minutely, sparsely hispid. Leaves subopposite or alternate, 11–35 mm. long, 2–3(7) mm. wide, oblong-spathulate, obtuse, glabrous occasionally sparsely short hispid on nerves beneath. Inflorescence with comparatively few flowers, clustered at ends of branches, lengthening only slightly to 2.5 cm. at fruiting. Bracts (2.5)4 mm. long, ovate-lanceolate, subacuminate, shortly hispid. Bracteoles 3–3.8 mm. long, 0.2–0.4 mm. wide, linear to narrowly linear-lanceolate, sparsely short ciliate to glabrous. Calyx 6–7(9) mm. long, obscurely 10-ribbed, 5-lobed, sparsely appressed hispid with forwardly directed hairs, becoming subglabrous; lobes 1.2–1.5 mm. long, linear-lanceolate, not spreading after flowering. Corolla mauve, blue-purple or rose-coloured; tube (6.5)9–10 mm. long, cylindric, straight, glabrous to shortly white pilose, occasionally sparsely intermixed with minute stipitate glands, throat densely long-pilose; limb (6.5)8–9 mm. in diam.; lobes 4.5–5 mm. long, 2.5–3.5 mm. wide, unequal, obovate. Capsule c. 4 mm. long, 2.3 mm. in diam., compressed ovoid-oblong.

Zambia. E: near Minga, c. 360 km. E. of Lusaka, fl. 26.viii.1929, *Burtt Davy* 939A/29 (K). **Zimbabwe**. N: Mwami (Miami) Distr., 1370 m., fl. & fr. 4.x.1947, *Wild* 1337 (K). E: Chimanimani Distr., Bundi Valley, 1585 m., fl. 27.ix.1966, *Plowes* 2795 (K; SRGH). **Mozambique**. MS: Báruè, Serra

da Chôa, 1200 m., fl. 17.ix.1942, *Mendonça* 308 (LISC).
Also known from Angola and possibly Zaire. High altitude grasslands; to over 2000 m.

29. **Buchnera hockii** De Wild. in Fedde, Rep. **13**: 199 (1914). Type from Zaire (Katanga).

Perennial herb 2.5–4.5(10) cm. tall; stems many arising from a short woody base, 1.5–4.5(10) cm. long, subterete, glabrous to sparsely hispid. Leaves: basal in dense rosettes, up to 6 × 2.5 cm., at times equalling stems, lanceolate to obovate-lanceolate, entire, 5-nerved, glabrous above, shortly appressed-hispid only on prominent major nerves beneath, not drying black; stem leaves 4–6.5 × 1.6–3 mm., elliptic or ovate-elliptic, shortly hispid, hispid-ciliate on margin, opposite or subopposite. Inflorescence spicate with terminal spikes, c. 1 cm. long, few-flowered increasing to c. 3 cm. long at flowering with flowers lax. Bracts 3.5–4.5(5) × 1.5–2 mm., ovate, 1-nerved, hispid on median nerve, ciliate-hispid. Bracteoles, 2.5–4 × 0.2–0.3 mm., linear-subulate, patent-hispid. Calyx 6–7 mm. long, 5-lobed, obscurely 10-nerved, hispid on nerves only or glabrous, lobes 1–1.5 mm. long, triangular, not spreading after flowering. Corolla dark blue to purple; tube 7–10 mm. long, cylindric, glabrous without, throat pilose; limb 5–7.5 mm. in diam.; lobes 2–5 × 1–1.8 mm., unequal, obovate. Capsule 5.5–6 × 1.5 mm., oblong in outline, subcompressed, remains of style persistent.

Zambia. W: Mwinilunga Distr., c. 19 km. W. of R. Kasingiko, fl. & fr. 4.viii.1930, *Milne-Redhead* 830 (K). S: Kafue Nat. Park, near Moshi Camp, fl. & fr. 30.vi.1964, *van Rensburg* KR52939 (K; SRGH).
Also known from Zaire. In grassland and open woodland, frequently on damp sandy soil.
This plant exhibits the peculiar character of the stems and other vegetative parts turning black on drying, while the basal rosette or cluster of large leaves dry pale brown. Very few collections have been seen and studied for this treatment, but it appears from the collectors' information that such plants appear to have been affected by burning and it is here considered possible that this may be responsible for this colour phenomenon on drying.

30. **Buchnera chisumpae** Philcox in Kew Bull. **42**: 384 (1987). TAB. **38**. Type: Zambia, Kasama Distr., Chishimba Falls, 40 km. W. of Kasama, 1610 m., 30.iv.1986, *Philcox, Pope & Chisumpa* 10219 (K, holotype; NDO; SRGH, isotypes).

Annual herb 35–45 cm. tall, erect; stems slender, much branched, glabrous to subglabrous above, sparsely minutely pubescent below, branches ascending, leafy throughout. Leaves opposite or alternate frequently with smaller leaves or short branches arising in axils, 30–40(55) × 0.5–1(2.25) mm., very slender to almost filamentous, mostly revolute, sparsely short-hispid, strict or occasionally irregularly contorted when dry, 1-nerved. Spikes up to 15 cm. long, slender, cylindrical; flowers laxly arranged, opposite or alternate, ascending, somewhat appressed to axis. Bracts (2)3–4(7) × c. 1 mm., lanceolate, glabrous except for short-ciliate margin and midrib. Bracteoles 1.6–2.8(4.3) mm. long, linear, ciliate. Calyx (5.5)6.8–8.5 mm. long, 5-lobed, 5-nerved, glabrous; lobes (0.5)1–1.8(2) mm. long, triangular, acute or obtuse, short-ciliate. Corolla cream, yellow, pink to mauve; tube often pale brown, 8–9 mm. long, glabrous, cylindrical, becoming arcuate after anthesis; limb 2.8–4.2 mm. in diam.; lobes 1.5–2.8 × 0.4–0.8 mm., lanceolate-oblong, obtuse, involute, appearing somewhat fleshy when dry, subequal. Capsule 9–10 × c. 1 mm., slender, linear-elliptic in outline tapering to acute, rostrate apex.

Zambia. N: near Kasama, fl. & fr. 31.iii.1955, *Exell, Mendonça & Wild* 1387 (BM; LISC; SRGH).
Known only from northern Zambia. Among rocks in dry, sandy areas; from 1260 to c. 1750 m.

31. **Buchnera lastii** Engl., Pflanzenw. Ost-Afr. **C**: 359 (1895). —Skan in F.T.A. **4**, 2: 392 (1906).
—Norlindh & Weimarck in Bot. Not. **98**: 121 (1951). Type: Mozambique, Namuli, Makua, *Last* s.n. (K, holotype).
Buchnera similis Skan in F.T.A. **4**, 2: 391 (1906). Type: Malawi, between Khondowe (Kondowe) & Karonga 600–1800 m., *Whyte* s.n. (K, syntype), Mt. Mulanje, *Whyte* s.n. (K, syntype; P, isosyntype).
Buchnera tuberosa Skan in F.T.A. **4**, 2: 391 (1906). Type: Malawi, Thuchila Plateau, Mt. Mulanje, 1800 m., *Purves* 21 (K, syntype; E, isosyntype).

Annual or perennial herbs, 12–40 cm. or more tall from woody rootstock, erect; stems simple or branched, with spreading branches, erect or arcuate, terete, slightly angular, minutely bifariously pubescent. Leaves mostly opposite, 7–35(50) × 1–4.5(8) mm., lanceolate or linear-lanceolate, sessile, acute, entire or minutely denticulate, glabrous above or sometimes minutely pubescent, glabrous or sparsely hispid-pubescent beneath especially on nerves, shortly ciliate, 1-nerved. Inflorescence of long slender spikes,

Tab. 38. BUCHNERA CHISUMPAE. 1, habit (× $\frac{2}{3}$); 2, flower (× 3); 3, corolla opened out showing gynoecium (× 3); 4, capsule with calyx removed (× 4), all from *Robertson* 634.

(2)3.5–16(30) cm. long, not quadrangular, most flowers imbricate, but the lower frequently interrupted and arranged more laxly on spike. Bracts (5.5)7.5–9 × (1.5)2–3.5 mm., lanceolate, acuminate, ciliate or finely pubescent. Bracteoles 3.5–5(8) × 0.35–1 mm., linear to narrowly lanceolate, ciliate or occasionally finely pubescent. Calyx (5)7–9.5(12.5) mm. long, 5-lobed, 10-nerved, glabrous; lobes (0.6)1.25–3 mm. long, unequal, lanceolate or narrowly triangular, glabrous or ciliate. Corolla white, cream or yellow, brown or reddish-purple, fragrant; tube 7–10(16) mm. long, cylindric, slender, glabrous or finely pubescent, gently curved throughout; limb (3)5.5–8 mm. in diam.; lobes 3.5–5(7) × 0.5–1.8 mm., obovate, sometimes thick with involute margins. Capsule 3.5–5 mm. or more long, 1.5–2 mm. broad, oblong.

Subsp. **lastii**

Corolla tube glabrous without; flower spikes 3.5–16 cm. long.

Zambia. N: Mbala Distr., 1.6 km. on Mbala–Kasama road, 1500 m., fl. 29.viii.1956, *Richards* 5978 (K). E: Nyika Plateau, fl. 23.xi.1955, *Lees* 57 (K). **Malawi**. N: Nkhata Bay Distr., c. 15 km. NE. of Chikangawa, c. 1760 m., fl. 11.xi.1977, *Phillips* 3048B (K; MO). S: Mt. Mulanje, NE. foot of Matambale, 2040 m., fl. 7.vi.1970, *Brummitt* 11345 (BR; K; SRGH). **Zimbabwe**. E: Chimanimani Distr., Mt. Pene, 2130 m., fl. & fr. 12.x.1908, *Swynnerton* 6150 (BM; K). **Mozambique**. Z: Namuli, Makua, fl. sine die, *Last* s.n. (K). MS: Báruè, Serra de Choa, c. 28 km. from Catandica (Vila Gouveia), 1600 m., fl. & fr. 9.xii.1965, *Torre & Correia* 13493 (LISC).

One herbarium specimen of *Walters* 2747 belonging to this subspecies, and located in Herb. Kew is purportedly collected from Harare, while another in the SRGH collection is labelled as from Melsetter (Chimanimani). The latter is considered correct by reason of the habitat and altitude favoured by this species.

Subsp. **pubiflora** Philcox in Bol. Soc. Brot., sér. 2, **60**: 269 (1987). Type: Zambia, Mbala Distr., above Ndundu, 1500 m., 16.ii.1957, *Richards* 8196 (K, holotype).

Corolla tube pubescent without; flower spikes 18–30 cm. long.

Zambia. N: Mbala Distr., above Ndundu, 1500 m., fl. 16.ii.1957, *Richards* 8196 (K). **Malawi**. N: Rumphi Distr., Nyika Plateau, c. 29 km. SE. on Kasaramba road, c. 2400 m., fl. 23.iv.1976, *Pawek* 11212 (K).

Also known from Tanzania. Hillsides and montane grasslands on rich or poor, rocky or sandy soils; 1500–2300 m.

With the exception of *Fanshawe* 4049 from Kawambwa, all other Zambian collections of subsp. *pubiflora* are from the immediate area around Mbala. The *Fanshawe* collection differs only in having a denser fine pubescence on the upper surface of the leaves.

Plants here referred to *B. lastii* have, in the past, been separated under three names by reason of inflorescence size and flower colour. Plants from Mt. Mulanje are recorded as, with one exception, having white, purple, red, pink or brown flowers and tend to have a compact inflorescence at flowering which elongates at fruiting. These have been widely considered to represent *B. similis*. Plants from the Nyika area of Malawi on the other hand are recorded as having cream or yellow flowers and tend to have elongate inflorescences with broader bracts, and have been considered to represent *B. tuberosa*.

Flower colour breaks down as a character separating *B. similis* from *B. tuberosa*, as is demonstrated by the following two examples: 1. The type specimen of *B. tuberosa* collected on Mt. Mulanje is brown-flowered while the flowers of those plants from the Nyika, which are referred to this species, are cream or yellow coloured. 2. Mt. Mulanje plants which are referred to *B. similis* are white, purple, red, pink or brown-flowered, however, a syntype of this species collected by Whyte on the Nyika is recorded as having yellow flowers.

These two elements have been distiguished from a third element, *B. lastii*, by reason of the density or laxity of the flowers on the inflorescence and the overall sturdiness of the plant. The laxity of the flowers, however, is seen to vary depending on the age of each spike, and in most cases the spikes elongate at maturity with the lower flowers becoming more distinct.

In view of the above, I consider that the three elements represent the same species and use *B. lastii* as the earliest valid name.

32. **Buchnera leptostachya** Benth. in DC., Prodr. **10**: 497 (1846). —Skan in F.T.A. **4**, 2: 394 (1906), excl. *B. mossambicensis* var. *usafuensis* Engl. —Hepper in F.W.T.A., ed. 2, **2**: 369 (1963). Types from West Africa and Pemba.
    Buchnera mossambicensis Klotsch in Peters, Reise Mossamb. Bot.: 224, t. 34 (1861). Type: Mozambique, Quirimba Island, *Peters* s.n. (not seen).

Annual or perennial herbs, 40–50 cm. tall, erect; stems simple or branched from middle or above, terete, subglabrous to minutely, finely appressed, antrorse hispid-pubescent, sometimes scabrid. Leaves opposite or occasionally alternate; upper 27–35(50) × 3–4(8) mm., linear-oblong to linear, obtuse or acute, shortly hispid-scabrid to scabrid, 1-nerved; lower leaves 20–75 × 4–12(25) mm., elliptic, oblong or obovate, 1–3-nerved, entire to crenate, indumentum as for upper. Inflorescence of long, slender, laxly-flowered spikes 7–25 cm. long; flowers shortly pedicellate on pedicels 0.5–1 mm. long, almost opposite above, alternate below. Bracts 2.5–3 × c. 1 mm., ovate to ovate-lanceolate, concave, acute to acuminate, sparsely hairy to glabrous except on margin. Bracteoles 1.5–3 × 0.3–0.75 mm., linear to linear-lanceolate, indumentum as for bracts. Calyx 4.5–6.5 mm. long, 5-lobed, 10-nerved, sparsely short hispid-strigose on nerves and lobes; lobes 1–1.5 mm. long, narrowly- to linear-triangular, acute. Corolla pink, purplish-blue to white; tube 5.5–6.5 mm. long, glabrous; limb up to 5.5 mm. in diam.; 1.5–3 × 0.5–2 mm., obovate. Capsule 4.5–6 × 1.75 mm., broad, narrowly oblong in outline, shortly apiculate.

Zambia. C: Chakwenga Headwaters, 100–129 km. E. of Lusaka, fl. & fr. 25.viii.1963, *Robinson* 5630 (K). **Zimbabwe**. N: Mwami (Miami), fl. iv.1926, *Rand* 60 (BM). **Malawi**. N: Chitipa (Fort Hill), Tanganyika Plateau, c. 1050–1200 m., fl. & fr. vii.1896, *Whyte* s.n. (K). **Mozambique**. N: Litunde, Lichinga (Vila Cabral), by Rio Lombala, 500 m., fl. & fr. 1.vii.1934, *Torre* 184 (COI).

Also known from Cameroon and Tanzania, Senegal, Gambia, Mali, Portuguese Guinea, Sierra Leone, Ghana and Nigeria. In wet grasslands, swamps and coastal bush, frequently with high salinity; from sea level to 1700 m.

33. **Buchnera geminiflora** Philcox in Kew Bull. **42**: 384 (1987). TAB. **39**. Type: Malawi, Mzimba Distr., S. of Mzambazi Mission, N. of Euthini, c. 1220 m., 18.iv.1974, *Pawek* 8369 (K, holotype; MO, isotype).

Annual herb, 10–25 cm. tall; stems slender, erect, sparsely short white, appressed pilose. Leaves: cauline 8–20 × 0.4–1 mm., minutely hispid-scabrid on both surfaces; basal leaves 8–14 × 3.5–5 mm., elliptic to oblong-elliptic, hispid-scabrid, 3-nerved. Inflorescence 4–6-flowered with 2 opposite flowers initially appearing at apex of stem, growth of axis continues before appearance of further pair of flowers giving rise to internodes c. 4.5 cm. long. Bracts 1.75–2.5 × 1–1.25 mm., ovate-lanceolate, glabrous but shortly hispid on main nerve and margin. Bracteoles 2.5 × 0.6 mm., linear, shortly hispid. Calyx c. 6.25 mm. long in flower, becoming 9.5 mm. long in fruit, 5-lobed, prominently 10-nerved, shortly hispid on nerves only; lobes broadly deltate, 1 × 1 mm. Corolla scarlet to red-orange; tube 6–6.5 mm. long, about equalling calyx, subglabrous or minutely glandular; limb c. 6.5 mm. in diam.; lobes 3–3.5 × 1.5–25 mm., obovate to obovate-oblong, entire to emarginate at apex. Capsule c. 5 × 2 mm., oblong in outline.

Zambia. S: Choma, c. 1300 m., fl. 19.iii.1958, *Robinson* 2808 (SRGH). **Malawi**. N: Mzimba Distr., 1.6 km. S. of Mzambazi Mission, c. 5 km. N. of Euthini, c. 1220 m., fl. & fr. 18.iv.1974, *Pawek* 8369 (K; MO).

Known only from Zambia and Malawi. Moist sandy soils; up to 1300 m.

34. **Buchnera subglabra** Philcox in Kew Bull. **42**: 208 (1987). TAB. **40**. Type: Zimbabwe, Mt. Peza, 1980 m., 15.x.1950, *Wild* 3599 (K, holotype; SRGH, isotype).

Annual herb 13–35 cm. tall, erect; stems glabrous, several arising from rootstock, simple, only very rarely branched, occasionally with a few long, patent, multicellular hairs arising at or near base. Leaves: cauline in 1–3 pairs, opposite or nearly opposite, somewhat appressed to stem, 3–8 × c. 1 mm., lanceolate to narrowly ovate-lanceolate, glabrous or shortly hirsute on slightly keeled midrib, ciliate; basal and lower cauline leaves 12–40 × 4–8(19) mm., obovate, obovate-oblong to oblong, glabrous to hispid on major nerves beneath, entire, 1–3-nerved with midrib occasionally pinnately divided, petiole-like base shortly hispid. Inflorescence spicate with spikes 10–15 mm. long, extending to c. 30 mm. at fruiting with lower flowers becoming remote. Bracts 2.8–4 × 2 mm., ovate, short acuminate, glabrous, ciliate. Bracteoles 2.8–3.5 × 1.2–1.8 mm., linear-lanceolate, ciliate. Calyx 4–5.25 mm. long, glabrous, 5-lobed, 10-nerved; lobes 2–3 mm. long, narrowly lanceolate, equalling or longer than calyx tube, ciliate. Corolla purple to deep violet; tube 7–7.5 mm. long, sparsely pubescent to subglabrous, straight, pilose at throat; limb (4)5.5–8 mm. in diam.; lobes 2.6–4 × 1–2.75 mm., irregularly uneven, obovate, rounded. Capsule c. 6 mm. long, 2.5–3 mm. broad, ovoid, subtruncate.

Zimbabwe. E: Chimanimani Mts., fl. 6.vi.1949, *Wild* 2886 (K; SRGH). **Mozambique**. MS: Chimanimani Mts., near Martin's Falls, c. 1500 m., fl. & fr. 6.ii.1958, *Hall* 389 (SRGH).

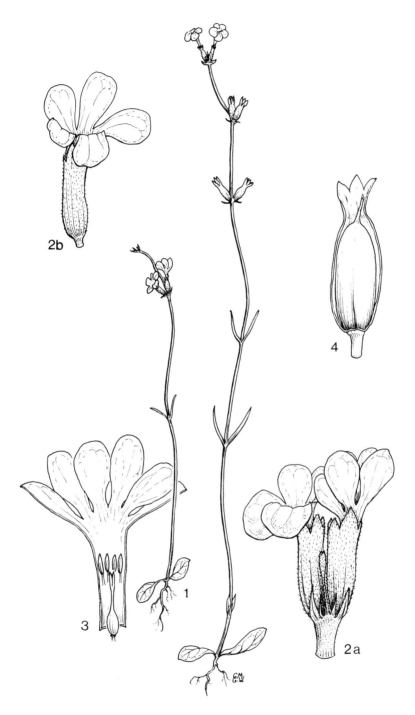

Tab. 39. BUCHNERA GEMINIFLORA. 1, habit (×⅔); 2a, flowering apex of inflorescence (× 4); 2b, flower (× 4); 3, corolla opened out showing gynoecium (× 4); 4, fruit (× 4), all from *Pawek* 8369.

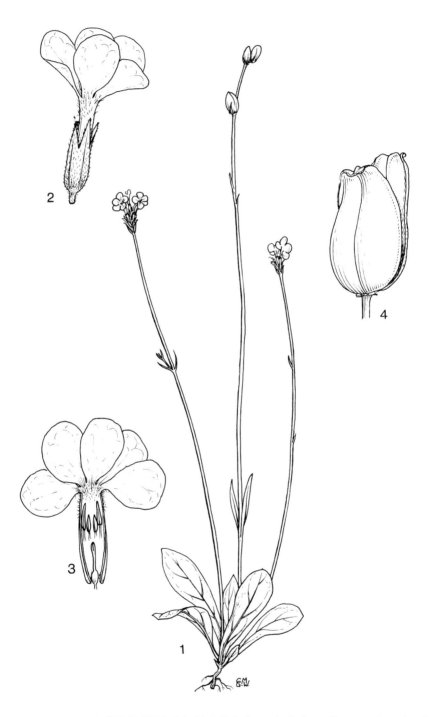

Tab. 40. BUCHNERA SUBGLABRA. 1, habit (× ⅔); 2, flower (× 4); 3, corolla opened out showing gynoecium (× 4); 4, dehisced capsule with calyx removed (× 6), all from *Phipps* 665.

Eastern Highlands of Zimbabwe and adjoining Mozambique only. In damp, marshy ground in the Chimanimani Mountains; from 1500–2450 m.

35. **Buchnera hispida** Buch.-Ham. ex D. Don, Prodr. Fl. Nep.: 91 (1825). —Benth. in DC., Prodr. **10**: 496 (1846). —A. Rich., Tent. Fl. Abyss. **2**: 128 (1850). —Hook. f., Fl. Brit. Ind. **4**: 298 (1884). —Skan in F.T.A. **4**, 2: 397 (1906). —Norlindh & Weimarck in Bot. Not. **98**: 124 (1951). —Hepper in F.W.T.A. ed. 2, **2**: 369 (1963). —Merxm. & Roessler in Merxm., Prodr. Fl. SW. Afr. **126**: 20 (1967). Type from Nepal.

   Striga schimperiana Hochst. in schaed. Schimperi Iter Abyss., Sectio Prima No. 23 (1840), Sectio Tertia No. 1516 (1844) nom. nud.

   Buchnera macrocarpa Hochst. in schaed. Schimperi Iter Abyss., nom. nud.

   Buchnera longifolia Klotzsch in Peters, Reise Mossamb., Bot.: 225 (1861). —Skan in F.T.A. **4**, 2: 398 (1906). —Hutch. & Dalz. in F.W.T.A. ed. 1, **2**: 225 (1931) non Kunth (1821).

Annual herb, up to 1 m. or more tall, erect; stems simple or much branched, terete above, usually long pilose below, hispid-pilose to scabrid above. Leaves: upper 15–25(40) × 1–2.5 mm., linear to linear-lanceolate, acute or obtuse, entire or occasionally remotely dentate, 1-nerved; lower leaves 45–55(75) × 9–12(25) mm., lanceolate, ovate-lanceolate, to broadly or narrowly oblong, obtuse, entire to coarsely irregularly dentate, 3-nerved with median nerve usually pinnately divided; basal leaves where present 22–35 × 15–20 mm., broadly elliptic to subcircular, entire, 3-nerved with a pinnately divided median nerve; all leaves appressed or antrorse ± hispid-scabrid. Flowers in terminal spikes, initially crowded at apex but soon becoming lax, spikes (2)6–25(40) cm. long; flowers sessile or with pedicels up to c. 0.5 mm. long. Bracts 4–5 × 0.8–1(3) mm., ovate to lanceolate, acute or acuminate, but bracts of flowers larger, more leaf-like, pilose hispid or glabrous with ciliate margins. Bracteoles 2.8–4 × 0.15–0.4(1) mm., linear-lanceolate to subulate. Calyx 4–8 mm. long, clearly to obscurely 10-nerved, 5-lobed, sparsely to variously densely hispid to hispid-short pilose, hairs mostly spreading; lobes 1–2 mm. long, linear-lanceolate to narrowly deltoid. Corolla blue, mauve to purple or white; tube 6–7.5 mm. long, cylindric, straight or rarely curved, glabrous to sparsely or slightly densely pilose; limb 3–5 mm. in diam.; lobes 2–2.8 × 1.75–2 mm., obovate to obovate-oblong, throat villous. Capsule 5–5.5 × 3 mm., ovoid.

Zambia. B: Senanga, fl. & fr. 27.vii.1962, *Fanshawe* 6969 (BR; K; SRGH). W: Kasempa Distr., Chezela (Chizera), 100 km. WNW. of Kasempa, fl. & fr. 7.vii.1963, *Robinson* 5560 (BR; K; LISC). C: Mkushi Distr., Lunsemfwa Gorge, Bell Point, Muchinga Escarpment, 850–1050 m., fl. & fr. 19.iv.1972, *Kornaś* 1589 (K). E: Katete, St. Francis Hospital, c. 1060 m., fl. & fr. 21.iv.1957, *Wright* 191 (K). S: Muckle Neuk, c. 19 km. N. of Choma, fl. & fr. 28.xi.1954, *Robinson* 584 (K). **Zimbabwe**. N: Guruve Distr., Dande, c. 450–600 m., fl. & fr. 16.v.1962, *Wild* 5758 (BR; K; SRGH). W: Hwange Distr., Deka R., c. 710 m., fl. & fr. 30.iii.1963, *Leach* 11619 (K; LISC). C: Harare Distr., Ngoma Kurira, Chinamora Reserve, c. 1650 m., fl. 25.iii.1952, *Wild* 3788 (K; SRGH). E: Mutare Distr., c. 21 km. S. of Mutare, 914 m., fl. & fr. 26.xii.1955, *Chase* 5924 (K). S: Mwenezi Distr., Rundi R. above Clarendon Cliffs, fl. & fr. 30.iv.1962, *Drummond* 7820 (BR; K; LISC; MO; SRGH). **Malawi**. N: Karonga Distr., Stevenson Road, c. 21 km. W. of Karonga, 700 m., fl. & fr. 22.iv.1975, *Pawek* 9497 (BR; K; MO; SRGH). C: Dowa Distr., Chimwere, 13 km. N. of Lombadzi towards Ntchisi, 1370 m., fl. 24.iii.1970, *Brummitt* 9353 (K). S: Blantyre Distr., 16 km. NW. of Blantyre towards Chileka, 910 m., fl. 13.iv.1970, *Brummitt* 9827 (K). **Mozambique**. N: between Cuamba and Mutali, by Riva Lurio, fl. & fr. 24.iv.1961, *Balsinhas & Marrime* 425 (BM; K; LISC; SRGH). Z: Mocuba Distr., Namagoa, 60 m., fl. & fr. vii.1945, *Faulkner* 229 (BR; COI; K; SRGH). T: Tete Distr., Chipera, fl. & fr. 11.iv.1972, *Macêdo* 5184 (K; SRGH). MS: Expedition Is., fr. viii.1858, *Kirk* s.n. (K). GI: Bazaruto Is., 1–92 m., fr. 28.x.1958, *Mogg* 28663 (K; LISC; SRGH).

Widespread in tropical and subtropical Africa, from West through Central to East Africa; southwards from Oman and the Yemen through the Sudan and Ethiopia; and in Angola and Namibia. Also in Madagascar and India.

In well-drained grassland on sandy soils, frequently in *Brachystegia* and other open woodland; from sea level to 1800 m. Apparently also a weed of cultivation in Casava fields in Zambia (*Loveridge* 946).

It is almost impossible to separate the material hitherto referred to *B. hispida* and *B. longifolia* Klotzsch. In the main, older fruiting material has been placed under *B. longifolia* and the younger flowering material under *B. hispida*. Here I have combined them under the earlier name on the following grounds: there is a great range in the size of the *B. hispida* material I have studied and it would appear that the species can vary from a small 7–8 cm. tall, precocious flowering plant to a coarse leafy plant of one metre or more in height. The basal rosette of apically rounded leaves is prominent in younger material but as the plant ages, the larger and more acute leaves of the central region of the plant become the prominent feature.

36. **Buchnera laxiflora** Philcox in Kew Bull. **42**: 208 (1987). Type: Zambia, banks of R. Kafue, 11 km. N. of Chingola, 1350 m., 4.v.1960, *Robinson* 3688 (K, holotype; SRGH, isotype).

Annual herb 35–55 cm. tall, erect, simple or branched; stems terete, soft appressed fine white pilose; branches (where present) ± opposite, erect-spreading. Leaves 25–30 × 2–8 mm., ± opposite, linear to linear-lanceolate above, obovate-lanceolate to subspathulate below, entire or more usually shallowly remotely dentate, acute, tapered at base, 1-nerved, finely hispid-scabrid on both surfaces. Inflorescences of widely interrupted spikes 5–17 cm. long, with flowers 1.5–3 cm. distant when ripening, mostly alternate or ± opposite. Bracts 3–4 × 1–1.5 mm., ovate-lanceolate, white pilose-hispid, long white ciliate. Bracteoles 2.75–3 × 0.4–0.6 mm., linear-lanceolate to narrowly lanceolate, long white ciliate. Calyx 6–9 mm. long, fine appressed, antrorse, short pilose, slightly recurved, 5-lobed, 10-nerved; lobes 1.5–2 mm. long, subulate to narrowly lanceolate. Corolla violet, purple or white; tube 9–10 mm. long, cylindrical, widening slightly at mouth, slightly curving, glabrous without, throat slightly pilose within; limb about 11.5 mm. in diam.; lobes c. 7 × 4.5 mm., obovate, subequal. Capsule 4.5–7.5 × 1.5–1.8 mm., oblong, obtuse; old style partly persistent.

Zambia. N: Mbala Distr., near Kalambo Falls, c. 1200 m., fl. 23.v.1967, *Robertson* 594 (K). W: Kasempa Distr., Chisela (Chizera), fl. 11.vi.1953, *Fanshawe* 83 (K). Known only from Zambia. Dambos; up to c. 1200–1400 m.

37. **Buchnera eylesii** S. Moore in Journ. Bot. **46**: 72 (1909). —Norlindh & Weimarck in Bot. Not. **98**: 124 (1951). Type: Zimbabwe, Mazowe (Mazoe), Iron Mask Hill, *Eyles* 334 (BM, holotype; SRGH, isotype).
    Buchnera mossambicensis var. *usafuensis* Engl., Bot. Jahrb. **30**: 404 (1901). Type from Tanzania.
    Buchnera usafuensis (Engl.) Melchior in Not. Bot. Gart. Berl. **11**: 680 (1932). Type as above.

Perennial up to 75 cm. tall, erect; stems branched, rarely simple, terete to obscurely quadrangular, subglabrous to somewhat sparsely hispid with short, stout, patent white hairs. Leaves opposite, entire; upper (10)15–40 × 1–3 mm., linear to narrowly linear-lanceolate, 1-nerved, hispid-scabrid; lower 40–60 × 4–6 mm., linear-lanceolate, 1–3-nerved, hispid-scabrid, basal leaves not seen; leaves often with clusters of small leaves or sterile branches in axils. Inflorescence of many long, spreading branches each of long laxly-flowered spikes 6.5–18(30) cm. long; flowers shortly pedicellate on pedicels 1.25–2.5 mm. long, opposite below to almost opposite or alternate towards apex. Bracts 2.5–4 × 0.5–1.25 mm., lanceolate, glabrous, scabrid-ciliate. Bracteoles 2.25–3(4) mm. long, c. 0.75 mm. wide, narrowly linear, scabrid on margin. Calyx 6–8 mm. long, 5-lobed, prominently 10-nerved, glabrous except for lobes; lobes 0.15–2 mm. long, lanceolate-oblong, acute, hispid short-ciliate, spreading to slightly reflexed in fruit. Corolla white; tube 7–10 mm. long, glabrous; limb 6–10(11.5) mm. in diam.; lobes 3–4.5(6) × 2–3(4) mm., obovate rounded. Capsule 5.5–6.5 × 3–3.5 mm., cylindric-ovoid.

Zimbabwe. N: Darwin Distr., Mavuradonha, c. 1450 m., fl. & fr. 12.v.1955, *Whellan* 925 (SRGH). C: Harare Distr., Ruwa, c. 1500 m., fl. & fr. vii.1957, *Miller* 4457 (K; SRGH). E: Mutare Distr., Nyachohwa Falls, fl. & fr. 20.vi.1948, *Fisher* 1607 (K; SRGH). S: Masvingo Distr., Kyle Nat. Park Game Res., Mutumushava Hill, fl. & fr. 21.v.1971, *Grosvenor* 516 (K; SRGH). **Malawi**. C: Ntchisi, 1350 m., fl. & fr. 8.ix.1946, *Brass* 17570 (K; MO; NY). **Mozambique**. N: Mecaloge (Macaloja), fl. & fr. 28.viii.1934, *Torre* 835 (COI).
Also known from Tanzania. Rocky and sandy terrain, *Brachystegia* open woodland; from 1000–1750 m.

38. **Buchnera speciosa** Skan in F.T.A. **4**, 2: 394 (1906). Type: Malawi, near Mt. Musisi (Masisi), 1200 m., ix.1902, *McClounie* 123 (K, lectoparatype); Masuku Plateau, c. 1900–2100 m., vi.1896, *Whyte* s.n. (K, lectotype, chosen here); Nyika Plateau, c. 1800–2100 m., vi.1896, *Whyte* s.n. (K, lectoparatype).

Perennial woody herb or undershrub, 60–200 cm. or more tall, erect; stems rigid, branched, terete, more often glabrous above, becoming bifariously hispid-scabrid-pubescent below, occasionally glabrescent leaving only thickened scabrid bases of hairs; branches opposite to subopposite, spreading. Leaves opposite to nearly opposite, (10)30–60 × (1)3–8.5(15) mm., lanceolate or oblanceolate, narrowed into petiole-like base, acute to obtuse at apex often with small apiculus present, entire, appressed white hispid-strigose on both surfaces, 1–3-nerved, frequently with tuft of small hairs in axils. Inflorescence of long interrupted spikes or racemes up to 30 cm. or more long; flowers opposite, subsessile or with pedicels up to 3 mm. long. Bracts 3.5–6(10) × 1.2–1.5 mm., ovate-lanceolate, shortly acuminate, sparsely hairy on upper surface, shortly ciliate. Bracteoles 2.5–5 × 0.5–0.7 mm., linear-lanceolate. Calyx 7.5–12 mm. long slightly curved, glabrous without or with few stiff hairs on nerves, 5-lobed, 10-nerved; lobes 2–4 mm. long, c. 1 mm. wide, ovate-lanceolate, shortly acuminate, short-ciliate, spreading after anthesis. Corolla white; tube 12–18 mm. long, cylindrical, widening somewhat above, straight or slightly curved, glabrous without, pilose at throat; limb 18–30 mm. in diam.; lobes 8–14 × 6–12.5 mm., subcircular. Capsule 6.5–9 × 2.5–3.8 mm., ellipsoid to cylindrical, narrower above, slightly curved, shortly apiculate.

Zambia. N: Mafinga Hills, fl. & fr. 31.x.1972, *Fanshawe* 11635 (BR; K). E: Nyika Plateau, c. 4 km. from Rest House, 2150 m., fl. & fr. 22.x.1958, *Robson* 257 (BM; K; LISC; SRGH). **Zimbabwe**. E: Mutare, fl. & fr. 22.ii.1960, *Head* 28 (BM). **Malawi**. N: Ntchisi Mt., 1400 m., fl. & fr. 25.vii.1946, *Brass* 16928 (BM; BR; K; MO; NY; SRGH). **Mozambique**. N: Kerimbas Is., *Peters* s.n. (ex lit.).
Also known from Tanzania. In grasslands (especially in wetter areas), montane and open woodland and swamp forests; from 975–2150 m.

39. **Buchnera candida** S. Moore in Journ. Bot. **64**: 306 (1926). Type: Zimbabwe, Mwami (Miami), *Rand* 158 (BM, holotype).

Annual herb, 15–70 cm. tall, erect; stems simple or rarely branched, terete, sparsely hispid-pilose with patent white hairs, becoming scabrid above, drying pale brown to greenish-fawn, not black. Leaves: cauline 7–28 × 0.3–3(4) mm., linear-oblong to linear or lanceolate to elliptic, obtuse, 1-nerved, hispid-scabrid especially beneath; basal leaves, where present, 5–20 × 3–20 mm., broadly ovate or obovate, entire, hispid-scabrid above mixed with small white glands, hispid beneath only on major nerves, 1-nerved with occasional secondary nerves. Inflorescence terminally spicate with spikes laxly flowered, 6–13(18) cm. long in fruit, internodes up to 2.5(4) cm. long. Bracts 3–4 × 1–1.5 mm., ovate-lanceolate, 1–3-nerved, hispid-scabrid especially on nerves, to subglabrous, shortly hispid-ciliate, somewhat keeled towards base. Bracteoles 2.5–4 × 0.4–0.6 mm., linear-lanceolate, 1-nerved, hispid-ciliate, keeled towards base. Calyx 4.5–7 mm. long, 5-lobed, 10-nerved, nerves prominent or not, where prominent sparsely hispid-scabrid; lobes 1.5–2 mm. long, narrowly lanceolate, shortly hispid on margin. Corolla blue or white; tube 5.5–6 mm. long, glabrous, subcylindrical below, c. 0.3 mm. in diam., widening to c. 0.8 mm. at throat, slightly arcuate; limb c. 2 mm. in diam.; lobes 0.8 × 0.5–0.6 mm. Capsule 4.5–5 × 1.8–2 mm., cylindric-ovoid, apex subtruncate, shortly rostrate.

Zambia. W: Luano, fl. & fr. 9.v.1967, *Mutimushi* 1871 (K; SRGH). C: near Mt. Makulu Res. St., c. 20 km. S. of Lusaka, fl. & fr. 2.vi.1956, *Angus* 1322 (K; SRGH). **Zimbabwe**. N: Mwami (Miami), fl. & fr. v.1926, *Rand* 158 (BM).
Also known from Angola. Open woodlands, sandy ground; up to 1150 m.

40. **Buchnera pulcherrima** R.E. Fries, Wiss. Ergebn. Schwed. Rhod.-Kongo-Exped. 1911–1912, **1**: 293 (1916). Type: Zambia, Luapula R., fl. & fr. 6.ix.1911, *Fries* 541 (K, isotype).

Annual or possibly perennial herbs, 25–30 cm. tall, erect or ascending; stem branched or simple, terete, sparsely to densely white, fine hispid intermixed with minute glands. Leaves: upper cauline opposite or alternate, 12–20 × c. 1 mm., narrowly linear, entire, obtuse, 1-nerved, hispid-scabrid especially beneath, margins somewhat involute, lower cauline to 30 × 5 mm., oblong, entire, obtuse; basal leaves c. 3 × 2 mm., obovate, 1-nerved, densely white hispid. Inflorescence spicate with spikes 5–9 mm. long at fruiting. Bracts 3–4 × 1–1.5 mm., ovate-lanceolate, acute, hirsute, ciliate. Bracteoles 2.5–3 × 0.5 mm., linear. Calyx 5.5–6 mm. long, laxly patent white hirsute, 5-lobed, 10-nerved; lobes 1–1.8 mm. long, narrowly triangular, acute. Corolla blue; tube 9.5–10.5 mm. long, glabrous, c. 1 mm. wide at base, widening to 2.5–3 mm. at throat; limb 9–14 mm. in diam.; lobes c. 6 × 4.5–5 mm., rotund-obovate. Capsule c. 4.8 × 2 mm., cylindric-ovoid, truncate, shortly rostrate.

Zambia. N: Luapula R., fl. & fr. 6.ix.1911, *Fries* 541 (K).
Known only from the type collection.

41. **Buchnera chimanimaniensis** Philcox in Kew Bull. **42**: 208 (1987). TAB. **41**. Type: Zimbabwe, Chimanimani (Melsetter) Distr., 1820 m., *Chase* 6906 (K, holotype; BM; BR; MO; SRGH, isotypes).

Annual or perennial herbs, up to 35 cm. tall, erect, decumbent or straggling, much branched with branches spreading; stems slender, minutely bifariously appressed fawn-pubescent. Leaves 10–30(40) × 0.5–1(2.5) mm., linear, entire, flexuous, patent, 1-nerved, sparsely minutely appressed hispid to subglabrous, frequently with secondary leaves or small axillary branches. Inflorescence with flowers clustered at ends of branches into

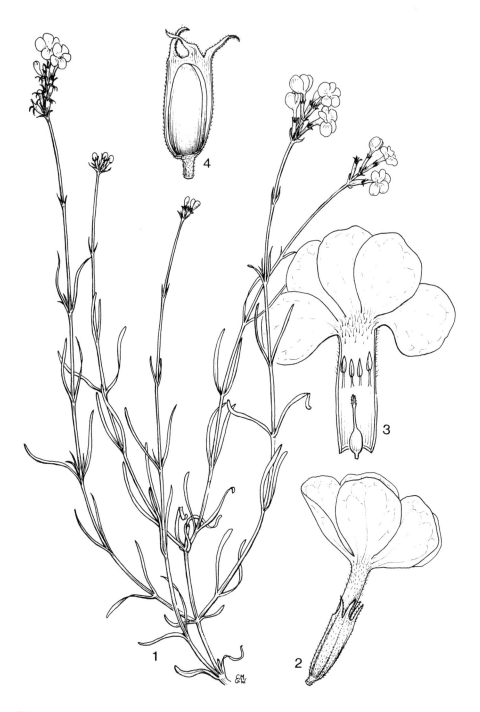

Tab. 41. BUCHNERA CHIMANIMANIENSIS. 1, flowering stem (× ⅔); 2, flower (× 4); 3, corolla opened out showing gynoecium (× 4); 4, capsule with part of calyx removed (× 6), all from *Grosvenor* 179.

short spikes 1–2 cm. long, increasing to rarely more than 4.5 cm. at fruiting; flowers sessile or shortly pedicellate with pedicels up to 0.6 mm. long. Bracts 2.5–3 × c. 1 mm., ovate-lanceolate, subglabrous. Bracteoles 1.75–2.5 × 0.3 mm., linear-oblong. Calyx (5)6.5–7.5 mm. long, 10-ribbed, 5-lobed, short appressed fawn pilose; lobes 1.5–2 mm. long, all perfect with none reduced in size, narrowly linear-lanceolate, widely spreading after flowering with apices tending to recurve. Corolla pale blue to mauve; tube 7.5–9 mm. long, widening towards mouth, long fawn pilose especially below limb; limb 9.5–14 mm. in diam.; lobes 4.5–7 × 2.5–4.5 mm., obovate, unequal. Capsule c. 5 × 2.25 mm., cylindric-ovoid, shortly rostrate.

Zimbabwe. E: Chimanimani Mts., Martin Forest Res., fl. 14.xi.1967, *Mavi* 611 (K; LISC; SRGH).
Mozambique. MS: Chimanimani Mts., Musapa Gap, c. 915 m., fl. 6.x.1950, *Munch* 337 (K; SRGH).
From Zimbabwe and Mozambique. Montane grasslands; 915–1850 m.
This plant has hitherto been confused with *B. henriquesii* Engl. from which it differs in being generally taller, more slender and bearing very fine, linear, flexuous leaves.

42. **Buchnera nervosa** Philcox in Kew Bull. **42**: 384 (1987). TAB. **42**. Type: Zambia, Kawamba, Ntumbacushi (Timnatushi) Falls, 1260 m., fl. 18.iv.1957, *Richards* 9313 (K, holotype).

Annual herb 12–36 cm. tall, erect, stems simple or branched with branches arising in axils of most cauline leaves, sparsely appressed hispid; branches mostly with terminal inflorescences, spreading. Leaves opposite: upper cauline (6)20–45 × 0.5–3.5 mm., linear, sparsely appressed hispid-scabrid, 1-nerved; lower cauline to sub-basal leaves 20–45 × (2.5)3.5–7 mm., linear-lanceolate to obovate-lanceolate, 1-nerved, entire or occasionally with 1–5 coarse teeth, slightly hispid especially beneath to subglabrous. Spikes 8–20 × 8–15 mm., with flowers compacted, lower flowers not becoming lax or remote on fruiting. Bracts 5–6 × 5.5–6 mm., rhombic to reniform-rhombic, shortly acuminate or not, glabrous, ciliate, prominently strongly parallel-nerved, concave. Bracteoles (3.5)5.5–6.5 × 2–3 mm., lanceolate-subobovate, long acuminate, somewhat inwardly curved, concave, keeled, glabrous, ciliate. Calyx 7–9 mm. long, 5-lobed, nerved, glabrous to subglabrous; lobes 1.75–2 mm. long, lanceolate, acute, short hispid-ciliate. Corolla purple or lilac; tube 10–10.5 mm. long, cylindric but widening at mouth, subglabrous to long-pilose abaxially without, throat long pilose; limb 6.5–9 mm. in diam.; lobes 3–6 × 2–4.5 mm., broadly obovate to oblong, very unequal. Capsule (immature) 3 × 1.75 mm., ellipsoid.

Zambia. N: Kawamba, 1340 m., fl. 21.vi.1957, *Robinson* 2330A (K).
Known only from Kawambwa in Zambia. Damp sandy areas, common in semi-open woodland; c. 1350 m.

43. **Buchnera henriquesii** Engl., Bot. Jahrb. **18**: 72 (1894). —Skan in F.T.A. **4**, 2: 396 (1906). —Norlindh & Weimarck in Bot. Not. **98**: 122 (1951). Type from Angola.
  Buchnera rhodesiana S. Moore in Journ. Bot. **38**: 468 (1900). Type: Zimbabwe, Harare, *Rand* 154 (BM, holotype).

Annual herb or apparently so, 6–20(35) cm. tall, erect or decumbent, much branched from thick woody base; stems terete, short hispid to hirsute. Leaves 15–45 × 1–3 mm., linear to narrowly linear-lanceolate, opposite, nearly opposite or occasionally alternate, frequently with groups of smaller leaves in axils, narrowed somewhat at base, more or less densely short-hispid above, densely so below, obtuse or acute, 1-nerved. Inflorescence with flowers crowded at apex, spike elongating with flowers becoming lax at fruiting, up to 10 cm. long; flowers sessile, nearly opposite. Bracts 2.5–5 × 1–1.75 mm., ovate to ovate-lanceolate, acuminate, sparsely hispid. Bracteoles 1.5–2.5 × 0.2–0.25 mm., linear. Calyx (4.75)6.5–8 mm. long, 10-ribbed, 5-lobed with anterior lobe frequently much reduced, sparsely to densely hispid-hairy with hairs spreading or forwardly directed; lobes 1.25–2 mm. long, linear-lanceolate to linear, spreading after flowering. Corolla blue, violet or mauve; tube (6.5)8–11 mm. long, cylindric, straight to slightly curved, appressed antrorse-pilose above, retrorse-pilose below to glabrous or subglabrous within calyx, throat densely pilose; limb 8.5–12 mm. in diam.; lobes 4–6 × 2–3.75 mm., unequal, obovate to obovate-oblong. Capsule 3.75–5.5 × 1.75–2.25 mm., oblong.

Zambia. B: Kaoma Distr., 65 km. E. of Kaoma (Mankoya), fl. 15.xi.1965, *Robinson* 6709 (K; MO; SRGH). N: Lake Kashiba, fl. & fr. 22.x.1957, *Fanshawe* 3794 (BR; K). W: Mwinilunga Distr., c. 25 km. W. of Kabompo R., fl. & fr. 11.ix.1930, *Milne-Redhead* 104 (K). C: Mumbwa Distr., Sanje, WSW. of Lusaka, 1050–1100 m., fl. & fr. 4.x.1972, *Strid* 2281 (K). S: Namwala Distr., between Lukomezi and

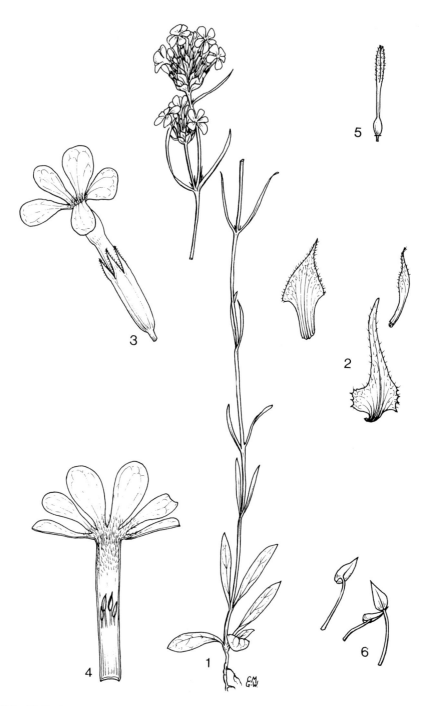

Tab. 42. BUCHNERA NERVOSA. 1, habit (× $\frac{2}{3}$); 2, bracts (× 3); 3, flower (× 3); 4, corolla opened showing stamens (× 3); 5, gynoecium (× 6); 6, stamens (× 6), all from *Lawton* 861.

Banga R., fl. 16.viii.1963, *van Rensburg* 2461 (K; SRGH). **Zimbabwe**. N: Mwami (Miami), fl. & fr. v.1926, *Rand* 157 (BM). W: Shangani, Gwampa Forest Reserve, fl. ii.1955, *Goldsmith* 78/55 (SRGH). C: Charter Flats, xi.1899, *Cecil* 83 (K). E: W. slopes of Mt. Chirinda, c. 1150 m., fl. 29.vii.1906, *Swynnerton* 303a (K). **Mozambique**. GI: Inhambane, c. 45 km. SW. of Panda, fl. & fr. 15.x.1968, *Balsinhas* 1391 (LISC).

Also known from Zaire and Angola. In grasslands and open woodlands; from 400–1800 m.

This species has long been confused with *B. longispicata* Schinz. *B. henriquesii* is generally a much smaller plant, rarely exceeding 20 cm. in height with very many short stems arising from a thickened woody base. The inflorescences in both species are spicate in the young stage with flowers crowded together at the apex. With maturity and towards fruiting those of *B. henriquesii* elongate to c. 10 cm. in length, while those of *B. longispicata* extend to at least 20 cm. or more long. The two species can be further and more clearly distinguished by the indumentum on the calyces. In *B. henriquesii* it is usually of fine to coarse pinkish to off-white short hairs which are either spreading or forwardly directed, while in *B. longispicata* it is of much denser, longer white hairs which are appressed and directed backwards. The indumentum on the corolla tube is quite clearly defined too, with the hairs on *B. longispicata* being totally appressed-retrorse, while in the other those above on the tube are clearly antrorse.

For the purpose of Flora Zambesiaca, I prefer to keep these two species separate.

44. **Buchnera longispicata** Schinz in Verh. Bot. Ver. Brand. **31**: 193 (1890). —Skan in F.T.A. **4**, 2: 395 (1906). —Merxm. & Roessler in Merxm., Prodr. Fl. SW. Afr. **126**: 20 (1967). Types from Namibia.

Annual herbs, 30–42 cm. or more tall, erect to decumbent, branched usually from base; stems subterete, strigose to short patent hirsute. Leaves opposite, almost opposite or alternate, frequently with groups of two or more smaller leaves in axils, 25–35 × 1.5–5 mm., linear to linear-lanceolate, obtuse to subacute, narrowed at base, densely short-hispid with hairs often thickened at base, 1-nerved, lowermost leaves often with somewhat thickened, revolute margins, scabrid. Inflorescence with flowers crowded at apex, developing into long spike 20–26 cm. long at maturity; flowers sessile, almost opposite. Bracts: lower leaf-like; upper 2–3 × 1–1.4 mm., ovate-lanceolate to lanceolate, acute or acuminate, hispid, slightly concave. Bracteoles 1.25–2 × 0.15–0.4 mm., linear-lanceolate, sparsely hispid. Calyx 5–6 mm. long, clearly 10-ribbed, 5-lobed, densely white pilose with downwardly directed hairs; lobes 1.2–1.5 mm. long, narrowly lanceolate to subulate, spreading after flowering. Corolla blue; tube 8–9 mm. long, funnel-shaped, oblique, slightly curved, appressed retrorse-pilose without, throat densely pilose within; limb 8.5–11 mm. in diam.; lobes 3.5–7 × 2–4 mm., unequal, oblong-obovate to obovate. Capsule 3.5–5 × 1.4–1.8 mm., broadly oblong in outline, subtruncate-emarginate.

Botswana. N: Chobe Distr., Kazuma Forest, c. 915 m., fl. xi.1966, *Mutakela* 151 (SRGH). **Zambia**. B: Sesheke Distr., c. 27 km. S. of Machili, fl. & fr. 21.xii.1952, *Angus* 992 (BR; K). S: 34 km. NE. of Livingstone towards Choma, fl. & fr. 10.vii.1930, *Hutchinson & Gillett* 3503 (BM; BR; COI; K; LISC; SRGH). **Zimbabwe**. N: Mutorashanga (Taoshanga) Pass, Umvukwe Mts., fl. & fr. 24–27.iv.1948, *Rodin* 4412 (K). W: Hwangi Distr., Kazuma Range, 1800 m., fl. & fr. 12.v.1972, *Gibbs Russell* 1981 (BR; K; LISC). C: Goromozi Distr., Bromley, c. 1600 m., fl. & fr. i.1917, *Walters* 2205 (K; SRGH). E: Chipinge Distr., 5 km. S. of Rusongo, 400 m., fl. 1.ii.1975, *Gibbs Russell* 2739 (K; SRGH). S: Masvingo Distr., c. 8 km. W. of Mashaba, near Tokwe R., fl. & fr. 12.i.1964, *Leach* 12061 (BR; K; LISC; P; SRGH).

Also known from Angola, Namibia and S. Africa. In sandy areas with open woodland or grasslands.

45. **Buchnera attenuata** Skan in F.T.A. **4**, 2: 383 (1906). Type from Angola.

Annual herbs, 35–65 cm. tall, erect; stem very slender, simple or branched within the lower half or immediately below inflorescence, terete, glabrous above becoming pilose below in area of leaves. Leaves 1 or 2 pairs on upper half to three-quarters of stem, small, 3–7 mm. long, up to 1 mm. wide, linear, those on lower part of stem (7)15–34 × 3–6.5(10) mm., opposite or nearly so, remote with internodes up to 7 cm. or more long, basal leaves crowded, internodes less than 1 cm. long, lamina narrowly obovate to broadly ovate, obtuse, rounded apex, narrowed at base, entire to crenate-dentate with 1–4 irregularly positioned teeth, pilose, scabrid. Spikes 7–12 × 5–8 mm., compact becoming lax after flowering, up to 70 mm. long in fruit. Bracts 3–3.5(4) × 1.5 mm., ovate, acuminate, pubescent, white ciliate. Bracteoles 3(4) mm. long, narrowly linear-lanceolate. Calyx 4.5–5.5 mm. long, 10-nerved, 5-lobed, pubescent; lobes 1–1.75 mm. long, lanceolate-subulate. Corolla pale blue, violet to purple or white; tube 6–8 mm. long, slender, hairy without, throat pilose; limb 4.5–6 mm. in diam.; lobes 2–3 mm. long, c. 1.5 mm. wide, oblong-obovate to obovate, entire. Capsule 4–5.25 × 1.5–2.5 mm., cylindric-ovoid, shortly rostrate.

Zambia. N: Mpika Distr., 57 km. SW. of Mpika, 1450 m., fl. & fr. 13.viii.1966, *Gillett* 17434 (K; LISC; SRGH). W: Mwinilunga Distr., SW. of Dobeka Bridge, fl. & fr. 3.i.1938, *Milne-Redhead* 3931 (BR; K). C: Mkushi Distr., Fiwila, 1220 m., fl. 5.i.1958, *Robinson* 2616 (K).
Also known from Angola. Bogs, marshes and wet grasslands; up to 1600 m.

46. **Buchnera pusilliflora** S. Moore in Journ. Bot. **46**: 310 (1908). Type: Zimbabwe, Mazowe (Mazoe), fl. & fr. 20.iv.1906, 1340 m., *Eyles* 367 (BM, holotype; K; SRGH, isotypes).

Annual herbs, 12–27 cm. tall, erect; stems simple, slender, quadrangular, patent-hispid to pilose or subglabrous. Leaves: cauline opposite or nearly so above, internodes often up to 5.5 cm. long, 10–35 × 1–3.5 mm., narrowly linear to linear-lanceolate below, entire or rarely with 1 or 2 blunt teeth below, 1-nerved, hispid-scabrid; basal leaves (18)25 × 8.5(12) mm., ovate, obtuse, 3-nerved, entire to sinuate-dentate or sinuate, hispid-scabrid. Spikes 8–10 × 5–7 mm., densely flowered, elongating to 50(75) mm. long with lower flowers becoming more remote at fruiting. Bracts: upper 2.5–3 × 1–1.25 mm., lanceolate, hispid-scabrid; lower up to 6.5 × 1 mm., involute, linear to linear-lanceolate, hispid-scabrid. Bracteoles 1.25–1.5 mm. long, linear. Calyx 4.5–4.8(5.5) mm. long, hispid, 5-lobed, 10-nerved; lobes 1–1.2(1.5) mm. long, narrowly lanceolate to subulate, subequal. Corolla blue to violet or purple; tube c. 5.5–6.5 mm. long, subcylindrical, strict, pilose without, throat pilose; limb 2.5–4.5 mm. in diam.; lobes 1–1.5(2.5) × 0.25–0.75(1) mm., oblong to obovate-oblong, sometimes involute on drying. Capsule 3.8–4 × 2–2.8 mm., ellipsoid to ovoid-ellipsoid, truncate to broadly rounded.

Zimbabwe. N: Mwami (Miami), fl. & fr. iv.1926, *Rand* 59 (BM). C: Harare Distr., Domboshawa Rock, fl. 23.iii.1963, *Loveridge* 627 (SRGH). W: Matobo Distr., Besna Kobila Farm, 1480 m., fl. 6.iii.1963, *Miller* 8398 (SRGH).
Known only from Zimbabwe. In open grasslands on well-drained soils; between 1300–1500 m.
B. pusilliflora has been little reported and some material I have seen has been confused with *B. randii* S. Moore. The latter usually is a much stronger, taller, branched and coarser plant with larger bracts and bracteoles, densely hispid-strigose within the inflorescence with rather long, stiff patent, spreading white callus-based hairs. There are, however, some collections which are shorter and simple-stemmed but still bear the indumentum and inflorescence characters. In the case of *B. pusilliflora*, the plants are always slender and simple stemmed, with much smaller and narrower bracts and bracteoles which are covered by a much finer indumentum of slender ± callus-based hairs. In the latter, the calyx lobes are always equal in length unlike those of *B. randii* which at most times can be seen to have four of the lobes of equal size with the fifth reduced to give the calyx the appearance of being 4-lobed.
Maybe future researchers will see reason to combine these two under one name after broadening the specific concept. For this treatment however, I prefer to keep them apart with the general calyx and inflorescence characters as diagnostic.

47. **Buchnera granitica** S. Moore in Journ. Bot. **57**: 218 (1919). Type: Zimbabwe, Harare, *Rand* 1431 (BM, holotype).

Annual herbs, up to c. 18 cm. tall; stems erect, subsimple, with small leafy branches arising in axils of cauline leaves, quadrangular, hispid-scabrid. Leaves: cauline mostly opposite, 25–45 × 3–4 mm., linear to linear-oblong, mucronate; basal leaves densely clustered, (20–30)35–45 × (5.5)7–12 mm., oblong to oblong-ovate, obtuse, mucronate, (2)3-nerved. Spikes immature, 10–12 × 10 mm., ovoid, not obviously with remote flowers at base, but occasionally with flowers appearing in axils of upper leaves. Bracts 4–5 mm. long, lanceolate, acute, hispid-scabrid. Bracteoles 2 mm. long, linear, acute. Calyx c. 8.5 mm. long, 9–10-nerved, hispid-scabrid, 5-lobed; lobes c. 1.5 mm. long, short-subulate. Corolla colour unknown; tube barely 9 mm. long, hispid above; limb. c. 5 mm. in diam.; lobes c. 3 mm. long, 2.5 mm. wide, obovate, obtuse. Capsule unknown.

Zimbabwe. C: Harare, fl. 9.iv.1909, *Rand* 1431 (BM).
Known only from the type collection. Granitic soil.

48. **Buchnera albiflora** Skan in F.T.A. **4**, 2: 387 (1906). Type: Malawi, Mt. Musisi (Masisi), *McClounie* 75 (K).

Annual herb to 40 cm. or more tall, erect; stem branched, mainly about middle, more or less densely white pilose; branches ascending, up to 20 cm. long. Basal leaves 20 × 6 mm., obovate, obtuse, hispid-scabrid, 1-nerved with the nerve pinnately-branching; upper leaves opposite or alternate, few, 18–45 × 1.5–3.5 mm., linear to narrowly linear-lanceolate, acute or obtuse, entire, scabrid with short white hairs thickened at the base,

1-nerved. Spike 8.5–10.5 cm. long overall, terminal, upper part densely-flowered, thick, 3–6 cm. long, lax below with flowers up to 40 mm. apart. Bracts 4–5.5 × 2–4 mm., broadly ovate-elliptic, acute, pubescent, ciliate, strongly nerved. Bracteoles 4–4.5 × 0.75 mm., linear-lanceolate, acute, sparsely hispid, short ciliate. Calyx 5–6 mm. long, strongly hispid-scabrid, 10-nerved, 5-lobed; lobes c. 1.5 mm. long, narrowly lanceolate, acute. Corolla white; tube c. 5 mm. long, slender, cylindric, sparsely pubescent; lobes 1–1.5 × c. 0.8 mm., obovate. Capsule 5–5.5 × 2 mm., cylindric-ovoid, shortly apiculate.

Malawi. N: Musisi (Masisi), 1370 m., ix.1902, *McClounie* 75 (K).
Known only from the type collection.

49. **Buchnera namuliensis** Skan in F.T.A. **4**, 2: 386 (1906). Type: Mozambique, Makua, Namuli Hills, *Last* s.n. (K, holotype).

Annual herb or apparently so, 30 cm. or more tall, simple below, branched above; branches quadrangular, long, slender, erect, appressed pubescent. Leaves: upper up to 35 × 3 mm., linear, sparsely hispid-pubescent, entire, obscurely 1-nerved. Spike terminal or lateral, 8–40 mm. long, densely flowered, becoming somewhat laxer at fruiting. Bracts c. 4 × 1 mm., lanceolate-acuminate, concave, shortly pilose. Bracteoles c. 2.5 × 0.25 mm., linear, acuminate, shortly pilose. Calyx 4.5–5.5 mm. long, 10-nerved, shortly hispid-pilose, 5-lobed; lobes up to 1.7 mm. long, triangular to linear-triangular. Corolla tube c. 10 mm. long, slender, cylindrical, very slightly curved, sparingly pubescent above; lobes c. 4 mm. long, 3 mm. wide, obovate. Flower colour unknown. Capsule 4 mm. long, ellipsoid, slightly curved, minutely apiculate.

Mozambique. Z: Namuli, Makua Country, 1887, *Last* s.n. (K). MS: Dondo, near Beira, xii.1899, *Cecil* 257 (K).
Known only from Mozambique. Lowland swamps.
Both collections are incomplete and likewise imperfect but there are enough characters of *Last* s.n. for me to tentatively agree with Skan that it is probably distinct. Correspondingly *Cecil* 257 shows enough similarities to ally it to the other.

32. STRIGA Lour.

By F.N. Hepper

Striga Lour., Fl. Cochinch.: 22 (1790).

Annual, rarely perennial, hirsute or scabrous root-parasitic herbs. Root system greatly condensed, adventitious roots arising from subterranean scales, these fine roots terminated by small (1–2 mm. in diam.) haustoria, in some species a large (to 5 cm.) primary haustorium present. Stems stiffly erect, green or grey, usually square in transverse section, often ridged. Leaves opposite or nearly so, sessile or subsessile, reduced to small scales near the base of the stem in most species. Inflorescence a spike, flowers in axils of leaves (bracts) or in dense heads. Bracts leaf-like or reduced, bracteoles 2. Calyx tubular, 5-lobed or with 5 (rarely 4) teeth, in some species intracostal veins occur. Corolla with a narrow tube and expanded bilabiate limb. Orifice of tube small, less than 1 mm. in diam. with hairs abundant at orifice. Upper lobes fused, erect, lower lobes 3, spreading. Stamens 4, didynamous, inserted in the tube below the orifice. Anthers unilocular, basifixed on short filaments. Pollen sparse, often sticky. Pistil tubular with numerous minute ovules. Style terete, elongate, stigma bifid. Nectary present at base of ovary. Capsule cylindric or subovoid, style usually persistent. Dehiscence loculicidal. Seeds minute ("dust seeds") with prominent encircling ridges. Embryo small. Germination hypogeal and cryptocotylar. Seedlings achlorophyllous, squamate.

A genus comprising some 40 species in the Old World tropics and subtropics with occasional introductions into the United States of America, semi-parasitic on grasses, cereal crops and a wider range of other hosts, causing great economic loss in Sudan and drier parts of West Africa.

1. Calyx with 4–5 ribs (nerves), each ending in a calyx tooth - - - - - 2
– Calyx with 10 or more ribs (nerves), at least one between each calyx tooth - - 7
2. Stems glabrous or slightly pubescent, ± succulent - - - - - 6. *gesnerioides*
– Stems conspicuously or harshly pubescent - - - - - - - - 3

3. Corolla tube conspicuously pubescent - - - - - - - - - - - 4
- Corolla tube almost glabrous - - - - - - - - - - - - 5
4. Indumentum of appressed grey hairs all over the plant, including the bracts   1. *linearifolia*
- Indumentum of hispid hairs, bracts ciliate-pubescent on margins and midrib   2. *bilabiata*
5. Flowers white (or yellowish); bracts leaf-like, those of lower flowers up to 5 times length of calyx, 1–4. mm. wide - - - - - - - - - - - - - - 3. *passargei*
- Flowers pink - - - - - - - - - - - - - - - - 6
6. Bracts lanceolate, 7–10 mm. long, margins ciliate - - - - - 5. *hermonthica*
- Bracts linear, c. 4 mm. long (lower bracts up to 10 mm.), not ciliate - - - 4. *aspera*
7. Calyx 10-ribbed i.e. with 1 extra rib to sinus between each calyx tooth; corolla scarlet, pink, yellow or white - - - - - - - - - - - - - - - - - - 8
- Calyx 15- or more ribbed i.e. with 2 extra ribs between each calyx tooth; corolla white   10
8. Leaves 3–9 mm. wide, coarsely dentate - - - - - - - - - - 9
- Leaves 1–2(3) mm. wide, ± not dentate - - - - - - - - - - 11
9. Flowers in dense heads, white; lower flowering bracts over 20 mm. long   7. *macrantha*
- Flowers axillary, salmon pink; lower flowering bracts c. 20 mm. long - - - 8. *forbesii*
10. Corolla tube c. 20 mm. long, limb 15–20 mm. across - - - - 9. *pubiflora*
- Corolla tube 12–15 mm. long, limb 10–12 mm. across - - - - 10. *angustifolia*
11. Flowers alternate usually 2 or 3 out at a time, corolla small - - - - 11. *asiatica*
- Flowers in opposite pairs, many out at a time, corolla large - - - 12. *elegans*

1. **Striga linearifolia** (Schum. & Thonn.) Hepper in Kew Bull. **14**: 416 (1960); in F.W.T.A. ed. 2, **2**: 372 (1963). —Musselman & Hepper in Kew Bull. **41**: 218, fig. 4B (1986). TAB. **44**, fig. B. Type from Ghana.
    Buchnera linearifolia Schum. & Thonn., Beskr. Guin. Pl.: 279 (1827). —Hepper, W. Afr. Herb. Isert & Thonning: 119 (1976). Type as above.
    Striga canescens Engl., Pflanzenw. Ost-Afr. **C**: 361 (1895). —Skan in F.T.A. **4**, 2: 406 (1906). Type from Uganda.
    Striga strictissima Skan in F.T.A. **4**, 2: 407 (1906). —F.W. Andr., Fl. Pl. Anglo-Egypt. Sudan **3**: 145 (1956). Type from Nigeria.

Slender erect stem 20–55 cm. high, usually single or 2 or 3 from base, often branched, branches erect, square in section, covered with appressed grey ascending hairs. Leaves inconspicuous, appressed to stem, opposite, linear-subulate, 4–10 mm. long. Bracts similar to leaves, slightly broader. Flowers opposite in terminal spike. Calyx tube 2 mm. long, subulate lobes 4 mm. long, 5-nerved, appressed-pubescent. Corolla whitish, tinged purple to bluish purple, with pink spot inside lower lip, densely appressed-pubescent outside, angled about middle, tube 6–11 mm. long, lobes 2–5 mm. long, narrow. Capsule included in calyx, acute with persistent style or base of style.

Malawi. N: Rumphi, fl. 5.ii.1977, *Pawek* 12317 (K; MAL; MO; SRGH; UC).
Also in East and West Africa. In grassland, both upland and lowland.

2. **Striga bilabiata** (Thunb.) Kuntze, Rev. Gen. Pl. **3**, 2: 240 (1898). —Hepper in Kew Bull. **14**: 413 (1960). Types from S. Africa.
    Buchnera bilabiata Thunb., Prodr. Pl. Cap.: 100 (1800). Types as above.
    Striga thunbergii Benth. in Hook., Comp. Bot. Mag. **1**: 363 (1836). —Hiern in F.C. **4**, 2: 380 (1904). —Skan in F.T.A. **4**, 2: 404 (1906). —Eyles in Trans. Roy. Soc. S. Afr. **5**, 4: 476 (1916). Type from S. Africa.

Semi-parasitic herb, stem erect (10)20–55 cm. high, usually simple, pubescent, hairs whitish. Leaves usually erect and close to stem, 5–20 mm. long, linear to linear-lanceolate, hairs thick-based, scabrid. Inflorescences dense terminal spikes, elongating in fruit, several flowers out at the same time. Bracts imbricated, lanceolate, ciliate-pubescent on margins and midrib; bracteoles linear. Calyx tube 2–3 mm. long, teeth 1–2 mm. long, subulate at apex, 5-nerved, pubescent, hyaline between nerves. Corolla pink, violet or whitish, tube up to 11 mm. long, angled just below lobes, ± pubescent; lower lobes c. 4 mm. long, all narrow. Capsules 2–3 × 1.5 mm.

Botswana. N: Ngamiland Distr., Xhere Lediba, fl. 15.ii.1973, *P.A. Smith* 390 (K; SRGH). SE: Southern Distr., Digkatlong Ranch 1030 m., fl. & fr. 2.ii.1977, *Hansen* 3016 (C; GAB; K; PRE; SRGH; UPS). **Zambia**. B: Sesheke Distr., Sichinga Forest, fl. & fr. 27.xii.1952, *Angus* 1066 (BR; FHO; K; P). N: Mbala Distr., L. Chila, fl. 31.xii.1956, *Richards* 7425 (K). W: Mwinilunga Distr., Dobeka, fl. 22.i.1975, *Brummitt, Chisumpa & Polhill* 13988 (K; NDO; SRGH). C: Kabwe Distr., Muka Mwanji Hills, Kalwelwe, 1200 m., fl. & fr. 25.ii.1973, *Kornaś* 3299 (K). S: Machili, 24.ix.1960, *Fanshawe* 5819 (K; SRGH). **Zimbabwe**. N: Makonde (Lomagundi), fl. x.1920, *Eyles* 2707 (K; PRE; SRGH). W: Matobo Distr., Besna Kobila, 1460 m., fl. ii.1955, *Miller* 2671 (K; SRGH). C: Macheke, 1524 m., xii.1919, *Eyles* 2015 (K; PRE; SRGH). E: Chimanimani (Melsetter) Distr., Chimanimani Mts., Martin Forest Res.,

14.xi.1968, *Mavi* 628 (K; SRGH). **Malawi**. C: Lilongwe, fl. 16.xi.1951, *Jackson* 649 (K). **Mozambique**. M: Namaacha, Mt. Ponduine, c. 800 m., fl. 20.xii.1978, *Schafer & Nuvunga* 6644 (BR; K; LMU).

The above description and distributions refer to var. *bilabiata* which also occurs in S. Africa (Transvaal, Swaziland, Natal, Cape Province) and Namibia, Angola and Zaire. The other varieties; *ledermannii* (Pilger) Hepper, *barteri* (Engl.) Hepper, *rowlandii* (Engl.) Hepper, *jaegeri* Hepper are distributed across East and West Africa. It is a species of grassland, often at considerable altitude.

3. **Striga passargei** Engl., Bot. Jahrb. **23**: 515, t. 12, figs. M, N (1897). —Skan in F.T.A. **4**, 2: 403, note under *S. aspera*, (1906). —Hepper in F.W.T.A. ed. 2, **2**: 372 (1963). —Musselman & Hepper in Kew Bull. **41**: 219 fig. 3B (1986). Type from Cameroon.

Plant annual 10–40 cm. tall, simple or branched, stems somewhat succulent, scabrid, slightly quadrangular. Leaves few on lower half of stem, c. 10 × 1 mm. Bracts much larger than leaves, up to 5 × 4 mm., often recurved. Flowers opposite or alternate, only two flowers at same node open per inflorescence branch. Bracteoles subulate, shorter than calyx tube. Calyx in flower c. 8 mm. long, 15 mm. in diam.; enlarging in fruit, subulate teeth up to half as long as the tube; lower portion connivent, upper spreading in fruit, nerves scabrid, hyaline between nerves. Corolla cream-white or pale yellow, 10–17 mm. long, finely glandular-pubescent outside, throat hairy. Upper lobe emarginate or bilobed, 2–5 × 2–4 mm., reflexed, lower lobes spreading 2–7 × 1.5–4 mm. Capsule ovoid, 8–13 × 2 mm.

Zambia. C: Luangwa Valley Game Reserve South, c. 609 m. fl. 21.iii.1967, *Prince* 401 (K).
Also in Tanzania, Sudan, SW. Arabia, Cameroon, Nigeria, Ghana and Senegal. Low Mopane woodland.

4. **Striga aspera** (Willd.) Benth. in Hook., Comp. Bot. Mag. **1**: 362 (1836). —Skan in F.T.A. **4**, 2: 403 (1906). —F.W. Andr., Fl. Pl. Anglo-Egypt. Sudan **3**: 145 (1956). —Hepper in F.W.T.A. ed. 2, **3**: 372 (1963). —Binns, First Check List Herb. Fl. Malawi: 97 (1968). —Musselman & Hepper in Kew Bull. **41**: 209 fig. 2A (1986). TAB. **43**. Type from Ghana.
Striga aspera var. *schweinfurthii* Skan in F.T.A. **4**, 2: 403 (1906). Type from Sudan.

Branched or simple annual to 30 cm. tall; stem slightly angled, hispid with upward pointed hairs. Leaves divergent, narrow, up to 6 cm. long, usually less than 3 mm. wide, scabrid. Bracts of lowest flowers 2.5 cm. long, exceeding the calyx, bracts of upper flowers as long as or only slightly exceeding the calyx, 0.25–0.5 mm. wide with hispidulous margins, bracteoles similar to bracts but greatly reduced, three quarter length of calyx, 0.2 mm. wide. Calyx 6–10 mm. long, including subulate teeth, teeth half as long as calyx tube. Corolla pink, tube c. 15 mm. long, densely puberulent, bent and inflated just below limb; upper lip of corolla emarginate, slightly recurved, lower lobes spreading, 5–8 mm. long. Capsule 4 mm. long, oblong in outline.

Malawi. N: Nkhata Bay Distr., Viphya Plateau, 1676 m., fl. & fr. 15.v.1976, *Pawek* 11298 (MO; SRGH; UC). C: Lilongwe Distr., fl. 6.iv.1978, *Pawek* 14332A (K; MO). S: Chiradzulu Mt., fl. & fr. 14.v.1970, *Brummitt & Banda* 9838 (K; MAL; SRGH; PRE; UPS; WAG).
Also in East Africa, Sudan and commonly found in West Africa. Among wild grasses but not in cultivated cereal fields.

5. **Striga hermonthica** (Del.) Benth. in Hook., Comp. Bot. Mag. **1**: 365 (1836). —Skan in F.T.A. **4**, 2: 407 (1906). —Hepper in F.W.T.A. ed. 2, **2**: 372, fig. 290 (1963). —Musselman & Hepper in Kew Bull. **41**: 216, fig. 1c (1986). Type from Egypt.
Buchnera hermonthica Del., Fl. d'Egypte: 245, t. 34 fig. 3 (1813). Type as above.

Stiffly erect, dark green, much branched herb, 20–50 cm. tall. Stem quadrangular with groove on each face, scabrous. Leaves rather thick, 6–9 × 1.1–1.5 cm., both surfaces scabrous, margins with hispid hairs at regular intervals. Flowers bright pink (rarely white), opposite or nearly so, 8–10 open at same time per inflorescence branch, variable in size, faint sweet fragrance in early morning. Bracts almost as long as calyx with prominent hispid hairs on margin, apex curved away from stem. Bracteole half length of calyx. Calyx 10–12 × 2.5 mm., intracostal portion whitish and translucent. Corolla tube 1–2 cm. long, bent just above calyx teeth, outside with very scattered hairs; upper lobe emarginate, erect, lower lobes 10–13 × 8–10 mm. Capsule 12–15 × 2–2.5 mm.

Mozambique. Without locality, *Forbes* s.n. (K).
Widespread in the drier parts of tropical Africa and Madagascar. Usually parasitising the roots of cultivated sorghum but also on some wild grasses. In Sudan the attacks on sorghum seriously reduce the yield.

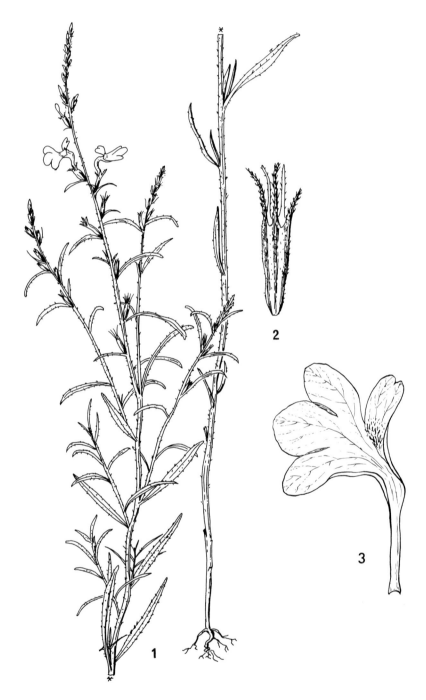

Tab. 43. STRIGA ASPERA. 1, habit (× $\frac{2}{3}$); 2, calyx (× 4); 3, corolla (× 4), all from *Schweinfurth* 1992. From Kew Bull.

6. **Striga gesnerioides** (Willd.) Vatke in Öst. Bot. Zeitschr. **25**: 11 (1875). —Wild, Rhod. Wild Fl.: 75 (1954); Common Rhod. Weeds, fig. 79 (1955). —F.W. Andr., Fl. Pl. Anglo-Egypt. **3**: 144, fig. 37 (1956). —Hepper in F.W.T.A. ed. 2, **2**: 373 (1963). —Binns, First Check List Herb. Fl. Malawi: 97 (1968). —Richards & Morony, Check List Fl. Mbala: 222 (1969). —Musselman & Hepper in Kew Bull. **41**: 213 fig. 1B (1986). Type from India.

Buchnera gesnerioides Willd., Sp. Pl. **3**: 338 (1800). Type as above.

Striga orobanchoides (R.Br.) Benth. in Hook., Comp. Bot. Mag. **1**: 361 (1836). —Hiern in F.C. **4**, 2: 380 (1904). —Skan in F.T.A. **4**, 2: 402 (1906). —Eyles in Trans. Roy. Soc. S. Afr. **5**, 4: 476 (1916). Type from Ethiopia.

Buchnera orobanchoides R.Br. in Endl., Öst. Bot. Zeitschr. **2**: 388, t. 2 (1832). Type as above.

Tufted, greenish-yellow, succulent herb, usually branching from base, 11–25(35) cm. tall. Single large primary haustorium c. 1–2 cm. in diam. usually present on each plant; adventitious roots abundant from subterranean scales. Stems square but only obtusely angled. Leaves scale-like, appressed to stem, 5–10 × 2–3 mm. Leaves and stems minutely puberulent with upward pointing hairs to almost glabrous. Flowers opposite or alternate, usually 2, rarely 3 per node, not fragrant. Bracts usually same length and width as calyx, acuminate. Bracteoles minute, three quarters the length of calyx. Calyx 4–6 mm. long (including teeth), 2 mm. wide. Corolla 1.2–1.5 cm. long, light blue or dark purple; bend in tube just below limb. Upper lobes 2–2.5 mm. long, sharply recurved, lower 3–3.2 mm. long. Capsules 1–2 × 3 mm. Pollen mass usually persistent on stigma.

Botswana. N: Maun, near airport, fl. & fr. ii.1967, *Lambrecht* 31 (K; SRGH). SW: Ghanzi to Lobatsi road, 950 m., fl. 1.ii.1970, *R.C. Brown* 8282 (K; SRGH). SE: Gaborone, Aedume Park, 1050 m., 15.iii.1978, *Hansen* 3375 (C; GAB; K; PRE; SRGH). **Zambia**. B: Mongu Distr., Mongu, 11.ii.1952, *White* 2048 (BR; FHO; K). N: Kumbula (Mbulu) Isl., L. Tanganyika, 760 m., 17.ii.1955, *Richards* 4512 (K). C: Serenje, 18.ii.1955, *Fanshawe* 2083 (K; NDO). E: Katete, 1300 m., 2.iii.1973, *Kornaś* 3364 (K). S: Choma, 1280 m., fl. 23.iii.1957, *Robinson* 2176 (K; SRGH). **Zimbabwe**. N: Umvukwe Mts., Darwendale, 20.iv.1948, *Rodin* 4339 (K; UC). W: Hwange, fl. ii.1955, *Levy* 1177 (K; PRE; SRGH). C: Charter Distr., Wiltshire Estate between Chivhu (Enkeldoorn) and Buhera, fl. 30.iv.1969, *Plowes* 3203 (K; SRGH). E: Odzi Distr., 1370 m., iii.1961, *Davies* 2890 (K, SRGH). S: Masvingo Distr., Kyle Nat. Park, 1065 m., fl. 10.iv.1971, *Basera* 343 (K; SRGH). **Malawi**. N: Karonga Distr., Chilumba, 470 m., fl. 20.iv.1922, *Pawek* 5135 (K; SRGH). C: Chitedze, Chirikanda Estate, 1220 m., 6.ii.1959, *Robson* 1487 (BM; K; SRGH). S: Mangochi Distr., 32 km. N. of Mangochi, fl. 20.ii.1982, *Hepper* 7377 (K; P). **Mozambique**. N: mouth of Malu R., ii.1912, *Allen* 124 (K). T: Tete (Tette), fl. & fr. ii.1859, *Kirk* s.n. (K). MS: Chibababa (Chibabava) R., Buzi, 5.i.1907, *Swynnerton* 1413 (BM; K).

Widespread in tropical Africa and Asia. In Mopane woodland, rocky grassland and cultivated ground. Some parasitic on legumes such as *Indigofera*, *Tephrosia*, *Vigna* and *Arachis*, also *Merremia*, *Jacquemontia*, and other genera; a form with a swollen haustorium grows on succulent *Euphorbia* species.

7. **Striga macrantha** (Benth.) Benth. in DC. Prodr. **10**: 503 (1846). —Skan in F.T.A. **4**, 2: 413 (1906). —Hepper in F.W.T.A. ed. 2, **2**: 371 (1963). —Musselman & Hepper in Kew Bull. **41**: 218, fig. 4A (1986). TAB. **44**, fig. A. Type from Sierra Leone.

Buchnera macrantha Benth. in Hook., Comp. Bot. Mag. **1**: 366 (1836). Type as above.

Coarse erect herb 50–100 cm. or more high, branched in upper part or simple, stems rounded-quadrangular, roughly hairy. Leaves opposite, 3–11 × 6–17 mm., linear-lanceolate to lanceolate, stiff, 3-nerved, nerves deeply impressed, margin coarsely toothed, roughly scabrid. Inflorescence very dense, terminal spike, 3–10 cm. long or up to 15 cm. long in fruit, closely glandular-hairy. Bracteoles broadly ovate, 4–6 mm. long. Calyx 5-toothed, 10-nerved. Corolla white, orifice yellow, tube c. 2 cm. long, angled towards lobes, upper half densely white-pilose; lobes broadly obovate, rounded, 6–9 mm. long, ± pilose outside. Capsules 8 × 4 mm., closely arranged, styles persistent.

Zambia. S: Mapanza, 28.iii.1954, *Robinson* 637 (BR; K; SRGH).

Also in Angola, Sudan (Jebel Marra) and frequent in West Africa but not recorded from E. Africa. In grassy, rocky places normally in hilly savanna country.

8. **Striga forbesii** Benth. in Hook., Comp. Bot. Mag. **1**: 364 (1836). —Hiern in F.C. **4**, 2: 384 (1904). —Skan in F.T.A. **4**, 2: 410 (1906). —Eyles in Trans. Roy. Soc. S. Afr. **5**, 4: 476 (1916). —Wild, Common Rhod. Weeds fig. 98 (1955). —Hepper in F.W.T.A. ed. 2, **2**: 371 (1963). —Binns, First Check List Herb. Fl. Malawi: 97 (1968). —Richards & Morony, Check List Fl. Mbala: 222 (1969). —Musselman & Hepper in Kew Bull. **41**: 213, fig. 4C (1986). TAB. **44**, fig. C. Type: Mozambique, *Forbes* s.n. (K, holotype).

Stiffly erect annual herb, dark green, sparsely branched, to 75 cm. high, usually much shorter. Young parts glandular-pubescent, becoming scabrous with age. Leaves sessile,

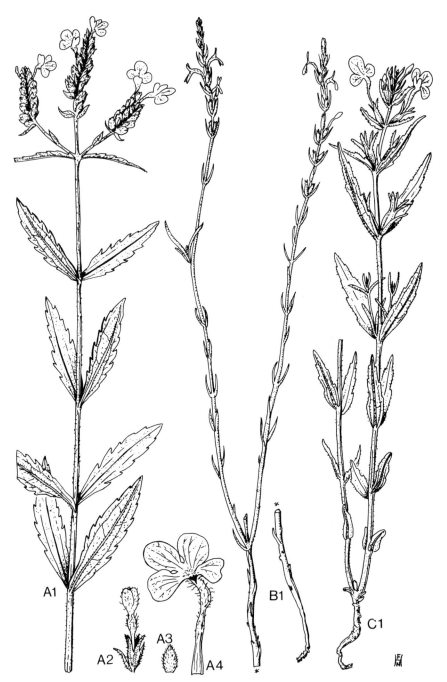

Tab. 44. A.—STRIGA MACRANTHA. A1, flowering branch (×⅔), from *Dalziel* 711; A2, flower bud (×¼); A3, bract (×¼); A4, corolla (×¼), A2–4 from *Rowland* s.n. B.—STRIGA LINEARIFOLIA. B1, habit (×⅔), from *Gilbert* 1250. C.—STRIGA FORBESII. Cl. habit (×⅔), from *Mackintosh* 1094. From Kew Bull.

opposite, 3-nerved, (1)2–4(7) × 3–7(10) cm., coarsely toothed each tooth the termination of one of the secondary veins, margins recurved, scabrid-pubescent. Bracts leaf-like, usually more than 2 cm. long. Flowers opposite to alternate. Calyx tube 5–9 mm. long, lobes lanceolate as long as tube, nerves scabrid hairy; bracteoles linear, shorter than calyx. Corolla salmon pink, densely glandular-pubescent, no obvious scent, tube c. 2 cm. long, angled near top, middle lower lobe 7–9(11) mm. wide, upper lobes smaller. Capsule shorter than calyx tube, flattened, rounded at apex.

Botswana. N: Okavango, Taoghe (Toakhe) R. Swamp, fl. 21.iii.1961, *Richards* 14824 (K). **Zambia**. N: Kafukula, Lunzua Valley, fl. & fr. 5.iii.1953, *Richards* 4780 (K). W: Kitwe, fl. 10.ii.1967, *Fanshawe* 10003 (K; NDO; SRGH). C: Chakwenga Headwaters, fl. 5.iii.1965, *Robinson* 6403 (K; SRGH). S: Mazabuka Distr., Siamambo stream, fl. 3.iii.1960, *White* 7574 (FHO; K; SRGH). **Zimbabwe**. N: Mazowe (Mazoe) Distr., Shamva near Lions Heads, fl. 18.ii.1963, *Wild* 5994 (K; SRGH). W: Hwange Distr., Victoria Falls, 880 m., fl. 7.iii.1979, *Mshasha* 182 (K; SRGH). C: Chegutu (Hartley) Distr., Poole, fl. 21.iii.1948, *Hornby* in GHS 20114 (K; SRGH). E: Mutare (Umtali) Distr., Battery Spruit, fl. & fr. 17.iv.1962, *Whiteside* in GHS 30476 (K; SRGH). **Malawi**. N: Rumphi Distr., c. 14 km. S. of Lura junction, fl. 26.ii.1978, *Pawek* 13913 (K; MA; MO; SRGH; UC). C: c. 3 km. S. of Kasungu, 1075 m, fl. 15.i.1959, *Robson & Jackson* 1199 (BM; K; SRGH). S: Chiromo, fr. 26.v.1949, *Wiehe* N/118 (K). **Mozambique**. Morrumbala (Moramballa), fl. 15.i.1863, *Forbes* s.n. (K). M: Maputo Distr., Vila Luiza, fl. 2.x.1957, *Barbosa & Lemos* 7913 (K).

Widespread throughout tropical Africa, also in S. Africa and Madagascar. Among short grasses in flood plains on cracking clay, occasional on maize and sorghum.

9. **Striga pubiflora** Klotzsch in Peters, Reise Mossamb., Bot.: 227 (1861). —Skan in F.T.A. **4**, 2: 412 (1906). —Moriarty, Wild Fl. Malawi: t. 39, fig. 3 (1975). Type: Mozambique, *Peters* s.n. (B†).
   Striga zanzibarensis Vatke in Linnaea **43**: 310 (1882). Type from Kenya.

Erect perennial herb with mauve tuberous roots. Stems slender 30–65 cm. tall, usually several simple stems from a woody base, sometimes branched above, shortly pubescent. Leaves opposite below, alternate above, 15–38 × c. 2 mm., linear, held vertically, shortly scabrid pubescent. Flowers 4–14, alternate, shortly pedicellate in loose terminal inflorescences. Bracts and bracteoles subulate, usually shorter than calyx. Calyx tubular, strongly 15-ribbed; tube 8–11 mm. long; teeth 5, 3–8 mm. long, linear. Corolla white with green tube; tube c. 20 mm. long, sharply bent above, densely pubescent outside; upper lip broadly truncate-ovate, undivided; lower lip broadly 3-lobed, lobes finely pubescent on margins and outside, central lobe up to 15 × 10 mm., obovate. Capsule 8 × 2.5 mm., valves recurved at apex after dehiscence.

Malawi. C: Kasungu, fl. 22.xii.1970, *Pawek* 4112 (K; SRGH). **Mozambique**. N: Cuamba (Nova Freixo) to Mandimba, fl. & fr. 25.v.1961, *Leach & Rutherford-Smith* 11011 (K; SRGH). Z: Quelimane Distr., Nawagre, fl. & fr. xii.194?, *Faulkner* 540 (K). MS: c. 40 km. S. of Muda, fl. 28.viii.1961, *Leach* 11232 (K; SRGH). GI: W. of Inhassoro, *Leach & Bayliss* 11866 (K; SRGH).

Also in Kenya and Tanzania. In open woodlands and damp grassy places.

10. **Striga angustifolia** (Don) Saldanha in Bull. Bot. Surv. India **57**, 1: 70 (1963); in Fl. Hassan Distr.: 526 (1976). —Hepper in Taxon **35**: 391 (1986). Type from India.
    Striga euphrasioides sensu auctt. non *Buchnera euphrasioides* Vahl (1794). —Skan in F.T.A. **4**, 2: 412 (1906). —Binns, First Check List Herb. Fl. Malawi: 97 (1968).

Erect annual or biennial herb 15–50 cm. high, usually simple or erect-branched, stems stiff, ± pubescent to densely covered with very short hispid hairs. Leaves opposite to nearly so, alternate above, linear to linear-lanceolate, 1–4 mm. wide, acute, finely pubescent. Flowers axillary, solitary, alternate, forming long lax terminal spikes; lower bracts leaf-like, the upper linear to subulate; bracteoles c. 4 mm. long, subulate. Calyx 10–12 mm. long, tubular, very prominently 15-ribbed, 5-toothed, finely pubescent; teeth 3–5 mm. long, linear-lanceolate, elongating in fruit. Corolla white or cream with greenish tube, pubescent outside; tube abruptly curved just above calyx teeth, inflated above; upper lip obovate, emarginate or truncate; lobes of lower lip 6–8 × 5 mm., obovate, rounded. Capsule ovoid, shorter than calyx, apiculate, valves sharply recurved after dehiscence.

Zambia. S: Choma, 1280 m., 11.vii.1930, *Hutchinson & Gillett* 3537 (BM; K). **Zimbabwe**. C: Marondera Distr., Membwe E., fl. 25.iii.1971, *Cleghorn* 2357 (K; SRGH). E: Nyanga Distr., Cheshire, fl. & fr. 5.ii.1931, *Norlindh & Weimarck* 4878 (K; ?LU). **Malawi**. N: Chitipa (Fort Hill), Tanganyika Plateau, 1066–1220 m., fl. & fr. vii.1896, *Whyte* s.n. (K). C: Tamanda, 1200 m., 8.i.1959, *Robson* 1094 (BM; K; SRGH). S: Mulanje Distr., above Likhubula, 1500 m., fl. & fr. ii.1982, *Hepper* 7368 (K). **Mozambique**. Z: Milanje Distr., Mt. Mongoe, fl. 27.i.1971, *Hilliard & Burtt* 6336 (E).

Apart from the Flora Zambesiaca area it occurs only in Tanzania in Africa; also in Oman and India. On seasonally wet rocks with water seepage among small sedges; up to 1500 m.

11. **Striga asiatica** (L.) Kuntze, Rev. Gen. Pl. **2**: 466 (1891). —Wild, Rhod. Wild Fl.: 74, pl. 27 (1954); Common Rhod. Weeds fig. 97 (1955). —Hepper in F.W.T.A. ed. 2, **2**: 372 (1963). —Binns, First Check List Herb. Fl. Malawi: 97 (1968). —Richards & Morony, Check List Fl. Mbala: 221 (1969). —Hepper in Rhodora **76**: 45–47 (1974). —Tredgold & Biegel, Rhod. Wild Fl.: 50, pl. 327 (1979). —Musselman & Hepper in Kew Bull. **41**: 207 fig. 1A (1986). Type from Comoro Islands.
   Buchnera asiatica L., Sp. Pl.: 630 (1753). Type as above.
   Striga lutea Lour., Fl. Cochinch.: 22 (1790). —Hiern in F.C. **4**, 2: 382 (1904). —Skan in F.T.A. **4**, 2: 409 (1906). —Eyles in Trans. Roy. Soc. S. Afr. **5**, 4: 476 (1916). —Saldanha in Bull. Bot. Surv. India **5**: 70 (1963). —Binns, First Check List Herb. Fl. Malawi: 97 (1968). Type from China (Canton).
   Striga lutea var. *bicolor* Kuntze, Rev. Gen. Pl. **3**, 2: 240 (1898). Type: Mozambique, 5.iv.1894, *Kuntze* (B†; K; NY).

Sparsely branched herb, stems 15–20 cm. tall, branching from above the middle, densely hispid, square in cross section. Leaves ascending, 8–16 mm. long, less than 1 mm. wide. Flowers opposite or alternate, only 2 flowers open at a time per inflorescence branch. Bracts 9–12 mm. long, longer than calyx; bracteoles 2 mm. long or less. Calyx 6 mm. long, including teeth (teeth 2 mm.), 2.5 mm. wide. Corolla scarlet within, yellowish without, rarely all yellow, 13 mm. long; tube bent in upper part near limb, whole limb positioned at right angles to tube; upper lobe emarginate, 4 mm. wide, lower lobes spreading 4 × 1.5 mm. Style persistent, usually with white pollen mass on the stigma. Capsule 7 × 2 mm.

Botswana. N: Kwando R., fl. & fr. 5.iv.1976, *Williamson* 90 (BR; K; SRGH). SE: Lobatsi Distr., Mogobane, fl. & fr. 23.ii.1959, *Powell* 876 (K; SRGH). **Zambia**. B: Sesheke, fl. & fr. iv.1910, *Gairdner* 412 (K). N: Chishimba Falls, Luombe R., fl. & fr. 31.ii.1955, *Richards* 5230 (K). W: Mwinilunga Distr., Sinkabolo Dambo, fl. 21.xii.1937, *Milne-Redhead* 3756 (K). C: Kafue Basin Survey, Mumbwa, fl. 22.iii.1963, *van Rensburg* KBS 1782 (K; SRGH). E: Chipata Distr., Chinzombo, Luangwa Valley, 609 m., fl. & fr. 20.ii.1969, *Astle* 5507 (K; SRGH). S: Mapanza Mission, fl. & fr. 21.i.1953, *Robinson* 60 (K). **Zimbabwe**. N: Shamva, fl. 17.ii.1961, *Rutherford-Smith* 542 (K; SRGH). W: Bulawayo Distr., Waterford, fl. 21.i.1974, *Norrgrann* 459 (K; SRGH). C: Harare, fl. & fr. 30.iv.1962, *Wild* 5718 (K; SRGH). E: Mutare Distr., Odzani R. Valley 1914, *Teague* 84 (BOL; K). **Malawi**. N: Karonga Distr., Vinthukutu Forest Res., fl. & fr. 26.iv.1975, *Pawek* 9575 (K; UC). C: Lilongwe Distr., Bunda Agric. Coll., fl. & fr. 19.ii.1970, *Brummitt* 8661 (K; SRGH). S: Zomba, fl. 12.ii.1984, *Hepper* 7332 (K; MO; P). **Mozambique**. N: Cabo Delgado, Mocimboa de Praia, fl. 7.iii.1983, *Jansen* 8158 (K). Z: Lugela Distr., Nawagre, fl. iii.194?, *Faulkner* s.n. (K). GI: Inhambane, Govuro, fl. & fr. 21.iii.1974, *Correia & Marques* 4142 (K). M: Namaacha, fl. ii.1931, *Gomes e Sousa* 432 (K).
   Throughout most of tropical and southern Africa, Madagascar, Arabia, across India to SE. Asia and China, introduced into N. America. Semiparasitic on wild grasses and certain crops such as maize in some areas.

12. **Striga elegans** Benth. in Hook., Comp. Bot. Mag. **1**: 363 (1836); in DC., Prodr. **10**: 502 (1846). —Hiern, Cat. Afr. Pl. Welw. **1**: 779 (1898); in F.C. **4**, 2: 382 (1904). —Skan in F.T.A. **4**, 2: 408 (1906). —Eyles in Trans. Roy. Soc. S. Afr. **5**, 4: 476 (1916). —Binns, First Check List Herb. Fl. Malawi: 97 (1968). Type from S. Africa.

Erect annual herb 11–36 cm. high, stems usually simple, densely strigose, when fresh yellowish green, drying greenish. Leaves mostly opposite, erect, 8–30(45) × 1–3 mm., linear, densely strigose. Inflorescences terminal, 3–11(19) cm. long, usually rather dense or interrupted with several opposite pairs of flowers out at the same time. Lower bracts longer than calyx, upper ones shorter; bracteoles subulate, shorter than calyx. Calyx 8–9 mm. long, including teeth, strigose. Corolla bright scarlet, yellowish outside, fragrant; tube c. 16 mm. long; limb upper lip bilobed, lower lip very deeply 3-lobed, about 1 cm. long.

Botswana. N: 18°07.8'S, 23°11.8'E, fl. 27.i.1978, *P.A. Smith* 2259 (K; SRGH). **Zambia**. B: Sesheke Distr., Masese For. Stn., 1050 m., fl. 2.ii.1975, *Brummitt, Chisumpa & Polhill* 14239 (K; NDO). C: between Undaunda and Rufunsa, fl. 6.iv.1972, *Kornaś* 1508 (K). S: Mazabuka Distr., Siamambo Forest Res., 11.i.1960, *White* 6213 (FHO; K). **Zimbabwe**. E: Nyanga (Inyanga), fl. 19.i.1948, *Chase* 666 (BM; K; SRGH). S: Mberengwa Distr., Mt. Buhwa, 10.xii.1953, *Wild* 4315 (K; SRGH). **Malawi**. C: Kasungu, Chipata (Chipala) Hill, fl. 14.i.1959, *Robson & Jackson* 1179 (BM; K; SRGH). S: Machinga Distr., Kawinga, 16.ii.179, *Blackmore, Brummitt & Banda* 453 (K; MAL).
   Also in Angola, S. Africa, Tanzania and Kenya. Among grasses and sedges in dambos and in *Brachystegia* woodland.

IMPERFECTLY KNOWN SPECIES

Striga diversifolia P. Lima in Broteria **1922**: 206. (1922).

Annual herb, glabrescent to 37 cm. high, with a rosette of oblong leaves. Spike lax terminal or lateral. Calyx 10-ribbed; corolla violet or pink.

Mozambique. ?Palma, 4.viii.1916, *Lima* 27 (type not seen).

33. RHAMPHICARPA Benth.

Rhamphicarpa Benth. in Hook., Comp. Bot. Mag. **1**: 368 (1836). —emend. Engl., Pflanzenw. Ost-Afr. **C**: 361 (1895). —O.J. Hansen in Bot. Tidsskr. **70**: 103–125 (1975).

Erect, slender, annual herbs, simple-stemmed to branched, glabrous, smooth. Leaves opposite or nearly so, entire, filiform or pinnatisect with filiform segments. Inflorescence racemose, often obscurely so, with opposite, leaf-like bracts. Flowers pedicellate, bibracteolate. Calyx 5-lobed. Corolla 5-lobed, long tubular, lobes equal or subequal. Stamens 4, didynamous, similar; filaments short, bearded at base; anthers unithecal without evidence of second theca. Ovary bilocular, ovoid or globose; ovules numerous. Fruit capsular, more or less distinctly rostrate, vertically somewhat compressed. Seeds numerous, cylindric or obovoid.

A genus of some 30 species occurring from Africa and Madagascar to India and Australia.

Pedicels (5)9–20(30) mm. long, as long as or longer than capsule; capsules straight or only slightly oblique, usually strongly winged along both sutures - - - - - 1. *fistulosa*
Pedicels 1–6 mm. long, shorter than capsules; capsules strongly oblique, strongly winged mainly on upper suture - - - - - - - - - - - 2. *brevipedicellata*

1. **Rhamphicarpa fistulosa** (Hochst.) Benth. in DC., Prodr. **10**: 504 (1846) excl. *Macrosiphon elongatus* Hochst. —Hiern in F.C. **4**, 2: 398 (1904) excl. *Macrosiphon elongatus*. —Hemsl. & Skan in F.T.A. **4**, 2: 419 (1906) excl. *Macrosiphon elongatus*. TAB. **45**. Type from the Sudan.
    Macrosiphon fistulosus Hochst. in Flora **1841**, 1: 374 (1841). Type as above.

Plants (3)5–50(120) cm. tall; stems simple or much branched. Leaves (5)10–60(100) mm. long, pinnatisect or rarely some simple; segments (0)2–7, 2–20(35) mm. long, under 1 mm. wide, filiform. Flowers solitary in axils of, or just above, bracts. Pedicels (5)9–20(30) mm. long. Bracteoles 1–5(7) mm. long, inserted near midway on pedicel. Calyx 3.5–9(13) mm. long; tube 0.5–2 mm. long, 1–4 mm. in diam.; lobes ovate at base with revolute margins, 2–5(8) mm. long, filiform. Corolla white, cream, pale pink or pale blue, externally stipitate-glandular; tube (22)25–30(35) mm. long, 1 mm. broad below, 1.5–2.5(4) mm. broad above, straight or curved; lobes 6–9 × 5–8 mm., circular to spathulate, rounded; corolla long persisting around the maturing capsule. Stamens inserted. Anthers 2–3.5 mm. long. Ovary 3–4 × 1.5–2 mm. Style 17–30 mm. long, slightly exserted. Capsule (6)7–10(15) mm. long to apex of beak, 4–7 mm. broad, winged along sutures.

Botswana. N: Boteti (Botetle) R., 9 km. E. of Makalamabedi, fl. & fr. 21.iii.1965, *Wild & Drummond* 7200 (K; LISC; SRGH). **Zambia**. W: Mwinilunga Distr., Kalenda Plain, fl. 23.i.1938, *Milne-Redhead* 4289 (BR; K). C: 16 km. NE. of Mfuwe, c. 610 m., fl. & fr. 26.iii.1969, *Astle* 5658 (K; SRGH). S: Mapanza, Choma, c. 1060 m., fl. & fr. 8.iii.1958, *Robinson* 2784 (K; SRGH). **Zimbabwe**. N: Binga Distr., Mwenda Res. Sta., fr. iv.1966, *Jarman* A6 (SRGH). W: Bulalima, Mangwe Distr., c. 10 km. N. of Marula, fr. 20.iv.1972, *Grosvenor* 744 (K; LISC; SRGH). C: Gweru Distr., Mlezu, fl. & fr. 6.iv.1965, *Molife* 138 (SRGH). E: Mutare Distr., Marange Res., 760 m., fl. 11.ii.1953, *Chase* 4788 (BM). S: Bikita Distr., W. bank of Turgwe R. at Dafana R. confluence, 1100 m., fl. & fr. 5.v.1969, *Biegel* 3021 (K; LISC; SRGH). **Malawi**. N: Nyika Plateau, fr. iii.1903, *McClounie* 96 (K). S: near Chiradzulu, fl. & fr. 21.x.1905, *Cameron* 100 (K). **Mozambique**. N: c. 45 km. E. of Ribáuè, c. 610 m., fl. & fr. 23.v.1961, *Leach & Rutherford-Smith* 10983 (SRGH). Z: Cundine, fl. & fr. 30.x.1902, *Le Testu* 509 (BM; BR). M: Maputo, Goba Fronteira, Fonte do Passo, fl. & fr. 21.iv.1955, *Mendonça* 4525 (LISC).

Widely spread from West Africa, Chad and Sudan, through Central and East Africa south to S. Africa; also in Madagascar, New Guinea and Australia. On peaty soils over rock substratum, on or between rocks in shallow, slow running rivers or streams, but more frequently near rivers and in grassy swamps; 600–1750 m.

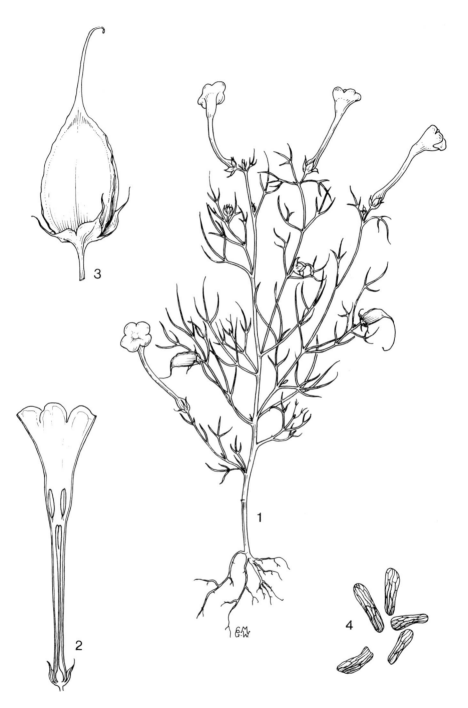

Tab. 45. RHAMPHICARPA FISTULOSA. 1, habit (× ⅔); 2, flower, longitudinal section (× 2); 3, capsule (× 2); 4, seeds (× 48). All from *Biegel* 3021.

2. **Rhamphicarpa brevipedicellata** O.J. Hansen in Bot. Tidsskr. **70**, 2–3: 119 (1975). Type from S. Africa.

Plants (5)10–40(70) cm. tall; stems simple or branched. Leaves 10–80 mm. long, pinnatisect, or rarely a few simple; segments (0)2–7, filiform, 3–20(35) mm. long, under 1 mm. wide. Flowers in axils of, or just above, bracts. Pedicels 1–6 mm. long. Pedicels 2–7 mm. long, inserted near middle of pedicel, to just below calyx. Calyx 5–11 mm. long; tube 1–3 × 2–4 mm.; lobes ovate at base with revolute margins, 2–5 mm. long, filiform. Corolla white to pink, externally stipitate-glandular; tube (20)25–35(40) mm. long, 1–2 mm. broad below, 2–3(5) mm. broad above, straight or curved; lobes 5–8 × 5–7 mm., subcircular to spathulate, rounded; corolla long persisting around the maturing capsule. Stamens inserted. Anthers 2–3 mm. long. Ovary 3–4 × 1.5–2 mm. Style 17–35 mm. long. Capsule 7–15 mm. long to apex of beak, 5–7 mm. broad, winged mainly on upper suture, often set at right angle to pedicel.

Botswana. N: Dobe catchment area, Quangwa watercourse, fl. & fr. 28.iv.1980, *P.A. Smith* 3471 (SRGH). **Zambia**. B: Masese, fl. 9.i.1961, *Fanshawe* 6097 (K; SRGH). **Zimbabwe**. N: Murewa Distr., Shawanoe R., fl. 14.iii.1954, *Whellan* 783 (SRGH). W: Matobo Distr., fr. iii.1902, *Eyles* 1034 (K; SRGH). C: Harare Distr., Cranbourne, fr. 31.iii.1946, *Wild* 1028 (K; SRGH). S: Masvingo Distr., Makoholi, fr. 17.iii.1978, *Senderayi* 296 (K; SRGH).

Also known from Namibia and S. Africa. Vleis and pan margins, often on sandy soil.

34. CYCNIUM E. Mey. ex Benth.

Cycnium E. Mey. ex Benth. in Hook., Comp. Bot. Mag. **1**: 368 (1836). —emend. Engl., Pflanzenw. Ost-Afr. **C**: 360 (1895). —O.J. Hansen in Dansk Bot. Arkiv. **32**, 3: 7–72 (1978).

Annual or perennial herbs or small shrubs with tuberous or elongated rhizomes; stems procumbent, ascending or erect. Leaves opposite, nearly opposite, occasionally alternate or in whorls of 3, pinnatisect or bipinnatisect or entire with variously incised or entire margins, sessile or shortly petiolate. Inflorescence terminally spicate or racemose, or flowers solitary-axillary or supra-axillary, pedicellate to sessile or subsessile. Bracts leaf-like. Bracteoles present or absent, where present inserted on pedicel or adnate to base of calyx tube. Calyx 4–5 lobed; tube ridged; lobes equal or subequal. Corolla 5-lobed, salver-form; tube curved, bent or straight, often covered with stipitate glands, throat bearded; limb bilabiate, zygomorphic, lower lip 3-lobed. Stamens 4, didynamous; filaments inserted in corolla tube, unilaterally bearded; anthers unithecal. Ovary bilocular, variable, compressed or not; ovules numerous. Fruit a capsule or berry, variable in shape. Seeds numerous.

A genus of 15 species occurring from East Africa and southwards to S. Africa.

1. Leaves pinnatisect or bipinnatisect - - - - - - - - 1. *recurvum*
– Leaves entire or variously toothed or lobed - - - - - - - - - 2
2. Leaves lanceolate, oblanceolate, elliptic, oblong or ovate, never linear, variously and regularly toothed, never lobed or irregularly toothed; fruit an indehiscent, fleshy berry 4. *adonense*
– Leaves linear or more rarely narrowly ovate, variously and irregularly toothed or lobed, or entire, never regularly toothed; fruit a dehiscent capsule - - - - - - - 3
3. Corolla tube gibbous above - - - - - - - - - 2. *filicalyx*
– Corolla tube cylindrical throughout, not gibbous above - - - - - 4
4. Plant annual; flowers sessile or subsessile; pedicels 0–1 mm. long; leaves narrowly linear, entire or very rarely lobed; filaments of short anthers 1.5–2 mm. long, filaments of long anthers 3–4 mm. long; anthers 1.8–2 mm. long - - - - - - - - 2. *filicalyx*
– Plant perennial; flowers pedicellate; pedicels 2–50 mm. long; leaves linear to ovate, irregularly toothed or lobed; filaments of short anthers 3–6 mm. long, filaments of long anthers 5–10 mm. long; anthers 2–4.5 mm. long - - - - - - - - - 3. *tubulosum*

1. **Cycnium recurvum** (Oliv.) Engl., Pflanzenw. Ost-Afr. **C**: 361 (1895). —O.J. Hansen in Dansk Bot. Arkiv **32**, 3: 25 (1978). Type from Tanzania.
    Rhamphicarpa recurva Oliv. in Trans. Linn. Soc. **29**: 122, t.87a (1875). —Hemsl. & Skan in F.T.A. **4**, 2: 420 (1906). Type as above.

Erect annual herb 20–80 cm. tall; stems simple to much branched; branches usually opposite, ascending. Leaves 10–70 cm. long, bi- or pinnatisect; segments 0.5–1.5(2) mm. wide. Inflorescence spicate or racemose. Pedicels 0–2 mm. long. Bracteoles 2.5–8 mm.

long, subulate to linear, adnate to base of calyx tube. Calyx 3–13 mm. long, campanulate, densely strigose to glabrous without; tube 2–5 mm. long; lobes five, 1–8 mm. long, triangular to narrowly triangular or linear, reflexed or spreading, erect in fruit, subequal or with upper reduced in size, furnished with short, somewhat appressed hairs, mixed with subsessile glands. Corolla white or pink; tube 8–18 mm. long, curved, somewhat gibbous above middle, glandular without; limb 6–16 mm. in diam. Filaments of shorter stamens 2–3 mm. long, of longer 3–7 mm. long. Anthers 1.5–2 mm. long. Ovary 1–2 mm. long and broad, subspherical laterally compressed. Style 4–6 mm. long including compressed stigma. Capsule 7–14 mm. long including beak, winged and dehiscent along upper suture only.

Malawi. N: Mzimba Distr., Phopo Hill, NE. end of Lake Kazuni, 1100–1150 m., fl. & fr. 20.v.1970, *Brummitt* 10947 (K).
Also known from Sudan, Rwanda and East Africa. Dry woodlands; 1100–1150 m.

2. **Cycnium filicalyx** (E.A. Bruce) O.J. Hansen in Dansk Bot. Arkiv **32**, 3: 29 (1978). Type from Tanzania.
    Rhamphicarpa filicalyx E.A. Bruce in Bull. Misc. Inf., Kew **1933**: 475 (1933). Type as above.

Erect annual herb 10–50 cm. tall; stems simple to much branched, stems glabrescent or strigose to hispid, more or less verrucose. Leaves up to 8 × 2 mm., linear, lower rudimentary, entire or rarely with two remote teeth. Inflorescence spicate or racemose. Pedicels 0–1 mm. long. Bracteoles usually absent, if present 1–3 mm. long, filiform, inserted on pedicels. Calyx 13–26 mm. long, campanulate; tube 2–3 mm. long, sparsely hispid, glabrescent; lobes (4)5, narrowly triangular to linear, 11–23 mm. long, scabrous on margins and midribs. Corolla white or cream or pale blue-pink; tube 15–22 mm. long, gibbous or not, glandular without; limb 8–16 mm. in diam. Filaments of shorter stamens 1.5–2 mm. long, of longer 3–4 mm. long. Anthers 1.8–2 mm. long. Ovary 2–3 mm. long, compressed ovoid. Style 5–7 mm. long including compressed stigma. Capsule 7–14 mm. long including beak, winged, dehiscent along upper suture only.

Botswana. N: near Tsimanemeha-Movombe track, fl. 28.i.1978, *P.A. Smith* 2299 (K; SRGH).
Zambia. S: Choma Distr., c. 20 km. N. of Choma, fl. & fr. 23.iii.1957, *Robinson* 2175 (K; SRGH).
Zimbabwe. W: Matobo Distr., Besna Kobila Farm, c. 1450 m., fl. i.1961, *Miller* 7663 (SRGH). C: Chegutu Distr., Avondale Farm Dam, fl. & fr. 25.ii.1968, *Mavi* 982 (K; LISC; SRGH). S: Masvingo Distr., fl. 1909–1912, *Monro* 1790 (BM).
Also known from Tanzania and Namibia. Seemingly restricted to grassy dambos and pans and wet areas surrounding them; 1000–1500 m.

3. **Cycnium tubulosum** (L. f.) Engl., Pflanzenw. Ost-Afr. **C**: 361 (1895). —O.J. Hansen in Dansk Bot. Arkiv **32**, 3: 33, fig. 13a–c (1978). TAB. **46**. Type from S. Africa.
    Gerardia tubulosa L. f., Suppl. Pl.: 279 (1781). Type as above.
    Rhamphicarpa tubulosa (L. f.) Benth. in Hook., Comp. Bot. Mag. **1**: 368 (1836). —Benth in DC., Prodr. **10**: 504 (1846). —Wettst. in Engl. & Prantl., Nat. Pflanzenfam. **4**, 3b: 95 (1891). —Hiern in F.C. **4**, 2: 399 (1904). —Skan in F.T.A. **4**, 2: 428 (1906). Type as above.
    Rhamphicarpa curviflora Benth. in Hook., Comp. Bot. Mag., **1**: 368 (1836); in DC., Prodr. **10**: 504 (1846). Type: Mozambique, *Forbes* s.n. (K); Madagascar.
    ?*Rhamphicarpa serrata* Klotzsch in Peters, Reise Mossamb., Bot.: 228 (1862). Type: Mozambique, *Peters* s.n. (B†).
    Rhamphicarpa montana N.E. Br. in Bull. Misc. Inf., Kew **1901**: 129 (1901). —Hiern in F.C. **4**, 2: 400 (1904). —Skan in F.T.A. **4**, 2: 427 (1906). Type: Zimbabwe, Matabeleland, *Elliot* s.n. (K, lectotype vide *Hansen* (1978)).
    Rhamphicarpa tubulosa var. *curviflora* (Benth.) Chiov., Fl. Somal. **2**: 337, fig. 193 (1932). Type as for *Rhamphicarpa curviflora*.
    Cycnium tubulosum subsp. *montanum* (N.E. Br.) O.J. Hansen in Dansk. Bot. Arkiv **32**, 3: 35, fig. 13d–g (1978). Type as for *Rhamphicarpa montana*.

Erect, decumbent, ascending or straggling perennial herb or more rarely, annual, branched or simple with a woody tuberous rhizome; stems 5–40(70) cm. long; whole plant, except corolla, glabrescent, hispid or pubescent, smooth or verrucose. Leaves 70(100) mm. long, linear to ovate, sometimes rudimentary below, apex acute or obtuse, entire or irregularly toothed or lobed, veins 1–3 somewhat parallel, pinnately nerved in broader leaves. Inflorescence racemose or flowers solitary. Pedicels 2–50(90) mm. long. Bracteoles absent or present, 0–10(20) mm. long, varied in shape, variously inserted on pedicels or adnate to base of calyx. Calyx 5–40 mm. long, campanulate or shortly tubular, becoming campanulate to broadly campanulate in fruit; tube 3–10(23) mm. long, ribbed

Tab. 46. CYCNIUM TUBULOSUM. 1, flowering branch (×⅔), from *Allen* 34; 2, basal leaf (× 1), from *Jackson* 2091; 3, flower, longitudinal section (× 1), from *Allen* 34; 4, habit of procumbent plant; 5, habit of erect plant, 4–5 after *Hansen,* fig. 13.

or not; lobes (4)5, equal, 2–12(18) mm. long, ovate to filiform, erect, reflexed or recoiled. Corolla white or pink; tube (12)14–55 mm. long, curved or straight, glandular without; limb (9)15–70 mm. in diam. Filaments of shorter stamens 3–6 mm. long, of the larger 5–10 mm. long. Anthers 2–4.5 mm. long. Ovary 2.5–4 mm. long, compressed-ovoid. Style 4–11 mm. long including compressed stigma. Capsule 4–15 mm. long including beak, winged, dehiscent mainly along upper suture.

Caprivi Strip. Kwando R., fl. 24.iv.1975, *Williamson* 43 (BR; SRGH). **Botswana**. N: behind Mutsoi Camp, E. of Nokaneng, fl. i.vi.1967, *Lambrecht* 212 (K; SRGH). SE: 3.2 km. S. of Lobatsi, fl. 17.i.1960, *Leach et al.* 160 (K; SRGH). **Zambia**. B: Mongu Distr., Barotse flood-plain between Mongu and Lealui, c. 1000 m., fl. 22.v.1972, *Strid* 2374 (K). N: Chinsali Distr., Lake Young, Shiwa Ngandu, 1350 m., fl. & fr. 17.i.1959, *Richards* 10711 (BR; K). W: Chingola Distr., fl. 18.iv.1954, *Fanshawe* 1105 (BR; K). C: Chilanga, 32 km. N. of Kafue, 915 m., fl. 10.x.1909, *Rogers* 5076 (K). E: Petauke-Sesare road, 800 m., fl. 4.xii.1958, *Robson* 820 (BM; BR; K; LISC; SRGH). S: Namwala, Mulela plain near Mbuilo fishing camp, fls. 17.x.1963, *van Rensburg* 2548 (BR; K; SRGH). **Zimbabwe**. N: Umvukwe Distr., fl. xii.1937, *Hopkins* in GHS 6758 (SRGH). W: Hwange Distr., Victoria Falls, Cataract and Livingstone Islands, fl. viii.1909, *Rogers* 5277 (K). C: Shurugwi Distr., 24 km. S. of Shurugwi (Selukwe), c. 1150 m., fl. 24.xii.1959, *Leach* 9667 (K). S: Bikita Distr., fl. & fr. 15.xii.1953, *Wild* 4390 (BR; SRGH). **Malawi**. N: Nkhata Bay Distr., Limpasa R., c. 670 m., fl. 26.xii.1976, *Pawek* 12049 (BR; SRGH). C: Livulezi R., near Ntcheu, 1075 m., fl. 29.i.1959, *Robson* 1333 (BM; K; LISC; SRGH). S: Blantyre Distr., near Njuli, N. of Blantyre, fl. & fr. 20.xii.1970, *Pawek* 4104 (K). **Mozambique**. Z: Mocuba Region, Namagoa, 200 km. inland from Quelimane, fl. Sept., *Faulkner* 169 (BR; K; SRGH). T: Angónia Distr., Ulongue, fl. 21.xi.1980, *Macúacua* 1292 (K). MS: Vila Machado, Pungwe Flats, fl. 13.x.1935, *Lea* 71 (K; SRGH). GI: Bilene, fl. 9.xii.1940, *Torre* 2266 (LISC).

Also known from West, Central and East Africa; Sudan, Madagascar and S. Africa; Angola and Namibia. Damp meadows and grassy plains, in mud of river banks, marshes and swamps, or in shallow water up to 1 m. deep; 10–1540 m.

For Flora Zambesiaca I feel that Hansen's reasons for the separation of two subspecies from this area are even weaker than those he gives for *Cycnium adonense*. The characters he uses show much flexibility for both subspecies and he agrees that there is evidence of much integration with resultant intermediates occurring. There are no clear-cut criteria for his subspecific division and too little consideration seems to have been given to the overall acknowledged habitats of the two. Plants grown in open montane areas generally assume the prostrate habit with the flowers arising in opposite positions on the stem but appearing unilateral, or with only one flower of the pair developed. By nature of their habitat, these plants tend to be comparatively much hairier and coarser with the leaves more strongly lobed. On the other hand, those assigned to the typical subspecies and occurring in the more lush and wetter situations of dense grassland, riversides and the like are found as more erect plants competing with the taller erect surrounding vegetation becoming themselves more so. In this situation the hairy indumentum is reduced in amount, the leaves are more elongated and the flowers arranged in a more well defined raceme.

4. **Cycnium adonense** E. Mey. ex Benth. in Hook., Comp. Bot. Mag. 1: 368 (1836). —Benth. in DC., Prodr. **10**: 505 (1846). —Hiern, Cat. Afr. Pl. Welw. **1**, 3: 777 (1898). —Engl., Bot. Jahrb. **28**: 479 (1900). —Hiern in F.C. **4**, 2: 395 (1904). —Hemsl. & Skan in F.T.A. **4**, 2: 431 (1906). —O.J. Hansen in Dansk Bot. Arkiv **32**, 3: 48 (1978). Type from S. Africa.
  Cycnium buchneri Engl., Bot. Jahrb. **18**: 73 (1893). Type from Angola.
  Cycnium camporum Engl., Bot. Jahrb. **18**: 73 (1893). Type from Sudan.
  Cycnium decumbens Gandoger in Bull. Soc. Bot. Fr. **66**: 217 (1919). Type from S. Africa.
  Cycnium pentheri Gandoger in Bull. Soc. Bot. Fr. **66**: 217 (1919). Type from S. Africa.
  Cyncium rectum Gandoger in Bull. Soc. Bot. Fr. **66**: 216 (1919). Type from S. Africa.
  Cycnium adonense subsp. *camporum* (Engl.) O.J. Hansen in Dansk Bot. Arkiv. **32**, 3: 53 (1978).

Perennial herb with woody tuber; stems 5–35 cm. long, prostrate, decumbent or erect; stems, leaves and calyx hispid, scabrous or pilose, at times additionally stipitate-glandular. Leaves up to 8(12) cm. long, oblanceolate, lanceolate, elliptic, oblong or ovate, apex acute or obtuse, cuneate at base, variously toothed. Flowers solitary or in racemes. Pedicels 2–30(50) mm. long, axillary or supra-axillary. Bracteoles (1)2–15 mm. long, subulate, filiform, linear or lanceolate, inserted on pedicel close to calyx or sometimes adnate to base of calyx tube. Calyx 8–75 mm. long, short- or long-tubular, rarely campanulate in flower, campanulate to urceolate in fruit; tube 5–60 mm. long, 3–13 mm. broad, ribbed, at times subplicate between ribs; lobes, 5, equal or unequal, 3–15 mm. long, narrowly triangular, ovate or lanceolate. Corolla white or pink; tube 30–95 mm. long, straight, occasionally slightly gibbous above middle, glandular without; limb 25–70(90) mm. in diam. Filaments of shorter stamens 3–13 mm. long, of the longer 6–17 mm. long. Anthers 4–7 mm. long. Ovary 2–5 mm. long, ovoid, somewhat compressed. Style 6–30 mm. long including compressed stigma. Berry 12–24 × 10–12 mm., globose to ovoid.

Botswana. SE: Lobatsi, c. 1200 m., fl. x.1913, *Rogers* 6236 (SRGH). **Zambia**. B: between Luampa R. (at Sikelenge) and Kaoma (Mankoya), fl. 19.x.1959, *Drummond & Cookson* 6638 (K; SRGH). N: Mbala Distr., NE. of L. Chila, 1590 m., fl. 25.xi.1964, *Richards* 19274 (BR; K; LISC). W: Kabompo Distr., 48 km. N. of Kabompo, fl. 5.x.1952, *White* 3458 (BR; K; P). C: Lusaka Distr., between Kasisi and Constantia, 1130 m., fl. 24.x.1972, *Kornaś* 2441 (K). E: Katete, 1060 m., fl. 17.xii.1957, *Wright* 206 (K). S: Mapanza, Choma, fl. 15.xi.1958, *Robinson* 2927 (K; SRGH). **Zimbabwe**. N: Chinhoyi (Sinoia), fl. xi.1926, *Rand* 316 (BM). W: Gwanda Distr., Mtsheleli Valley [? Mtshabezi], fl. 29.xi.1951, *Plowes* 1351 (NY; SRGH). C: South Marondera (Marandellas), fl. xii.1931, *Myres* 124 (K). E: Mutare Distr., Vumba Mts., c. 1700 m., fl. 10.v.1956, *Chase* 6111 (BM; K; SRGH). S: Masvingo Distr., Great Zimbabwe Nat. Park, fl. 23.xii.1970, *Chiparawasha* 252 (SRGH). **Malawi**. N: Chitipa Distr., Nyika Plateau, 7 km. NW. of Lake Kaulime, 2135 m., fl. 16.v.1970, *Brummitt* 10816 (K). C: Dedza Distr., Chongoni, 1700 m., fl. 19.i.1959, *Robson* 1258 (BM; K; LISC; SRGH). S: Thyolo (Cholo) Mt., 1200 m., fl. 26.ix.1946, *Brass* 17823 (BR; K; NY; SRGH). **Mozambique**. N: Plateaux de Lichinga, fl. xii.1932, *Sousa* 1043 (K). Z: Lugela Distr., Mocuba Region, Namagoa, 200 km. inland from Quelimane, fl. xii.1943, *Faulkner* 105 (BM; BR; K; NY; P; SRGH). T: Angónia Distr., Domue near N'tchide, fl. 20.xi.1980, *Macúacua* 1272 (K). MS: near Zembe, S. of Chimoio (Vila Pery), c. 610 m., fl. vii.1959, *Leach* 9206 (K; SRGH). M: Maputo, *Gomes e Sousa* 396 (LISC).

Occurring from West Africa through Central to East Africa; also in Angola. Montane grasslands and open woodlands, forest edges and occasionally streamsides; 610–2400 m.

In his treatment of the genus, Hansen distinguished two subspecies, the typical one and subsp. *camporum*. His criteria for this separation, I feel are far too weak and indecisive. He has investigated all aspects of the species from calyx, corolla, stamen and shape and size of pistil, habit and geographical distribution and has used his findings to present apparently good and valid reasons for this separation. However, from all the material from the Flora Zambesiaca area I have studied, I have found far too many examples of intermediates between the subspecies, even as identified by Hansen, to feel that the distinction is clear cut.

Here, for the convenience of Flora Zambesiaca I have decided not to recognise Hansen's two subspecies.

35. BUTTONIA McKen ex Benth.

Buttonia McKen ex Benth. in Hook., Ic. Pl. **11**: 63, t. 1080 (1871). —Hiern in F.C. **4**, 2: 384 (1904).

Perennial, slender, climbing shrub. Leaves opposite, pinnatisect, petiolate. Flowers solitary, axillary, pedicellate, bracteate; bracts deciduous. Calyx (4)5-lobed, campanulate, inflated in fruit; lobes ovate, shorter than tube. Corolla 5-lobed; lobes rounded, subequal; tube funnel-shaped with wide throat, somewhat curved. Stamens 4, didynamous, included; filaments linear; anthers connivent in pairs, bithecal with one empty theca or empty theca absent on lower stamens. Ovary bilocular. Ovules numerous. Capsule globose, enclosed in persistent calyx. Seeds numerous, truncate at both ends, conical-cylindric.

A genus of 2 species native to tropical east and S. Africa.

Buttonia natalensis McKen ex Benth. in Hook., Ic. Pl. **11**: 63, t. 1080 (1871). —Hiern in F.C. **4**, 2: 385 (1904). TAB. **47**. Type from S. Africa.
  Buttonia hildebrandtii Engl., Bot. Jahrb. **23**: 509 (1897). —Hemsl. & Skan in F.T.A. **4**, 2: 439 (1906). Type from Kenya.

Stems somewhat woody, many-branched, glabrous or occasionally glabrescent with few minute hairs. Leaves with petiole 3.0–6.5 cm. long and 1.5–4.0 cm. wide, glabrous to rarely shortly pubescent; lower lobes ovate to broadly elliptic, obtuse, narrowed at base, entire to 1- or 2-toothed; central lobes broadly lanceolate to rhombic, variously toothed or laciniate. Flowers carmine, rose-lilac to violet, throat darker, fragrant, supra-axillary, solitary; pedicels 2.5–3.7 cm. long, stout, spreading, glabrous, becoming recurved in fruit. Bracteoles 2, opposite, subsessile immediately below calyx, subcircular to reniform, 8.5–16 mm. in diam., glabrous or sparsely furfuraceous. Calyx 15–20 mm. long in flower, campanulate to urceolate, longer in fruit, 10-nerved without, shortly 4–5 lobed; lobes reticulately veined within. Corolla tube about 3 cm. long, curving upward, inflated above middle; limb 3–7 cm. in diam., spreading; lobes 1.5–2 × 2–3 cm. Stamens glabrous. Capsule (dried) 1.5–2.5 cm. in diam. Seeds 2–3.25 mm. long.

Zimbabwe. E: Chimanimani Distr., Umvumvumvu R. Gorge above east bridge, 700 m., fr. 20.i.1957, *Chase* 6297 (K; SRGH). S: Chibi Distr., near Rundi R. Bridge, Chibi Tribal Trust Lands, fl.

142

Tab. 47. BUTTONIA NATALENSIS. 1, flowering and fruiting branch (×⅔); 2, flower, longitudinal
section (× 1); 3, stamens (× 2), 1–3 from *Balsinhas* 689; 4, fruiting calyx (×⅔); 5, fruit with part of
calyx removed (× 1), 4–5 from *Chase* 8271.

3.ii.1973, *Ngoni* 189 (K; SRGH). **Mozambique**. GI: Inhambane Distr., 20 km. N. of Zavala, fl. & fr. iv.1938, *Gomes e Sousa* 2089 (K). M: Maputo Distr., Maputo region, fl. & fr. 21.ii.1948, *Gomes e Sousa* 3686 (COI; K).

Also in Kenya, Tanzania and S. Africa. Open woodlands, among rocks and on riverbanks; from 30–1200 m.

Hitherto it has been considered that *Buttonia natalensis* McKen ex Benth. and *B. hildebrandtii* Engl. represented two distinct taxa, but the material now available leaves little doubt that they are conspecific. Some of the Kenyan material named *B. natalensis* studied has leaves and flowers similar in size to some of the South African *B. hildebrandtii* and it would seem that the material used for the latter description was of a depauperate plant due to much drier and poorer local conditions. Here they are considered together under the earlier name.

36. SOPUBIA Buch.-Ham. ex D. Don

Sopubia Buch.-Ham. ex D. Don, Prodr. Fl. Nep.: 88 (1825).

Annual or perennial herbs or undershrubs, usually erect, branched, glabrous, scabrid, woolly or tomentose. Leaves opposite or verticillate, or upper alternate, linear to linear-lanceolate, entire, or pinnatifid with linear or filiform segments. Inflorescence terminal, racemose or spicate; flowers bracteate, pedicellate or not, bi-bracteolate. Calyx campanulate, 5-lobed; lobes valvate, linear-triangular to broadly deltate. Corolla tube usually short, at times exserted, enlarged at throat; limb spreading, lobes 5 broad subequal entire. Stamens 4, didymanous, slightly included; anthers bithecal, all coherent or coherent in pairs; one cell of each anther perfect, ovoid or ellipsoid, often subapiculate, the other cell much smaller, linear or clavate, stipitate, quite empty or nearly so. Style elongated, thickened or flattened at apex. Capsule ovoid, ellipsoid or subglobose, often compressed above, retuse, emarginate or rounded at apex, loculicidal. Seeds numerous, oblong to obovoid or at times narrowly cylindric.

A genus of about 40 species, from tropical and southern Africa and from the Himalayas to Indo-China and Formosa; 1 Australian species.

1. Calyx not densely woolly without, glabrous or hairy only on margin of lobes - - 2
 - Calyx densely woolly without - - - - - - - - - - - 6
2. Stems many-ribbed; leaves few, undivided, more or less appressed to stem 1. *simplex*
 - Stems few-ribbed; leaves many, undivided or trifid to multifid, spreading, not appressed to stem - - - - - - - - - - - - - - - 3
3. Plants perennial, robust; calyx lobes with densely woolly margins - - - 4
 - Plants annual, delicate; calyx lobes with margins only slightly woolly - - - 5
4. Midrib of leaves not prominent, often obscure; leaves linear to filiform, undivided to multifid; pedicels 2–24 mm. long - - - - - - - - - - - 2. *mannii*
 - Midrib of leaves prominent beneath; leaves linear to lanceolate, undivided or trifid; pedicels less than 12 mm. long - - - - - - - - - - - 3. *ramosa*
5. Upper leaves undivided with the lower trifid or pentafid - - - - 4. *eminii*
 - Leaves undivided throughout - - - - - - - - - 5. *parviflora*
6. Inflorescence compact with axis not visible between whorls or pairs of flowers - - - - - - - - - - - - 6. *lanata* var. *densiflora*
 - Inflorescence lax with axis visible between whorls or pairs of flowers, at least below 7
7. Plant below inflorescence pubescent to tomentose or at the most sparsely woolly; leaves undivided or tri- or pentafid, leaves or segments c. 1 mm. wide - - - - 8
 - Plant covered with a dense woolly or appressed silver-grey indumentum; leaves simple, 0.5–5 mm. wide - - - - - - - - - - - - - - 9
8. Plant usually white silky tomentose throughout; leaves entire; bracteoles 2.5–5 mm. long; calyx lobes 2 mm. long, deltate; ovary pilose - - - - - - 7. *angolensis*
 - Plant puberulent to glabrescent; leaves entire, tri- or pentafid; bracteoles 6.5–11 mm. long; calyx lobes 3–13 mm. long, narrowly triangular; ovary glabrous - - - 8. *karaguensis*
9. Leaves 0.5–1 mm. wide; bracteoles 2.5–3.5 mm. long, narrowly linear; corolla limb spreading to 7–11 mm. in diam.; capsule oblong, more or less densely pilose - - 7. *angolensis*
 - Leaves 1–5 mm. wide; bracteoles 4–10 mm. long, linear to linear-lanceolate; corolla limb spreading to 10–16 mm. in diam.; capsule shortly oblong to subglobose, sparsely pilose - - - - - - - - - - - - 6. *lanata* var. *lanata*

1. **Sopubia simplex** (Hochst.) Hochst. in Flora **27**: 27 (1844). —Skan in F.T.A. **4**, 2: 450 (1906).
—Hepper in F.W.T.A. ed. 2, **2**: 369 (1963). —O.J. Hansen in Kew Bull. **30**: 550 (1975). Type from
S. Africa.

    Raphidophyllum simplex Hochst. in Flora **24**: 667 (1841). Type as above.

    Sopubia dregeana Benth. in DC., Prodr. **10**: 522 (1846) nom illegit. (see Hepper in Kew Bull.
14: 409 (1960)).

Perennial herb, erect up to 45 cm. tall; stems several arising from a woody base, simple
or branched above, markedly angled, many ribbed, glabrous to minutely sparingly
pubescent, sparsely leafy. Leaves alternate to whorled, undivided, (5.5)12–20 × (0.2)0.5–1
mm., narrowly linear, usually appressed, revolute, subglabrous to scabrid on margins and
midrib beneath. Inflorescence loose to somewhat clustered, terminal, racemose up to
20(30) cm. long; flowers numerous, alternate to whorled, pink to lilac or pale mauve,
rarely white, with dark purple throat, pedicellate. Pedicels 6–13 mm. long, rather slender,
subglabrous to shortly hispid-pubescent. Bracteoles 1.5–2 × c. 0.15 mm., linear to subulate,
subglabrous to minutely scabrid. Calyx 4–6 mm. long, brownish-green when fresh,
verrucose to hispid or minutely pubescent without; lobes 1–3 mm. long, triangular, woolly
at margin and within. Corolla rotate, limb (10)14–15 mm. in diam., caducous. Fertile
anther cells 2.5–2.8 mm. long, more or less ellipsoid. Capsule 4.5–5 mm. long, 2.5–3.5 mm.
in diam., oblong to ovate- or obovate-obong in outline, retuse, bearing persistent base of
style.

    Zambia. B: Kaoma Distr., 56 km. W. of Kaoma (Mankoya) on Mongu road, fl. 8.xi.1959,
Drummond & Cookson 6245 (K; LISC; SRGH). N: Kawambwa Distr., 16 km. from Kawambwa towards
Nchelenge, 1290 m., fl. & fr. 29.xi.1961, *Richards* 15445 (K; SRGH). W: Mwinilunga Distr., 16 km. W.
of R. Lunga, 64 km. S. of Mwinilunga (Boma), fl. 12.viii.1930, *Milne-Redhead* 872 (K). C: c. 30 km.
NNE. of Lusaka, Chongwe R., N. of Kasisi, 1150 m., fl. & fr. 29.ix.1972, *Strid* 2226 (K). E: Nyika
Plateau, 2100 m., fl. 27.xi.1955, *Lees* 103 (K). **Zimbabwe**. N: Mazowe (Mazoe), 1300 m., fl. xii.1906,
Eyles 481 (BM; SRGH). W: Matobo Distr., Besna Kobila Farm, c. 1450 m., fl. iv.1961, *Miller* 7862
(BR). C: Marondera (Marandellas), fr. January, *Myers* 54 (K; SRGH). E: Martin Forest Reserve,
Chimanimani Mts., fl. 15.xi.1967, *Mavi* 636 (K; LISC; SRGH). S: Masvingo Distr., fl. & fr. 1909, *Monro*
585A (BM). **Malawi**. N: Nkhata Bay Distr., Viphya, c. 37 km. SW. of Mzuzu, c. 1600 m., fl. 15.xi.1970,
Pawek 4000 (K). C: Dedza Distr., Chongoni, base of Chiwawo Hill, 1650 m., fl. 4.xi.1959, *Robson* 1444
(BM; K; SRGH). S: Mulanje Mt., Luchenya Plateau, c. 1890 m., fl. & fr. 18.ix.1970, *Pawek* 3878 (K).
Mozambique. N: Mandimba (Mandumba), fl. & fr. 25.xi.1941, *Hornby* 3488 (K). Z: Namuli, Makua,
fl. *Last* s.n. (1887) (K). MS: between Skeleton Pass and The Plateau, Chimanimani Mts., fl.
27.ix.1966, *Grosvenor* 199 (K; LISC; SRGH). M: Maputo, Zitundo, Matutuine (Bela Vista), fl.
10.xii.1961, *Lemos & Balsinhas* 279 (BM; K; LISC; SRGH).

    Also known from West Africa, East and S. Africa. Mostly in burnt or moist grasslands, riversides,
swamps or marshes; from 600–2280 m.

2. **Sopubia mannii** Skan in F.T.A. **4**, 2: 450 (1906). —O.J. Hansen in Kew Bull. **30**: 545 (1975). Type
    from Cameroon.

Erect, rigid, scabrid, branched herb, 20–150 cm. tall; stems very leafy, numerous arising
from a thick root-stock. Leaves opposite or verticillate, undivided or multifid, closely or
loosely arranged, often with leafy branches in axils, 15–55 × 0.3–1 mm., narrowly linear,
acute, subscabrid on margins, margins reflexed or not, mid-vein indistinct. Inflorescences
of usually terminal racemes, up to 10 cm. long; flowers usually numerous in whorls of 3,
pink, mauve-pink to mauve or purple (or yellow fide *Phillips* 1374). Bracteoles 2–4 ×
0.15–0.8 mm., arising immediately beneath or up to 4 mm. below calyx. Pedicels 4–26 mm.
long. Calyx 3.5–7 mm. long, variously hirsute within and without. Corolla limb 10–22 mm.
in diam. Anthers 2–3.5 × 0.7–1 mm. Capsule 3.5–8 × 2.5–4.5 mm.

Pedicels less than 10 mm. long; leaves undivided, closely arranged on stem; corolla limb rarely
    exceeding 12 mm. in diam. - - - - - - - - - - - var. *mannii*
Pedicels 12–18(26) mm. long; leaves undivided or 3–5-fid, more openly arranged on stem; corolla
    limb 14–18(22) mm. in diam. - - - - - - - - - - var. *tenuifolia*

Var. **mannii** —Hepper in Kew Bull. **14**: 409 (1960); in F.W.T.A. ed. 2, **2**: 369 (1963). —O.J. Hansen in
    Kew Bull. **30**: 545 (1975).

Perennial herbs. Leaves undivided, 0.75–1 mm. wide, closely arranged on stem. Flowers
closely arranged on inflorescence axis. Pedicels 4–8.5(10) mm. long. Calyx 3.5–5.5 mm.
long, smooth, glabrous or subglabrous without, tomentose on margin. Corolla limb
10–12(13.5) mm. in diam.

Zambia. B: Kwando R., fl. ix.1959, *Guy* in GHS 98850 (SRGH). N: c. 72 km. E. of Mbala (Abercorn) on road to Mbosi, c. 1600 m., fl. 31.iii.1932, *Thompson* 1110 (K). **Zimbabwe**. C: St. Triashill, ii.1917, *Mundy* 3167 (K). W: Matobo Distr., Matopos Hills, fl. xi.1902, *Eyles* 1102 (SRGH). E: Nyanga Distr., Inyangani Mt., 300 m. below summit ridge, fl. 25.i.1979, *Müller* 3584 (K; SRGH). **Malawi**. N: Chitipa Distr., Nyika Plateau, Nganda Peak, 2600 m., fl. & fr. 10.iv.1969, *Pawek* 2074 (K). C: Mchinji, on road to Kasungu, 1215 m., fl. 27.iv.1970, *Brummitt* 10214 (K; SRGH). S: Mulanje Distr., Lichenya Plateau, c. 1830 m., fl. 11.ii.1958, *Chapman* 483 (SRGH). **Mozambique**. MS: Beira Distr., Gorongosa Mt., summit area, 1700–1860 m., fl. 12.iii.1972, *Tinley* 2439 (LISC; SRGH). .

Also recorded from Uganda, Tanzania; Ivory Coast and Cameroon. Grasslands; 1500–2600 m.

Var. **tenuifolia** (Engl. & Gilg) Hepper in Kew Bull. **14**: 410 (1960); in F.W.T.A. ed. 2, **2**: 369 (1963). —O.J. Hansen in Kew Bull. **30**: 546 (1975). Type from Angola.
　　Sopubia dregeana var. *tenuifolia* Engl. & Gilg in Warb., Kunene-Samb. Exped. Baum: 365 (1903). Type as above.
　　Sopubia trifida forma *humilis* Engl. & Gilg in Warb., Kunene-Samb. Exped. Baum: 365 (1903) excl. *S. decumbens* Hiern. Type from Angola.

Perennial herbs. Leaves undivided, or rarely 3–5-fid, 0.3–0.5(0.75) mm. wide, more loosely arranged on stem. Flowers loosely arranged on inflorescence axis. Pedicels (10)12–18(26) mm. long. Calyx 4–5.5(7) mm. long, smooth, glabrous without, tomentose on margin. Corolla limb (12)14–18(22) mm. in diam.

Caprivi Strip. Singalamwe, c. 1000 m., fl. & fr. 31.xii.1958, *Killick & Leistner* 3208 (SRGH). **Botswana**. N: Ngamiland Distr., Okavango, near Xudum R., 930 m., fl. 16.iii.1961, *Richards* 14744 (BR; K). **Zambia**. B: 10 km. E. of Mongu, fl. 24.x.1965, *Robinson* 6688 (K; SRGH). N: Mbala Distr., Kambole Escarpment, c. 1900 m., fl. & fr. 22.iv.1969, *Sanane* 628 (K). W: Mwinilunga Distr., West Lunga R., 8 km. N. of Mwinilunga, 1300 m., fl. & fr. 23.i.1975, *Brummitt, Chisumpa & Polhill* 14022 (BR; K; P; SRGH). **Zimbabwe**. N: Sebungwe Distr., fl. & fr. 8.xii.1956, *Lovemore* 522 (K; SRGH). W: near Victoria Falls, fl. 7.vii.1930, *Hutchinson & Gillett* 3481 (BM; K; SRGH). C: Marondera (Marandellas), c. 1650 m. fl. April, *Eyles* 7039 (K; SRGH). E: Nyanga Distr., below Inyangani Summit, fl. 3.iii.1956, *Whellan & Davies* 994 (K; SRGH). S: Masvingo Distr., 6.4 km. S. of Great Zimbabwe Ruins, fl. & fr. 18.xii.1970, *Müller & Pope* 1732 (K; SRGH). **Malawi**. N: Nyika Plateau, 25.6 km. E. of Rest House, 2438 m., fl. 5.vi.1957, *Boughey* 1635 (K). S: Mulanje Distr., Lichenya Plateau, Chilemba Peak, 2310 m., fl. & fr. 17.v.1962, *Richards* 16565 (K). **Mozambique**. MS: Gorongosa Mt., Gogogo Summit, fl. & fr. 21.iv.1971, *Tinley* 2103 (BR; K; LISC; SRGH).

Also known from Sierra Leone, Kenya, Uganda, Tanzania and Angola. In wet and dry grasslands, bordering rivers and among rocks. Mostly from above 1000 m. and up to 2500 m.

3. **Sopubia ramosa** (Hochst.) Hochst. in Flora **27**: 27 (1844). —Skan in F.T.A. **4**, 2: 449 (1906). —Hepper in F.W.T.A. ed. 2, **2**: 369 (1963). —O.J. Hansen in Kew Bull. **30**: 546 (1975). Type from Ethiopia.
　　Raphidophyllum ramosum Hochst. in Flora **24**: 668 (1841). Type as above.
　　Sopubia trifida var. *ramosa* (Hochst.) Engl., Bot. Jahrb. **18**: 65 (1893); Pflanzenw. Ost-Afr. **C**: 359 (1895). Type as above.
　　Sopubia similis Skan in F.T.A. **4**, 2: 447 (1906). Type from Ethiopia.
　　Sopubia laxior S. Moore in Journ. Bot. **49**: 186 (1911). Type from Angola.

Erect, perennial, much branched herb or undershrub, 45–100(150 or more) cm. tall, scabrid; stems usually branched above with branches ascending, densely pubescent with pubescence usually in 3 or 4 rows alternating with leaf insertions. Leaves usually in whorls of 3–4, 15–30(35) × 1–3.5 mm., linear to linear-lanceolate, undivided or rarely 3-fid, scabrid, margin revolute, strigose; midrib very prominent. Flowers pink to lilac or lavender, or occasionally purplish, in whorls of 3–4 in long terminal racemes. Pedicels 1.5–4.5 mm. long, increasing to 6.5 mm. in fruit. Bracteoles 2.5–5 mm. long, linear to linear-lanceolate, arising 0.25–0.5 mm. below base of calyx, long pubescent to hispid-strigose or subglabrous. Calyx 3.5–4 mm. long, increasing to 6.5 mm. in fruit, 10-ribbed, shortly scabrid-pubescent without; lobes c. 1.5 mm. long, deltate, densely tomentose at margin and within. Corolla 7–8 mm. long, limb 9–13.5 mm. in diam. Fertile anther thecae 2.8–3 × c. 1 mm., oblong. Capsule 4–5 mm. long, 3–3.5 mm. in diam., broadly ovoid to subglobose, emarginate, glabrous.

Zambia. N: Nkolemfumu, 48 km. S. of Kasama, on Mpika road, fl. & fr. 6.v.1958, *Lawton* 363 (K). W: Solwezi Distr., c. 20 km. NW. of Kansanshi on border, fl. 19.iii.1961, *Drummond & Rutherford-Smith* 7071 (SRGH). C: Mkushi Distr., 22 km. E. of Mkushi, Chilongoma Hills, 1400 m., fl. 25.iii.1973, *Kornaś* 3549 (K). E: Lundazi Distr., Nyika Plateau, Kangampande Mt., 2130 m., fl. & fr. 7.v.1952, *White* 2743 (BR; K). S: Mazabuka Distr., Mazabuka, fl. 8.iv.1963, *Astle* 2354 (SRGH). **Zimbabwe**. N: near Mutoroshanga (Toroshanga) Pass, Umvukwe Mts., fl. & fr. 24–27.iv.1948, *Rodin* 4411 (K;

SRGH). C: Harare Distr., Umwindisi Valley near Newmarch Farm, fl. iv.1974, *West* 7727 (K; SRGH). E: Mutasa Distr., Honde Valley, 800 m., fl. & fr. 18.iv.1958, *Phipps* 1124 (K; SRGH). S: Bikita Distr., c. 9 km. SE. of Silveira Mission, fl. & fr. 7.v.1969, *Biegel* 3050 (K; LISC; SRGH). **Malawi**. N: Mzimba Distr., South Viphya Plateau, 11 km. S. of Chikangawa on road to Luwawa, 1740 m., fl. & fr. 8.v.1970, *Brummitt* 10469 (K; SRGH). C: Dedza Distr., Mtendere, 22.4 km. E. of Dedza-Lilongwe road, c. 1350 m., fl. & fr. 25.iv.1971, *Pawek* 4672 (K). S: Zomba Distr., Zomba Plateau, 1430 m., fl. & fr. 30.v.1946, *Brass* 16082 (BM; BR; K; NY; SRGH). **Mozambique**. N: c. 29 km. N. of Mandimba, c. 950 m., fl. 26.v.1961, *Leach & Rutherford-Smith* 11014 (SRGH). Z: Quelimane Distr., Namagoa Estate, fl. & fr. viii.1944, *Faulkner* 177 (BR; K; SRGH). T: between Zóbuè and Vila Monzinho, c. 5.9 km. from Zóbuè, fl. & fr. 20.vii.1949, *Barbosa* 21062 (K). MS: Mafusi, 1065 m., fl. & fr. 22.ii.1907, *Johnson* 141 (K).

Also known from West and Central Africa, Sudan and Ethiopia. Forest borders, open woodland to grassy hillsides and plains; from almost sea-level in Mozambique to over 2200 m.

4. **Sopubia eminii** Engl., Pflanzenw. Ost-Afr. **C**: 359 (1895). —Skan in F.T.A. **4**, 2: 447 (1906). —O.J. Hansen in Kew Bull. **30**: 544 (1975). Types from Tanzania and Kenya.
   Sopubia trifida sensu Skan in F.T.A. **4**, 2: 446 (1906) non Buch.-Ham. ex D. Don (1825).

Annual, erect to slightly erect, scabrid herb, 30–60 cm. tall; stems simple to much branched especially near base, obscurely angled, branches spreading to slightly erect, scabrid, frequently with minute retrorse bristly hairs. Leaves up to 4 cm. long, opposite to almost opposite, occasionally in whorls of 3, usually trifid particularly below with narrowly linear lobes, becoming entire above, bracteate within inflorescence, often subtending small axillary leafy branches. Inflorescence loose, terminal, racemose; flowers pedicellate, yellow or white with dark purple centre, opposite. Pedicels 5–9 mm. long, slender, scabrid. Bracteoles 3–4 × 0.4–0.6 mm., narrowly linear to linear-lanceolate, scabrid, arising towards apex of pedicel. Calyx 3.5–4.5 mm. long, subglabrous to minutely scabrid without, subglabrous within, not densely woolly, lobes c. 1.25 mm. long, broadly triangular, slightly to barely tomentose at margin, sparsely to more or less densely orange to black glandular-punctate when dried. Corolla 6.5–7 mm. long, limb 8–10 mm. in diam. Fertile anther thecae c. 2 mm. long cylindric-ellipsoid. Capsule 3.5–5 mm. long, c. 3 mm. in diam., broadly cylindric-ellipsoid, glabrous or sparsely reddish glandular-punctate when dried.

Zambia. N: Mbala Distr., Saisi Valley, 1500 m., fl. & fr. 20.v.1968, *Richards* 23289 (BR; K; P). **Zimbabwe**. W: Matobo Distr., fl. & fr. v.1954, *Garley* 1028 (SRGH). E: Nyanga Distr., Nyanga, 1825 m., fl. v–vi.1922, *Barnes* in GHS 3569 (SRGH). S: Masvingo Distr., fl. & fr. 1909, *Monro* 1028 (BM; SRGH). **Malawi**. N: Rumphi Distr., Nyika Plateau, S. of Chelinda above Mwenyenyesi Dam, 2200–2300 m., fl. & fr. 10.iii.1977, *Grosvenor & Renz* 1126 (K; SRGH).

Also known from Kenya, Uganda and Tanzania. Woodlands and grassy places and hillsides; up to 2300 m.

5. **Sopubia parviflora** Engl., Bot. Jahrb. **18**: 65 (1893). —Hiern, Cat. Afr. Pl. Welw. **1**: 772 (1898). —Skan in F.T.A. **4**, 2: 452 (1906). Type from the Sudan.

Annual, erect herb (15)30–60 cm. tall; stems slender, slightly quadrangular, much branched, branches erect-spreading, up to 27 cm. long, subglabrous to sparingly bifariously pubescent. Leaves opposite, (10)20–45 × (0.2)0.5–1.5 mm., narrowly linear to filiform, entire, margins revolute or thickened-scabrid, scabrid on midrib beneath, strict or occasionally flexuous. Inflorescence (6.5)15–30 cm. long, loose, terminal, racemose; flowers numerous, pale yellow or cream to pale orange or bluish (fide *Mutimushi* 3252) with purple throat, long or short pedicellate. Pedicels 3–11 mm. long, slender, glabrous to shortly pubescent. Bracteoles 1.25–2.5 mm. long, linear, subglabrous. Calyx 2.75–4 mm. long, glabrous to very shortly pubescent without; lobes 0.75–1.75 mm. long, deltate to narrowly deltate, acute to obtuse, clearly to obscurely pilose at margin, sometimes within. Corolla c. 6 mm. long, limb (6)7–9 mm. in diam. Fertile anther thecae c. 1.5 mm. long, ellipsoid. Capsule 3 × 2.5–3.5 mm., subglobose.

Zambia. N: Mporokoso Distr., 8 km. N. of Muzombwe, fl. & fr. 16.iv.1961, *Phipps & Vesey-FitzGerald* 3228 (K; SRGH). W: Mwinilunga Distr., c. 14 km. N. of Kalene Hill on Salujinga road, c. 1250 m., fl. 21.ii.1975, *Hooper & Townsend* 273 (K). C: Mumbwa Distr., between Landless Corner and Mumbwa, fl. & fr. 19.iii.1963, *van Rensburg* 1703 (K; SRGH). **Zimbabwe**. N: Zwipani, Hurungwe (Urungwe) Native Reserve, fl. & fr. 25.iii.1958, *Goodier* 558 (K; SRGH). C: Marondera Distr., fl. & fr. 10.iv.1942, *Dehn* 27 (SRGH). E: Nyanga Distr., World's View, Troutbeck, c. 2320 m., fl. & fr. 15.i.1959, *Lennon* 76 (SRGH).

Also known from Angola, Sudan, Zaire and West Africa. In moist grassland and open woodlands, also on poorer soils over laterite and in old cultivated areas; 1050–1500 m.

This species, as noted by Hansen (Kew Bull. **30**: 545 (1975)) is very closely related to *S. eminii* Engl., appearing to differ only in having simple leaves compared with trifid or pentafid leaves of that species. For the convenience of Flora Zambesiaca I have decided to keep these two elements separate, though later study may find it advisable to treat them as subspecies.

6. **Sopubia lanata** Engl., Bot. Jahrb. **18**: 67 (1894). —Skan in F.T.A. **4**, 2: 454 (1906). —O.J. Hansen in Kew Bull. **30**: 553 (1975). TAB. **48**. Type from Angola.

Erect, robust perennial or undershrub to 1 m. or more tall, densely grey to silvery, appressed lanate almost throughout; stems many from a woody base, up to 6.5 mm. thick below, simple or branched usually above, densely leafy. Leaves subverticillate, 15–40 × 1–2.5(5) mm., linear to linear-lanceolate, spreading, strict or laterally arcuate, with tufts of smaller leaves in axils, entire, margins slightly revolute, midrib prominent beneath. Inflorescence many-flowered, 4–12(16) × 1.5–2.5 cm., spicate or racemose with flowers closely and tightly arranged so as to obscure the inflorescence, axis and base of flowers, or more loosely arranged showing main axis and base of at least lower flowers. Flowers pink to lilac or mauve-pink with darker pink or mauve throat, pedicellate or not. Pedicels 0–10 mm. long. Bracteoles 4–10 mm. long, linear or linear-lanceolate. Calyx 6–10 mm. long; lobes 3–5.5 mm. long, linear-triangular. Corolla limb 10–16 mm. in diam. Fertile anther thecae 2–3 mm. long, cylindric or ellipsoid. Capsule 4–5 × c. 4 mm., shortly cylindric to subglobose, rounded at apex, sparsely pilose.

Var. **lanata**
  Sopubia carsonii Skan in F.T.A. **4**, 2: 455 (1906). Type: Zambia, Fwambo, *Carson* 77 (K).

Flowers loosely arranged showing inflorescence axis, calyx and pedicels at least of lower flowers at flowering time.

Zambia. N: Mbala Distr., Lake Chila, 1875 m., fl. 25.iv.1986, *Philcox, Pope & Chisumpa* 10129 (K). Known also from Zaire and Angola. In moist grasslands and open woodlands on sandy soil; 1500–1900 m.

Var. **densiflora** (Skan) O.J. Hansen in Kew Bull. **30**: 554 (1975). Type from Tanzania.
  Sopubia densiflora Skan in F.T.A. **4**, 2: 454 (1906). Type as above.

Flowers closely and tightly arranged so that the inflorescence axis, calyx, base of flowers, and pedicels if present, are not visible at flowering time.

Zambia. N: Nchelenge Distr., Lake Mweru, 32 km. S. of Chiengi, c. 915 m., fl. 18.ix.1957, *Whellan* 1402 (K; SRGH). W: Mwinilunga Distr., Lisombo R., fl. 8.vi.1963, *Loveridge* 864 (BR; K; SRGH). C: Chakwenga Headwaters, 100–129 km. E. of Lusaka, fl. 27.iii.1965, *Robinson* 6477 (K). **Malawi.** N: Mzimba Distr., Viphya, c. 18 km. SW. of Mzuzu, 1524 m., fl. 10.v.1970, *Pawek* 3463 (K).
Known also from Uganda, Tanzania, Rwanda, Zaire and Angola. Marshes, swamps, riversides, wet grasslands and rainforest borders; 900–2500 m.

7. **Sopubia angolensis** Engl., Bot. Jahrb. **18**: 67 (1893). —Skan in F.T.A. **4**, 2: 453 (1906). Type from Angola.

Erect perennial herb or undershrub, 35–65 cm. tall, usually white silky-tomentose throughout; stems terete, one to several arising from a woody base, simple or branching mostly above. Leaves opposite or apparently in whorls, 15–40 × 0.5–1(1.5) mm., linear, acute, entire, midrib prominent beneath, margin obscurely revolute. Inflorescence 5–20 cm. long, terminal, loose-racemose; flowers many, pink to pinkish-mauve, pedicellate. Pedicels (3.5)5–7 mm. long. Bracteoles 2.5–3.5(5) mm. long, narrowly linear. Calyx 5–5.5 mm. long; lobes 2 mm. long, deltate, acute. Corolla 7–8 mm. long, limb 7–11 mm. in diam. Fertile anther thecae 2–2.5 mm. long, ellipsoid. Capsule 3.5–4.5 × 2–2.75 mm., oblong in outline, rounded to subtruncate or shallowly emarginate at apex, pilose to a varying degree.

Zimbabwe. N: Mazowe (Mazoe), c. 1300 m., fl. iv.1906, *Eyles* 322 (BM; SRGH). W: Matobo Distr., Besna Kobila Farm, c. 1450 m., fl. & fr. v.1957, *Miller* 4375 (K; SRGH). C: Marondera Distr., fl. & fr. 25.iii.1931, *Brain* 3216 (K). S: Masvingo Distr., Kyle Nat. Park, fl. 31.i.1975, *Lightfoot* 43 (K; SRGH).

8. **Sopubia karaguensis** Oliv. in Trans. Linn. Soc. **29**: 123, t. 87, fig. B (1875). —De Wild. & Durand, Études Fl. Katanga: 126 (1903). —S. Moore in Journ. Linn. Soc., Bot. **37**: 191 (1905). —Skan in F.T.A. **4**, 2: 448 (1906). —O.J. Hansen in Kew Bull. **30**: 552 (1975). Type from Uganda.

148

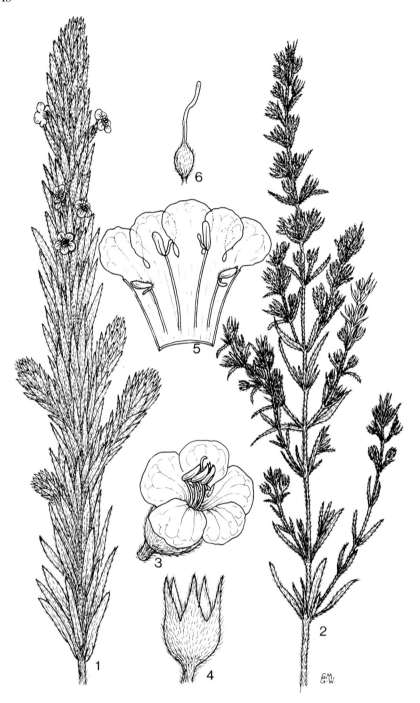

Tab. 48. SOPUBIA LANATA. 1, flowering stem (var. *densiflora*) (×⅔), from *Sanane* 476; 2, fruiting branch (var. *lanata*) (×⅔), from *Richards* 11362; 3, flower (× 3); 4, calyx (× 3); 5, corolla opened showing stamens (× 3); 6, (× 3), 3–6 from *Sanane* 476.

Sopubia welwitschii var. *micrantha* Engl., Pflanzenw. Ost-Afr. **C**: 359 (1895). Type from Tanzania.

Sopubia fastigiata Hiern in F.C. **4**, 2: 387 (1904). Type from Swaziland.

Erect, robust perennial or undershrub to 1 m. tall; stem to 6 mm. in diam. at base, branched or occasionally simple, pubescent to long-pilose or glabrescent, branches woody, ascending. Leaves opposite to subverticillate, 15–40 mm. long, undivided to trifid or pentafid often bearing smaller leaves in their axils, simple leaves or segments 1–1.5 mm. wide, margins somewhat thickened, pilose to shortly pubescent, slightly scabrid. Inflorescence 5–20 cm. long, loosely spicate or racemose with clearly defined internodes, simple to much branched; flowers many, pink, mauve to purple, sessile or pedicellate. Pedicels 0–6.5 mm. long. Bracteoles 6.5–11 mm. long, narrowly linear, acute, villous. Calyx 6–15 × 4–11 mm., densely lanate-tomentose within and without; lobes 3–13.5 mm. long, narrowly triangular. Corolla limb 15–22 mm. in diam. Fertile anther thecae 3.5–4.25 mm. long, cylindric. Capsule 5–5.5 × 3.8–5 mm., glabrous, subglobose to broadly ovoid.

Var. **karaguensis**.

Calyx 6–11 × c. 4–5 mm.; lobes 3–9.5 mm. long, calyx not apparently inflated or enlarged.

Zambia. N: 70 km. from Nakonde on Mbala road, 1420 m., fl. 23.iv.1986, *Philcox, Pope & Chisumpa* 10094 (K). S: Choma Distr., Mochipapa, near Choma, fl. & fr. 9.iii.1962, *Astle* 1439 (K; SRGH). **Zimbabwe**. W: Plumtree, c. 1390 m., fl. & fr. 8.iii.1930, *Brain* 604 (K; SRGH). **Malawi**. N: Chitipa Distr., Nyika Plateau, near base of Nganda Mt., c. 2350 m., fl. 12.iii.1977, *Grosvenor & Renz* 1155 (K; SRGH). C: Kasungu Distr., near Chasato Estate, fl. 1.iii.1979, *Salubeni, Banda & Tawakali* 2557 (SRGH). **Mozambique**. M: Libombos, near Namaacha, Mt. Mponduine (Mpondium), 800 m, fl. 22.ii.1955, *Exell, Mendonça & Wild* 503 (BM; SRGH).

Also known from Kenya, Uganda and Tanzania. Damp ground; up to 1450 m.

Var. **macrocalyx** O.J. Hansen in Kew Bull. **30**: 553 (1975). Type: Zimbabwe, Gwampa Forest Res., fl. & fr. iii.1956, *Goldsmith* 93/56 (K, holotype).

Calyx 11–15 mm. long or more, 7.5–10 mm. wide; lobes 10–13.5 mm. long, whole calyx appearing inflated and enlarged.

Zambia. C: N. of Serenje, near Livingstone Memorial, 1180 m. fl. & fr. 6.v.1972, *Kornaś* 1694 (K). S: Namwala Distr., Ngoma, Kafue Nat. Park, fl. 20.iii.1963, *Mitchell* 19/22 (P; SRGH). **Zimbabwe**. N: Hurungwe (Urungwe) Reserve, 1060–1210 m., fl. iv.1956, *Davies* 1888 (K). W: Gwampa Forest Res., fl. & fr. iii.1956, *Goldsmith* 93/56 (K). C: Harare Distr., Mukuwisi, fl. iii.1982, *Hall* 782 (SRGH). E: Mutare Distr., Odzani R. Valley, fl. & fr. 1915, *Teague* 380 (K). S: near Rundi R., fl. 30.vi.1930, *Hutchinson & Gillett* 3281 (BM; K). **Malawi**. C: Kasungu Distr., c. 6 km. N. of Kasungu, fl. 5.iii.1977, *Grosvenor & Renz* 1041 (K; SRGH). **Mozambique**. ?Jagersberg Mt., fl. 7.v.1948, *Munch* 57 (K; SRGH).

Also known from Tanzania. Grasslands, swamps and mixed woodland; 1050–1250 m.

Hansen (Kew Bull. **30**: 553 (1975)) in his treatment of the genus for East Africa, included var. *welwitschii* (Engl.) O.J. Hansen in *S. karaguensis*. This he separates from the typical variety by reason only of the calyx length which never exceeds 11 mm. in length. This variety, although recorded from Angola and Zaire has so far not been recorded from the Flora Zambesiaca area.

There is one collection, *Fanshawe* 8771 from Kasama, Zambia, in which the calyx is almost totally glabrous. I have chosen not to include this character in my description above, nor do I feel it warrants infraspecific ranking, and am therefore allying it to the typical variety. The discovery of more examples in the future may be reason for reconsideration of this point.

37. GRADERIA Benth.

Graderia Benth. in DC., Prodr. **10**: 521 (1846).

Perennial undershrub, many-stemmed from a woody rootstock. Leaves opposite or alternate, pinnatisect, dentate to entire. Flowers solitary, axillary. Calyx 5-lobed, campanulate. Corolla tubular, 5-lobed, limb spreading, lobes almost entire. Stamens 4, didynamous, included; filaments filiform, pilose at least towards base, inserted below middle of corolla tube; anthers bithecal, free, thecae divergent, oblong, curved, mucronate at base. Ovary bilocular, globular, compressed; ovules numerous. Style slender, glabrous, exserted, incurved above. Capsule compressed perpendicular to septum, loculicidal. Seeds numerous, obovoid-cylindrical.

150

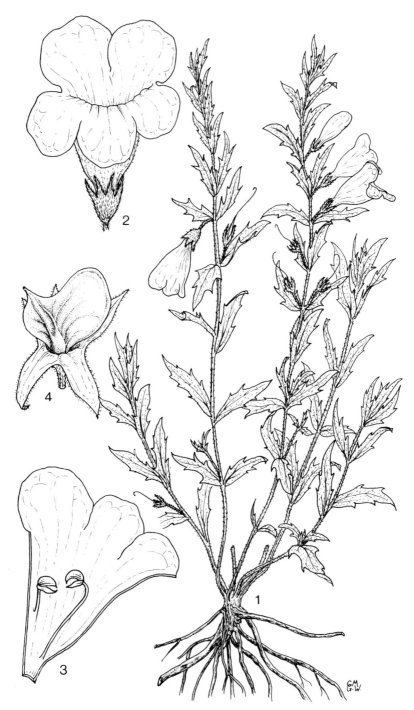

Tab. 49. GRADERIA SCABRA. 1, habit (×⅔), from *Methuen* 10; 2, flower (× 2); 3, corolla, longitudinal section showing stamens (× 2), 2–3 from *Richards* 22573; 4, fruit (× 3), from *Fries et al.* 3611.

Genus of 4 species; 3 from Southern Africa and 1 from Socotra.

Graderia scabra (L. f.) Benth. in DC., Prodr. **10**: 521 (1846). TAB. **49**. Type from S. Africa.
    Gerardia scabra L. f., Suppl.: 279 (1781). Type as above.
    Sopubia scabra (L. f.) G. Don, Gen. Syst. **4**: 560 (1837). Type as above.
    Melasma zeyheri Hook., Ic. Pl., t. 255 (1840). Type from S. Africa.
    Bopusia scabra (L. f.) Presl. in Abh. Böhm. Ges. Wiss. **3**: 521 (1845). Type as for *Graderia scabra*.

Erect, ascending or rarely procumbent perennial suffrutex, 7–30 cm. tall, from a woody rootstock, usually branched at or near the base; stems obtusely quadrangular above, indumentum of large multicellular pilose hairs, leafy, slightly herbaceous (at least when young). Leaves opposite or nearly so or alternate, 10–22(30) × 4–17 mm., ovate elliptic or more usually lanceolate, acute, subcuneate to subsessile at base, margins incised-pinnatifid with acute teeth or rarely subentire, scabrid-hispid, 3–5-nerved with nervation prominent beneath. Flowers purple mauve or pink, in axils of upper leaves, subsessile to shortly pedicellate and bibracteolate. Pedicels up to c. 2 mm. long. Bracteoles 5–8.5 × 0.6–1 mm., sublanceolate to linear, opposite, 1-nerved, inserted at base of calyx. Calyx 7–10 mm. long, campanulate, deeply 5-lobed, 10-nerved, scabrid hispid or pubescent especially on nerves; lobes 4–6 mm. long, triangular lanceolate, acute. Corolla funnel-shaped; tube 10–15 mm. long, sparsely pubescent without; limb 13–16(22) mm. in diam.; lobes 4–9 mm. long, broadly rounded, spreading. Style filiform to 14 mm. long, persistent in young fruit. Capsule 7 mm. long, 6–7 mm. wide, strongly longitudinally compressed, appearing broadly unilaterally winged, included in persistent calyx.

Zimbabwe. E: Chimanimani (Melsetter) Distr., Chimanimani Mts., Bundi Plain, fl. 26.x.1959, *Goodier & Phipps* 282 (K; SRGH). **Malawi**. N: Nyika Plateau, Kasaramba, 2250 m., fl. & fr. 17.xi.1967, *Richards* 22573 (K). C: Dedza Distr., Chongoni Forest Res., fl. 11.ix.1967, *Salubeni* 830 (K; SRGH). **Mozambique**. MS: Manica, Serra Zuira, Tsetserra, 2100 m., fl. 4.ii.1965, *Torre & Pereira* 12625 (LISC).
Also known from S. Africa, Swaziland and Tanzania. Montane grassland; from 1500–2400 m.

Two collections from Maputo in Mozambique, *Moura* 131 (LISC) and *Myre & Carvalho* 1424 (LISC) appear to be very close to this species but differ in lacking the large multicellular, pilose indumentum on the stem and with leaves lacking the very prominent venation. I would not consider them as representing a new taxon until further material is available, I have chosen only to record them here.

38. MICRARGERIA Benth.

Micrargeria Benth. in DC., Prodr. **10**: 509 (1846).

Annual, rigid, erect, branched herbs. Leaves opposite to alternate, linear, entire or trifid. Flowers small, solitary-axillary or terminally racemose, bi-bracteolate. Calyx campanulate, 5-lobed; lobes ovate to lanceolate, acute or obtuse. Corolla limb subequally 5-lobed; lobes entire; tube enlarged above, sometimes incurved. Stamens 4, somewhat didynamous, included; anthers free, bithecal; thecae arched or parallel, distinct, apically attached. Ovary 2–4-locular. Ovules numerous. Capsule globose to subglobose. Seeds numerous, obovoid.

A genus of 4 or 5 species from tropical west, east and southern Africa with 1 occurring in India.

Micrargeria filiformis (Schum. & Thonn.) Hutch. & Dalz. in F.W.T.A. **2**: 223 (1931). —Hepper in F.W.T.A. ed. 2, **2**: 366 (1963). TAB. **50**. Type from Guinea.
    Gerardia filiformis Schum. & Thonn., Beskr. Guin. Pl.: 272 (1827). Type as above.
    Sopubia filiformis (Schum. & Thonn.) G. Don, Gen. Syst. **4**: 560 (1837). —Hiern, Cat. Afr. Pl. Welw. **1**: 772 (1898) nom. illegit.
    Gerardianella scopiformis Klotzsch in Peters, Reise Mossamb., Bot.: 229, t. 36 (1861). Type: Mozambique, Querimba, *Peters* s.n. (B†).
    Sopubia scopiformis (Klotzsch) Vatke in Linnaea **43**: 313 (1882). Type as above.
    Micrargeria scopiformis (Klotzsch) Engl., Pflanzenw. Ost-Afr. **C**: 359 (1895). —Hemsl. & Skan in F.T.A. **4**, 2: 457 (1906). Type as above.

Slender herb, 35–80 cm. tall; stem much-branched above rarely simple; minutely scabrid on all vegetative parts. Leaves opposite, 0.75–3.5(5) × 0.5–1.25 mm., filiform to narrowly linear, erect to spreading, strict to somewhat flexuous. Flowers usually opposite, in open racemes, each subtended by a reduced leaf-like bract, pedicellate, bi-bracteolate. Pedicels 2–4 mm. long. Bracteoles 1.3–3.2 mm. long, narrowly linear to subulate. Calyx

Tab. 50. MICRARGERIA FILIFORMIS. 1, habit (×⅔); 2, flower (× 4); 3, corolla opened out showing androecium (× 4); 4, part of calyx and bracts removed showing gynoecium (× 4); 5, fruit (× 8); 6, fruit with part of calyx removed (× 8). All from *Philcox, Drummond & Pope* 9086.

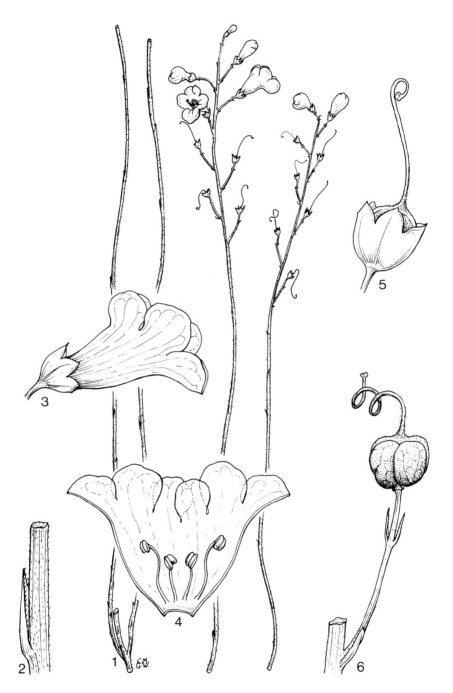

Tab. 51. MICRARGERIELLA APHYLLA. 1, flowering and fruiting branches (×⅔); 2, leaf (× 8), 1–2 from *Richards* 15448; 3, flower (× 2); 4, corolla opened out showing androecium (× 2), 3–4 from *Marks* 38; 5, fruiting calyx (× 4); 6, fruit with calyx removed (× 4), 5–6 from *Greenway & Brenan* 8230.

2.5–4.5 mm. long, lobes 0.8–1.4 mm. long, broadly deltate. Corolla 8.5–12 mm. long, campanulate, pinkish-white, lilac to reddish-purple, externally pubescent, narrowed at base into short tube, lobes rounded, finely ciliate. Anther thecae subequal. Capsule 2.5–3.5 × 2.8–3.5 mm., subglobose, slightly longitudinally compressed, glabrous, dark brown to black.

Zambia. N: Kaputa Distr., 24 km. SW. Bulaya, 1150 m., fl. & fr. 19.iv.1961, *Phipps & Vesey-FitzGerald* 3304 (K; SRGH). C: Chiwefwe, fl. & fr. 1.v.1957, *Fanshawe* 3241 (BR; K; SRGH). S: Siamsunda, 1.6 km. E. Mapanza, 1066 m., fl. & fr. 4.iv.1954, *Robinson* 654 (K; SRGH). **Zimbabwe**. N: Shamva Distr., c. 12 km. from Shamva road towards Nyagui R., 900 m., fl. & fr. 29.iii.1981, *Philcox & Drummond* 9056 (K; P; SRGH). W: Matobo Distr., 1460 m., fl. & fr. 28.iii.1963, *Miller* 8420 (K; SRGH). C: Harare Distr., Cranbourne, fl. & fr. 31.iii.1946, *Wild* 1023 (K; SRGH). E: Mutare Distr., 800 m., fl. & fr. 18.iii.1954, *Chase* 5208 (BM; K; SRGH). S: Masvingo Distr., fl. & fr. 17.iv.1946, *Greatrex* in GHS 14776 (K; SRGH). **Malawi**. N: Lower Plateau, N. Lake Malawi (Nyassa), fl. & fr. x.1880, *Thomson* s.n. (K). C: 6 km. N. Nkhota Kota, 490 m., fl. & fr. 16.vi.1970, *Brummitt* 11443 (K). S: Liwonde Nat. Park, fl. 28.xii.1985, *Dudley* 1802 (K). **Mozambique**. N: between Cuamba and Mutuali, near bridge over Rio Lurio, fl. & fr. 24.iv.1961, *Balsinhas & Marrime* 428 (BM; K; LISC). Z: Quelimane Distr., Mocuba, Namagoa, fl. & fr. ix.1946, *Faulkner* 59 (BM; COI; K; SRGH).

Also in West Africa, Sudan, East Africa and Angola. Swamps and damp areas; from 490–1800 m.

39. MICRARGERIELLA R.E. Fries

Micrargeriella R.E. Fries, Wiss. Ergebn. Schwed. Rhod.-Kongo-Exped. 1911–1912: 290 (1916).

Perennial, erect herb; stems simple or sparsely branched. Leaves alternate, scale-like, entire. Flowers medium-sized, terminally racemose, pedicellate, bi-bracteolate. Calyx campanulate, shortly 5-lobed. Corolla 5-lobed; lobes broad, entire, subequal, erect, spreading, tube enlarged above, somewhat incurved. Stamens 4, didynamous, included, glabrous; anthers free, thecae apically attached, becoming divergent at anthesis. Ovary bilocular. Ovules numerous.

A monotoypic genus presently known only from Zambia.

Micrargeriella aphylla R.E. Fries, Wiss. Ergebn. Schwed. Rhod.-Kongo-Exped. 1911–1912: 291 (1916). TAB. **51**. Type: Zambia, Kawendimusi to Lake Bangweulu, fl. 26.ix.1911, *Fries* 790 (UPS, holotype).

Slender herb up to 55 cm. tall; stems terete, sulcate, glabrous. Leaves 2–3 × c. 0.5 mm., subulate, acute, scabrid especially on margins towards the base. Inflorescence laxly many-flowered, 6–15 cm. long. Flowers bracteate; bracts 1.5–2.25 × 0.5 mm., subulate, glabrous. Pedicels (2.5)7–14(18) mm. long, glabrous. Bracteoles 0.75–1.3 mm. long, subulate. Calyx 2–3 mm. long, lobes 0.5–0.75 mm. long, subtruncate, apiculate, sparsely scabridulous. Corolla 15.5–20 mm. long, externally puberulent, white to very pale pink. Style 7–13 mm. long. Capsule c. 4 × 2.5 mm., dark brown.

Zambia. N: Kawambwa Distr., Nchelenge Road, 16 km. from Kawambwa, 1290 m., fl. & fr. 29.xi.1961, *Richards* 15448 (K).

Only known from northern Zambia. Swamps and dambos; 1290–1525 m.

40. GERARDIINA Engl.

Gerardiina Engl., Bot. Jahrb. **23**: 507, t.10 G–M (1897). —Hiern, Cat. Afr. Pl. Welw. **1**: 770 (1898); in F.C. **4**, 2: 378 (1904).

Annual or perennial herbs, erect. Stems simple. Leaves opposite, entire. Inflorescence racemose. Flowers opposite, bracteate, pedicellate. Calyx 5-lobed, campanulate, lobes subequal, shorter than tube. Corolla campanulate, oblique, narrow at base, dilated above, 5-lobed; lobes subequal, rounded. Stamens 4, didynamous with anterior pair about 2–2.5 times longer than posterior pair; anterior filaments long pilose above middle; anthers

Tab. 52. GERARDIINA ANGOLENSIS. 1, flowering and fruiting stem (× ⅔), from *Milne-Redhead* 3181; 2, flower (× 1); 3, flower, longitudinal section showing stamens (× 2); 4, gynoecium (× 2); 5, fruit (× 2), 2–5 from *Richards* 1148.

bithecal, thecae divergent, subequal. Ovary bilocular, ovoid, glabrous. Ovules numerous. Capsule ovoid, equalling to somewhat longer than calyx. Seeds numerous, narrowly cylindric usually tapering.

Monotypic genus from tropical Africa.

Gerardiina angolensis Engl., Bot. Jahrb. **23**: 507, t.10 G–M (1887). —Hiern, Cat. Afr. Pl. Welw. **1**: 770 (1898); in F.C. **4**, 2: 378 (1904). TAB. **52**. Type from Angola.

Herb 35–80 cm. tall. Stems subquadrangular, minutely hispid to subglabrous, sulcate. Leaves erect or appressed to stem (2.5)6–11 × 0.25–1.9 cm., linear-lanceolate, obtuse or rarely subacute at apex, sessile decurrent at base, 3–5-nerved, scabrid above, subglabrous beneath except scabrid on prominent nerves, much exceeding internodes. Racemes (5)10–20 cm. long; flowers numerous. Bracts 5–8.5 mm. long, cordate, broadly ovate, somewhat connate, glabrous. Pedicels 4–14 mm. long, glabrous. Calyx 5–8 mm. long, obscurely nerved, glandular-punctate; lobes 1.5–3 mm. long, broadly ovate-deltate, obtuse, glabrous. Corolla pink to purple or blue, or occasionally white, 18–30 mm. long, externally glabrous, pubescent within especially at base; lobes subequal, occasionally upper 2 clearly smaller. Style 16–18 mm. long, arcuate above, equalling or slightly exceeding longer stamens, base persistent in fruit. Capsule 7–8 × 5–6.5 mm.

Zambia. N: Mbala Distr., Lake Chila, c. 1520 m., fl. & fr. 7.iii.1952, *Richards* 1148 (K). W: Mwinilunga Distr., NE. of Dobeka Bridge, fl. 10.xi.1937, *Milne-Redhead* 3181 (BR; K). C: Serenje Distr., Kundalila Falls near Serenje, 1500 m., fl. 4.v.1972, *Kornaś* 1652 (K). **Zimbabwe**. E: Chimanimani Mts., c. 1700 m., fl. 20.iii.1981, *Philcox & Leppard* 9018 (K). **Malawi**. N: Nyika Plateau, Lake Kaulime, 2250 m., fl. & fr. 4.i.1959, *Richards* 10447 (BR; K). S: Limbe, c. 1200 m., fl. 14.iii.1948, *Goodwin* 97 (BM). **Mozambique**. Z: Gúrùe, Serra do Gúrùe, Chá Mozambique, c. 1700 m., fl. & fr. 4.i.1968, *Torre & Correia* 16876 (LISC). MS: Chimanimani Mts., fl. & fr. 4.vi.1948, *Munch* 94 (K; SRGH).

Also in Zaire, Burundi, Tanzania, Angola and S. Africa. Riversides, wet grasslands and open forests; 1200–2350 m.

41. HEDBERGIA Molau

Hedbergia Molau in Nord. Journ. Bot. **8**: 194 (1988).

Perennial herbs, sometimes shrubby, erect to scrambling, branched below, becoming tufted, pubescent. Leaves opposite, sessile, crenate or serrate, decreasing in size above. Flowers solitary, axillary, shortly pedicellate, ebracteolate. Calyx 4-lobed; lobes straight. Corolla subrotate, 5-lobed; limb weakly bilabiate but not separated into galea and lip. Stamens 4, didynamous; anther thecae 2, equal, parallel, mucronate. Style simple, filiform; stigma entire. Capsule ovoid. Seeds numerous, white, longitudinally winged.

Molau described this genus as distinct from *Bartsia* L. by reason of its 5-lobed, subrotate not markedly bilabiate and galeate corolla.
A monotypic genus of tropical Africa.

Hedbergia abyssinica (Hochst. ex Benth.) Molau in Nord. Journ. Bot. **8**: 195 (1988). TAB. **53**. Type from Ethiopia.

　　Bartsia abyssinica Hochst. ex Benth. in DC., Prodr. **10**: 545 (1846). —Engl., Hochgebirgsfl.: 384 (1892). —Hemsl., in F.T.A. **4**, 2: 460 (1906). —Hedberg in Bot. Not. **133**: 207 (1980). Type as above.

　　Alectra abyssinica (Benth.) A. Rich., Tent. Fl. Abyss. **2**: 118 (1851). Type as above.

　　Alectra petitiana A. Rich., Tent. Fl. Abyss. **2**: 118 (1851). Type from Ethiopia.

　　Bartsia mannii Hemsl. in F.T.A. **4**, 2: 459 (1906). Type from Cameroon.

　　Bartsia petitiana (A. Rich.) Hemsl. in F.T.A. **4**, 2: 460 (1906). Type as for *Alectra petitiana*.

　　Bartsia elgonensis R.E. Fries in Acta Hort. Berg. **8**: 67 (1924). Type from Kenya.

　　Bartsia nyikensis R.E. Fries in tom. cit. **8**: 66 (1924). Type: Malawi, Nyika Plateau, c. 2250 m., ix.1902, *McClounie* 60 (K, holotype).

　　Bartsia abyssinica var. *nyikensis* (R.E. Fries) O. Hedb. et al. in Bot. Not. **133**: 211 (1980). Type as above.

　　Bartsia abyssinica var. *petitiana* (A. Rich.) O. Hedb. et al. in Bot. Not. **133**: 211 (1980). Type as for *Alectra petitiana*.

Tab. 53. HEDBERGIA ABYSSINICA. 1, flowering and fruiting branch (× ⅔); 2, flower (× 3); 3, corolla opened out showing androecium (× 3); 4, gynoecium (× 3); 5, dehiscing fruit (× 3); 6, seeds (× 12), all from *White* 2748.

Perennial suffrutescent herb, erect, ascending or scrambling to 3 m. tall. Stems several arising from a woody rootstock, much branched especially within inflorescence, rarely almost simple, terete, usually more or less densely pubescent with straight or hooked, patent or retrorse, glandular hairs. Leaves sessile or subsessile, thick, rigid, spreading, 3–12(30) × 1–2(4) mm., elliptic to lanceolate, apex acute or obtuse, cuneate or rounded at base, minutely short-hispid to subglabrous, margin shallowly or deeply serrate or crenate, ± reflexed. Inflorescence a usually much branched many-flowered raceme. Flowers white to pale pink or at times purple, opposite to subopposite, pedicels 2–5 mm. long. Calyx 5–7.5 mm. long, shortly hispid; lobes equal to or slightly shorter than the tube, slightly bilabiate with lobes at times irregularly crenate-dentate. Corolla white to pink or magenta, obliquely campanulate; tube curved, shorter than limb; limb almost equally 5-lobed. Style 5.5–7 mm. long, minutely pubescent or glabrous. Capsule (5)6.5–13 × (4.5)5–6.5 mm., ovoid to subglobose, short to long pilose. Seeds slightly curved, longitudinally ridged with 6–14 ridges.

Zambia. E: Chama Distr., Nyika Plateau, near source of Chire R., c. 2135 m., fl. 7.v.1952, *White* 2748 (K). **Malawi**. N: Rumphi Distr., Nyika Plateau, below Sangule Kopje, 7 km. SW. of Chelinda Camp, 2255 m., fl. 15.v.1970, *Brummitt* 10759 (K; SRGH).
Also known from Nigeria, Cameroons, Ethiopia, Sudan, Zaire, Uganda, Kenya and Tanzania. Open grasslands or forest borders; from 2150–2440 m.
Hedberg and others (Bot. Not. **133**: 205 (1980)) in their treatment of the *Bartsia abyssinica* complex recognise three varieties, including the typical variety, under that name incorporating the five major elements involved, and base their varietal differentiation on four main characters. These are the branching of the inflorescence, pubescence of the style, the size and shape of the capsule and the number of ridges on the seeds. Having seen most of the specimens cited in their work over the whole range of occurrence of the species, I prefer to treat the Flora Zambesiaca material in the broad sense and not distinguish varieties. This is further accepted by Molau (1988).

42. CISTANCHE Hoffm. & Link

Cistanche Hoffm. & Link, Fl. Port. **1**: 318 (1809).

Root parasites totally lacking chlorophyll, fleshy. Stems succulent, simple, often thickened at base. Leaves reduced to fleshy scales. Inflorescence congested, spicate. Flowers simple or subsessile, bracteate, bibracteolate, rather large. Calyx gamosepalous, 4- or 5-lobed, persistent; lobes obtuse, rounded, equal or two posterior narrower. Corolla tubular below, funnel-shaped above, more or less curved or bent; limb 5-lobed, spreading, lobes broad, equal or subequal. Stamens 4, didynamous, more or less exserted, inserted deep in corolla tube; anthers usually bearded. Stigma globular; style apically decurved. Ovary unilocular, ovules many. Capsule bi-valved. Seeds numerous, minute.

A genus of 16 species from southern Mediterranean Europe and north Africa; Ethiopia to India and northwest China.

Cistanche tubulosa (Schenk) Hook. f., Fl. Brit. Ind. **4**, 2: 324 (1884) quoad nom. et syn. *Phelipaea tubulosa* Schenk et excl. descr. TAB. **54**. Type from Egypt.
    Phelipaea tubulosa Schenk, Pl. Sp. Aegypt. Arab. Syr.: 23 (1840). —Boiss., Fl. Or. **4**: 500 (1879). Type as above.

Stem erect, 10–20 cm. tall (in Flora Zambesiaca material), possibly taller, fleshy, swollen at base, yellow-brown to mauve-purple, glabrous. Scales up to 15 × 8 mm., broadly ovate-lanceolate, obtuse, scattered above, crowded below, glabrous. Inflorescence 7–10 × 7–8 cm. Bracts purplish, scale-like. Bracteoles up to c. 14 × 5 mm., subobovate to subspathulate. Calyx 1.6–1.8 mm. long; lobes rounded, c. 1 × 2.5 mm., slightly overlapping at base. Corolla cream-coloured, 4–5 cm. long, erect at first, becoming outwardly curved; lobes entire, broadly rounded. Stamens slightly exserted; anthers glabrous (in Flora Zambesiaca material), apically rounded to shortly mucronate. Style glabrous, slightly exserted. Capsule not seen.

Mozambique. N: Goa Isl., Mossuril Bay, c. 4.6 m., fl. 10.viii. 1964, *Leach* 12348 (K; SRGH).
Known only from the above collection in the Flora Zambesiaca area; also known from Morocco to Egypt, to India; Ethiopia to Kenya and Tanzania; Senegal and Socotra. Dry stony areas and often in saline areas in coastal vegetation (host not designated); up to 1500 m.

Tab. 54. CISTANCHE TUBULOSA. 1, habit (×⅔); 2, bract (× 2); 3, flower (× 1); 4, corolla opened out showing androecium (× 1); 5, gynoecium (× 1), all from *Leach* 12348.

160

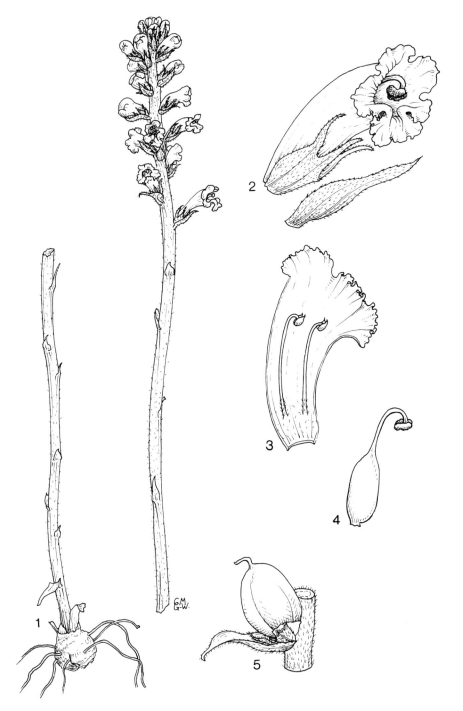

Tab. 55. OROBANCHE MINOR. 1, habit (× ⅔); 2, flower and bract (× 3); 3, corolla, longitudinal section showing androecium (× 3); 4, gynoecium (× 3), 1–4, from *Brummitt & Little* 9814; 5, fruit with calyx removed (× 3), from *Brass* 16410.

43. OROBANCHE L.

Orobanche L., Sp. Pl.: 632 (1753); Gen. Pl., ed. 5: 281 (1754). —Benth. & Hook.f., Gen. Pl. **2**,
    2: 984 (1876).

Parasitic perennial, biennial or annual herbs, lacking chlorophyll, usually covered with
gland-tipped hairs; stems erect, simple or branched, stout or slender, slightly succulent or
not, often thickened or more or less woody at the base. Leaves occurring as large scales.
Inflorescence spicate or racemose. Flowers sessile or shortly pedicellate, congested to
laxly arranged, bracteate; bracteoles 2 or absent. Calyx campanulate, 3–5-lobed, or
divided to base into two bifid or bilobed segments. Corolla tubular, usually curved, throat
somewhat widened, strongly bilabiate; upper lip entire, emarginate or bifid or bilobed,
with equal lobes; lower lip 3-lobed, lobes equal or unequal with prominent folds between.
Stamens 4, didynamous, included or nearly so, inserted below middle of tube; filaments
usually thickened towards base, glabrous to glandular-hairy; anthers often cohering,
thecae parallel or somewhat divergent, mucronate. Capsule bivalved, ovoid-globose or
ovoid-ellipsoid, loculicidally dehiscent. Seeds numerous, minute, subglobose.

A genus of about 150 species mostly from temperate or warm areas worldwide, with a few species
occurring in the subtropics and tropics.

Orobanche minor Smith in Sowerby & Smith, English Bot. **6**: t. 422 (1797). —Stapf in F.T.A. **4**, 2: 467
    (1906). —Graham in F.T.E.A., Orobanchaceae: 5, fig. 1, 3, 3a & b (1957). TAB. **55**. Type from
    England.

Erect herb (12)25–40(55) cm. tall; stem simple, yellowish, whole plant more or less
glandular-pilose. Scales 5–17 × 2–4.5 mm., ovate to ovate-lanceolate, acute or obtuse,
entire. Inflorescence terminal, rounded-compact, laxly elongated at maturity; flowers
sessile. Bracts similar to scales but more acuminate. Bracteoles absent. Calyx 5.5–11 mm.
long, subequally divided to base dorsally and ventrally in 2 segments, each ovate to
ovate-lanceolate with 2 long acuminate, entire teeth. Corolla 10–18 mm. long, varied in
colour from dull yellowish to violet-blue, with darker purplish-blue venation; upper lip
shallowly bilobed or emarginate, lobes rounded; lower lip subequally 3-lobed, lobes
rounded, central lobe smaller. Stamens inserted 2–3 mm. from base of tube; filaments
more or less hairy, at least below. Stigma bilobed, lobes 0.6 × 1 mm. Capsule 4.5–7 × 2.5–4
mm., ovoid-ellipsoid, valves remaining apically attached when mature.

Zambia. N: Mbala Distr., Chilongowelo, 1440 m., fl. 9.ii.1957, *Richards* 8133 (K). W: Mwinilunga
Distr., E. of R. Matonchi, fl. 29.i.1938, *Milne-Redhead* 4397 (K). C: Serenje Distr., Kundalila Falls, S. of
Kanona, 1300 m., fl. & fr. 13.iii.1975, *Hooper & Townsend* 721 (K). **Zimbabwe**. N: Darwin Distr., near
Musengezi Camp, 1200 m., fl. 11.v.1956, *Whellan* 900 (K; SRGH). C: Harare Distr., Cleveland Dam, fl.
& fr. 26.iii.1950, *Chase* 2115 (BM; K; SRGH). E: Mutare Distr., road to Mandambiri Mt., c. 1000 m., fl.
& fr. 2.iii.1958, *Chase* 6846 (LISC). **Malawi**. N: Mzimba Distr., Champira Forest, c. 1450 m., fl. & fr.
20.iv.1974, *Pawek* 8418B (K). C: Lilongwe, 1100 m., fl. 29.iii.1970, *Brummitt & Little* 9514
(K). S: Mulanje Distr., Mulanje Mt., W. slope, 1000 m., fl. & fr. 24.vi.1946, *Brass* 16410 (K).
Mozambique. N: N. of Serra de Ribáuè (Mepalué), c. 25 km. from Ribáuè, c. 600 m., fl. 27.i.1964,
Torre & Paiva 10261 (LISC). T: Moatize, E. of Mt. Zóbuè, c. 1000 m., fl. 11.iii.1964, *Torre & Paiva*
11141 (LISC).
Known throughout most of the northern hemisphere. A weed, usually of cultivated ground where
it parasitises a large number of host genera and species; also in grasslands, woodlands and forest
margins; up to c. 3000 m.

44. HEBENSTRETIA L.

Hebenstretia L., Sp. Pl.: 629 (1753); Gen. Pl., ed. 5 277 (1754).

Annual or perennial herbs or undershrubs. Leaves alternate, or the lower opposite, or
appearing fasciculate with smaller leaves clustered in axils, narrow, entire or toothed.
Flowers sessile in short or elongate spikes, spikes usually dense, frequently paniculate,
bracteate. Bracts usually imbricate, exceeding calyx, lowest often leaf-like. Calyx
spathaceous, entire or emarginate, membranous or hyaline, frequently binerved. Corolla
tube slender, divided almost to the base forming a large, flattened or concave 4(5)-lobed
limb. Stamens 4, didynamous, inserted on limb below lobes, filaments short, anthers

oblong or linear. Ovary bilocular; style entire. Capsule cylindrical-ovoid to broadly ovoid, subterete or compressed; at times separating into 2 cocci at maturity.

A genus of 24 species from tropical and southern Africa.

1. Plant annual - - - - - - - - - - - - - - - - - 2
– Plant perennial - - - - - - - - - - - - - - - - - 3
2. Leaves (15)35–55 × 0.5–1.25 mm.; corolla 10–12 mm. long; fruit 3–5 mm. long   1. *integrifolia*
– Leaves (10)20–35 × 0.2–0.5 mm.; corolla 7–8(9) mm. long; fruit 3.5–4 mm. long   2. *holubii*
3. Plant relatively small, mostly 20–30 cm. tall; stems simple, or branched from the base - - - - - - - - - - - - - - - - - 3. *comosa*
– Plant larger, usually 60–160(200) cm. tall; stems simple or branched throughout   4
4. Stems strongly branched; inflorescence open, of distinct spikes; spikes laxly paniculate; leaves entire or rarely obscurely toothed - - - - - - - - - 4. *angolensis*
– Stems simple; inflorescence of shorter spikes, in closely crowded panicles; leaves always shallowly or coarsely toothed - - - - - - - - - 5. *oatesii*

1. **Hebenstretia integrifolia** L., Sp. Pl.: 629 (1753); Sp. Pl., ed. 2: 878 (1763), as 'Hebenstreitia'. —Roessler in Mitt. Bot. Staatss. Münch. **15**: 51 (1979). Type from Ethiopia.
    Hebenstretia scabra Thunb., Prodr. Pl. Cap.: 103 (1800). Type from S. Africa.
    Hebenstretia aurea Andrews, Bot. Reposit., t. 252 (1802). Type from S. Africa.
    Hebenstretia tenuifolia Schrader & Reichenb., Hort. Bot. **2**: 13, t. 133 (1828). Type from S. Africa.
    Hebenstretia virgata E. Mey., Comment. Pl. Afr. Austr.: 249 (1837). Type origin uncertain.
    Hebenstretia watsonii Rolfe in F.C. **5**, 1: 103 (1901). Type from S. Africa.

Annual herb, up to c. 60 cm. tall. Stems erect, somewhat woody, especially at base, branched; branches minutely pubescent to subglabrous, many-leaved. Leaves 10–30(50) × 0.5–1.25 mm., narrowly linear, shortly hispid-bristly to subglabrous, usually entire but occasionally with one to few small indistinct teeth, frequently with smaller leaves in axils. Inflorescence spicate, up to 20 cm. long, longer in fruit, dense to somewhat more lax. Bracts 3–4.25 mm. long, up to 2 mm. wide, ovate, acuminate, glabrous. Calyx c. 2 × 1 mm., shorter than bracts, ovate-oblong with 2 slender, parallel nerves. Corolla 10–12 mm. long, white or pale cream with orange throat; tube slender, glabrous; lobes about 1 mm. long, oblong to rounded. Capsule 3–5 mm. long, narrowly oblong in outline.

Botswana. SW: W. Kalahari, Hanahai Hills, fl. & fr. 16.iii.1976, *Vahrmeijer* 3121 (K). SE: Kgatleng Distr., Masama Ranch, fl. & fr. 10.xi.1978, *Hansen* 3544 (K; SRGH). **Zimbabwe**. W: Bulawayo Distr., fl. & fr. 12.xi.1945, *Wild* 340 (SRGH). S: Masvingo Distr., 32 km. N. of Masvingo (Fort Victoria), fl. 4.v.1962, *Drummond* 7961 (LISC).
Also known from Namibia and S. Africa. Woodlands and well-drained sandy savannas.
This is a very variable species, as is apparent from Rolfe's (1901) more complete synonymy. For the purpose of Flora Zambesiaca I have omitted to cite this in full, but accept and follow his concept in this instance.

2. **Hebenstretia holubii** Rolfe in F.T.A. **5**, 1: 266 (1901). —Roessler in Mitt. Bot. Staatss. Münch. **15**: 56 (1979). Type: Zambia, Barotseland, Sesheke, xii.1875, *Holub* 366 (K, lectotype chosen by Roessler); *Holub* 388 (K, lectoparatype).

Annual herb, 20–90 cm. tall. Stems erect, somewhat woody especially at the base, branched; branches shortly pubescent to subglabrous, many-leaved. Leaves (10)20–35(40) × 0.2–0.5 mm., narrowly linear, subglabrous, entire, occasionally with smaller leaves in axils. Inflorescence spicate up to 28 cm. long in fruit, flowers compact on spike, imbricate, rarely laxly distributed on spike. Bracts (2.75)3–4(4.5) × (1.25)1.5–2 mm., ovate to ovate-lanceolate, acuminate, glabrous, margins hyaline. Calyx 2.5–3 mm. long, shorter than bracts, ovate-oblong, emarginate, obtuse with 2 slender, parallel nerves. Corolla (5)7–8 mm. long, white with orange throat; tube slender, glabrous; lobes about 1 mm. long, oblong to rounded. Capsule 3–4 mm. long, c. 1.25 mm. broadly cylindric-ovoid.

Caprivi Strip. Lisikili 24 km. E. of Katima Mulilo near lake, 975 m., fl. 17.vii.1952, *Codd* 7098 (BM; K; SRGH). **Botswana**. N: Chobe Nat. Park, Kasane, Chobe R., fl. & fr. 28.viii.1970, *Mavi* 1120 (K; LISC; SRGH). SW: Ghanzi and Kgalagadi distr., 4 km. N. Dondong borehole fl. & fr. 12.ii. 1977, *Skarpe* S-142 (K). **Zambia**. B: Kanda Lake, c. 11.5 km. NE. of Mongu, fl. & fr. 11.xi.1959, *Drummond & Cookson* 6351 (E; K; LISC). S: Livingstone Distr., Kazungula, 975 m., fl. & fr. 5.i.1957, *Gilges* 711 (K; SRGH). **Zimbabwe**. W: Islands at Victoria Falls, fl. & fr. ix.1905, *Gibbs* 116 (BM; K).
Known only from the Flora Zambesiaca area. Lake- and river-sides; up to 1000 m.
This species appears to be very close to *H. integrifolia* L. differing only in the size of the leaves and

corolla. In 1957, Hedberg (Afroalpine Vascular Plants, p. 319) measured much material under the name *H. holubii* along with three other species viz. *H. angolensis* Rolfe, *H. bequaertii* De Wild and *H. dentata* L. His findings were such that no clear and diagnostic character became evident to separate them, and he placed all under the earliest valid name, *H. dentata* L. He had not however considered *H. integrifolia* and it is here that stronger differences occur in both leaf and flower size. Maybe eventually a monographer will find that all five species are conspecific; however, for the purpose of Flora Zambesiaca and considering also their geographical distribution I follow Roessler and keep *H. holubii* and *H. integrifolia* as distinct species.

3. **Hebenstretia comosa** Hochst. in Flora **28**: 70 (1845), as '*Hebenstreitia*'. —Rolfe in F.C. **5**, 1: 99 (1901). —Roessler in Mitt. Bot. Staatss. Münch. **15**: 72 (1979). Type from S. Africa.
> *Hebenstretia elongata* Bolus ex Rolfe in F.C. **5**, 1: 99 (1901). Type from S. Africa.
> *Hebenstretia comosa* var. *integrifolia* Rolfe in F.C. **5**, 1: 99 (1901). Type from S. Africa.

Perennial herb, 15–60(80) cm. tall, woody especially at the base; stems erect, simple or branching from the woody rootstock, or prostrate with simple erect branches; branches longitudinally striate, shortly pubescent all over or sometimes only within furrows or subglabrous. Leaves 15–40 × (0.5)1–3.5(5) mm., narrowly oblong, elliptic-lanceolate to linear-lanceolate, acute to subacute, glabrous, entire to shallowly or coarsely dentate, with few teeth in upper third. Inflorescence spicate, with spikes 35–50(90) mm. long, dense. Bracts 5–8 mm. long, ovate-lanceolate, acuminate, glabrous. Calyx 4–4.5 mm. long, ovate-oblong, obtuse, binerved, somewhat hyaline at margin. Corolla 10–12 mm. long, white with a pale or deep orange throat; tube 3.5–4 mm. long, slender, glabrous; lobes 1–1.5 mm. long, subequal. Capsule 4–6 × 1 mm., oblong in outline.

Zambia. E: Nyika Plateau, near Chelinda, 2135 m., fl. & fr. 27.xi.1955, *Lees* 118 (K). **Zimbabwe**. E: Chimanimani (Melsetter) Distr., Chimanimani Mts., Bundi Plain, c. 1600 m., fl. & fr. 26.x.1959, *Goodier* 278 (K; SRGH). **Malawi**. N: Nyika Plateau, Mupopo Area, Nyika Game Park, 2250 m., fl. 18.xi.1967, *Richards* 22607 (K; P). **Mozambique**. MS: Manica, Rotanda, 1750 m., fl. & fr. 18.xi.1965, *Torre & Correia* 13096 (LISC). T: Zóbuè, fl. & fr. 3.x.1946, *Mendonça* 594 (LISC).

Also known from Lesotho and S. Africa; montane grassland; 1350–2450 m.

4. **Hebenstretia angolensis** Rolfe in Journ. Bot. **24**: 174 (1886), as '*Hebenstreitia*'. —Roessler in Mitt. Bot. Staatss. Münch. **15**: 77 (1979). TAB. **56**. Type from Angola.
> *Hebenstretia rariflora* A. Terrac. in Bull. Soc. Bot. Ital. **1892**: 424 (1892). Type from Somalia.
> *Hebenstretia bequaertii* De Wild. in Rev. Zool. Afr. **8**, Suppl. Bot.: 41 (1920). Type from Zaire.
> *Hebenstretia dentata* auctt. plur.

Perennial herb, 30–60(150) cm. tall. Stems erect, much branched; branches decurrent, densely leafy, minutely patent- or slightly retrorse-pubescent to subglabrous. Leaves 10–45(70) × (0.5)1–3(5) mm., linear or less frequently linear-lanceolate, entire or occasionally faintly toothed, rarely distinctly so throughout length, glabrous, subfasciculate with smaller leaves in axils. Inflorescence spicate with spikes (1.5)3–10(16) cm. long, not densely compacted-paniculate. Bracts 2.5–4.5 mm. long, ovate, acuminate or apiculate, glabrous. Calyx 3–5 mm. long, ovate, subobtuse, glabrous. Corolla 8–12(15) mm. long, white with an orange to reddish-brown throat, tube 2.5–4 mm. long, limb to 8 mm. long, slender, glabrous; lobes 1–1.6 mm. long, oblong. Capsule c. 4 mm. long, oblong in outline.

Zambia. N: Saisi Valley, Kalambo Farm, fl. & fr. 21.v.1952, *Richards* 1798 (K). E: Nyika Plateau, Chama Distr., nr. source of Chire R., fl. 3.v.1952, *White* 2564 (K). **Zimbabwe**. C: Marondera Distr., fl. & fr. iv.1955, *Corby* 802 (K; SRGH). E: Nyanga Distr., c. 1.5 km. S. of Nyanga (Inyanga) Village, fl. & fr. 19.iv.1966, *Biegel* 1135 (K; SRGH). S: Masvingo Distr., c. 32 km. N. of Masvingo (Victoria), fl. & fr. 4.v.1962, *Drummond* 7961 (K; LISC; SRGH). **Malawi**. N: Rumphi Distr., Nyika Plateau, fl. 23.v.1967, *Salubeni* 720 (K; LISC; SRGH). C: Nkhota Kota Distr., Ntchisi (Nchisi) Mt., 1600 m., fl. 26.vii.1946, *Brass* 16963 (K; NY; SRGH). S: Mulanje Mt., below Thuchila (Tuchila) Hut, 1940 m., fl. 5.iv.1970, *Brummitt* 9626 (K; SRGH). **Mozambique**. T: between Vila Mouzinho and Zóbuè, 59.2 km. from Vila Mouzinho, fl. 19.vii.1949, *Barbosa & Carvalho* 3685 (SRGH).

Also known from Ethiopia, Eritrea, Sudan and Somalia, East Africa, S. Africa, Lesotho, Zaire and Angola. High rainfall and submontane grasslands; up to 3000 m.

5. **Hebenstretia oatesii** Rolfe in Oates, Matabele Land and Victoria Falls, ed. 2: 406, t. 12 (1889), as '*Hebenstreitia*'. —Rolfe in F.C. **5**, 1: 98 (1901). —Roessler in Mitt. Bot. Staatss. Münch. **15**: 83 (1979). Type from S. Africa.
> *Hebenstretia polystachya* Harvey ex Rolfe in F.C. **5**, 1: 98 (1901). Type from S. Africa.

Perennial herb, 30–120 cm. or more tall. Stems erect, simple but branching profusely

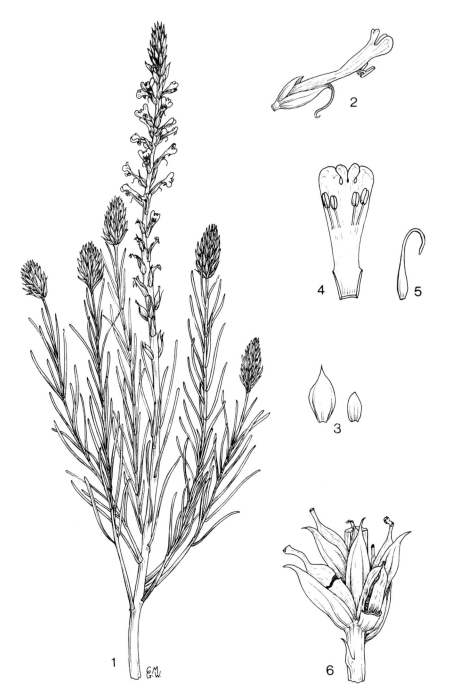

Tab. 56. HEBENSTRETIA ANGOLENSIS. 1, flowering and fruiting branch (× ⅔), from *Adamson* 374; 2, flower (× 3); 3, sepals (× 3); 4, corolla opened out showing androecium (× 3); 5, gynoecium (× 3), 2–5 from *Salubeni* 720; 6, ripe capsules (× 4), from *Brass* 16788.

within inflorescence, rarely so below, patent- or somewhat retrorse-pubescent or crisped pubescent, decurrent from leaf bases. Leaves 10–35(40) × 1.5–3.5(5) mm., narrowly lanceolate or linear, margins always finely or coarsely serrate throughout or at least in upper third, glabrous, fasciculate, frequently with smaller leaves in axils. Inflorescence compactly-paniculate, composed of many spikes 2.5–5.5(8) cm. long, with flowers crowded in spikes. Bracts 3.5–5.6 mm. long, ovate, acuminate, glabrous. Calyx 2–2.5(2.8) mm. long, ovate, obtuse, glabrous. Corolla 8–10.5 mm. long, white with yellow-orange throat, tube 2–2.5 mm. long; limb 5.5–8 mm. long, glabrous, lobes c. 1–1.25(2) mm. long, oblong. Capsule c. 4 mm. long, oblong in outline.

Subsp. **oatesii**

Leaves 40–75 × 4–8 mm., lanceolate or narrowly lanceolate, sharply toothed almost to the base. Bracts 5–7 mm. long.

Not occurring in Flora Zambesiaca area; known only from Swaziland and S. Africa.

Subsp. **rhodesiana** Roessler in Mitt. Bot. Staatss. Münch. **15**: 85 (1979). Type: Zimbabwe, Nyanga Distr., Mt. Inyangani, c. 2000 m., 14.ii.1931, *Norlindh & Weimarck* 5049 (PRE, holotype, n.v.).

Leaves 30–40 × 1–3(3.5) mm., very narrowly lanceolate to almost linear; margins shallowly, occasionally coarsely, toothed in upper third only. Bracts 3–4 mm. long. Corolla 8–10 mm. long.

Zimbabwe. E: Nyanga Distr., Nyanga (Inyanga) Downs, Upper Pungwe R., 1980 m., fl. 9.vi.1957, *Goodier & Phipps* 64 (K; LISC; SRGH).
Also known from S. Africa (Transvaal). Moist montane grasslands and forest margins; 1600–2000 m.

Subsp. **inyangana** Roessler in Mitt. Bot. Staatss. Münch. **15**: 87 (1979). Type: Zimbabwe, Nyanga Distr., Mt. Inyangani, 2130–2430 m., 16.ii.1964, *Plowes* 2430 (SRGH, holotype; K, isotype).

Leaves 20–35 × 1.5–3 mm., narrowly lanceolate to linear-lanceolate, margins toothed in upper third only. Bracts 4–5 mm. long. Corolla 11–15 mm. long.

Zimbabwe. E: Nyanga Distr., Mt. Inyangani summit ridge, 2560 m., fl. iv.1935, *Gilliland* 1902 (BM; K).
Only known from the eastern region of Zimbabwe. High montane grasslands; 2130–2600 m.
Roessler has tried to clarify this very difficult species and has partially succeeded by differentiating the three subspecies by, what appears to be a series of weak characters. Their weakness is more apparent when one considers the collections from S. Africa and Swaziland intermediate between our two subspecies. However, I feel that in the Flora Zambesiaca the material divides easily enough and have upheld Roessler's subspecies, though with more concise study, future workers may possibly unite them with the typical subspecies.

45. WALAFRIDA E. Mey.

Walafrida E. Mey., Comment. Pl. Afr. Austr.: 272 (1837).

Small, much-branched shrubs, undershrubs, woody perennials, or annual herbs. Leaves small to minute, narrow, alternate or axillary fasciculate, entire. Flowers sessile to shortly pedicellate, bracteate, in short terminal globose or elongate spikes, or racemes arranged in corymbs or panicles at ends of branches. Calyx 3-lobed with dorsal lobe smaller, sometimes minute or absent making calyx appear bilobed. Corolla tubular, 5-lobed; tube short or elongate, narrowed at base, wider at throat; lobes unequal to subequal, with the central anterior lobe considerably longer than the others. Stamens 4, didynamous, inserted in throat, more or less exserted; filaments filiform; anthers monothecal. Ovary bilocular; style slender, obtuse, subclavate to bidentate at apex. Capsule ovoid, globose or cylindric, often separating into two distinct cocci at maturity, included in calyx.

A genus of about 40 species from tropical and southern Africa.

1. Flowers in subglobose spikes or racemes arranged in lax compound panicles along the stems
   and branches   -   -   -   -   -   -   -   -   -   -   -   -   -   - 4. *paniculata*
 - Flowers in cylindrical or subglobose spikes or racemes arranged in dense terminal compound
   corymbs   -   -   -   -   -   -   -   -   -   -   -   -   -   -   -   -   -   2
2. Leaves 2–8 × 0.3–0.7 mm., narrowly linear   -   -   -   -   -   -   - 1. *swynnertonii*
 - Leaves 6–35 × 0.5–2.5(3) mm., linear to linear-lanceolate   -   -   -   -   -   -   3
3. Axillary fascicles of secondary leaves rarely present, but if so are produced just below the
   inflorescence and are much smaller than the main leaves; fruit clearly minutely
   muricate   -   -   -   -   -   -   -   -   -   -   -   -   -   -   2. *angolensis*
 - Axillary fascicles of leaves present throughout the length of the stem; fascicle-leaves half as long
   as or subequalling main leaves; fruit smooth, glabrous, rarely obscurely muricate   4
4. Bracts 2.5–4 × c. 1.25 mm. at base, usually longer than the flowers; spikes up to 60 mm. long,
   flowers sessile; calyx 0.8–1.25 mm. long; corolla white   -   -   -   - 5. *nachtigalii*
 - Bracts 1.75–2.5 × 0.75-1 mm. at base, shorter than the flowers; flowers minutely pedicillate in a
   spike-like arrangement; calyx 2–3.5 mm. long; corolla blue to purple   -   - 3. *goetzei*

1. **Walafrida swynnertonii** S. Moore in Journ. Linn. Soc., Bot. **40**: 165 (1911). Type: Zimbabwe,
   Nyahodi R., 1500 m., iv.1907, *Swynnerton* 2135 (BM, holotype; K, isotype).
       Selago swynnertonii (S. Moore) Eyles in Trans. Roy. Soc. S. Afr. **5**: 473 (1916). Type as above.

Perennial herbs, 18–40 cm. tall; stem slender, erect or scrambling, branches and leaves
scabrid-pubescent or rarely subglabrous to glabrous; branches ascending, slender, many-
leaved. Leaves 2–8(14) × 0.3–0.7(0.9) mm., alternate or fasciculate, narrowly linear, obtuse.
Inflorescence somewhat laxly branched, corymbose, spikes 3.5–7 × c. 3.5 mm.,
occasionally lengthening in fruit, at ends of branches, few-flowered, ovate, short. Pedicels
0–0.2 mm. long. Bracts c. 2 × 1.5 mm., boat-shaped, ciliate, subglabrous to minutely hispid.
Calyx 1–1.75 mm. long, (2)3-lobed, lobes subequal, linear-oblong, obtuse, shortly ciliate,
subglabrous. Corolla mauve to pinkish-mauve or white, or red (fide *Boughey* 458), 3–3.8
mm. long, sparsely short-pubescent without; tube 2–2.5 × c. 0.3 mm. at base, widening to c.
0.8 mm. at throat; lobes 0.8–1.3 × 0.5–0.7 mm., ovate-oblong, obtuse. Capsule c. 1.25 × 1.25
mm., of 2 cocci, attached but separating at maturity, glabrous.

Usually in submontane grassland or at lower altitudes in deciduous woodlands on gravelly or
sandy soils; up to 2100 m.

Var. **swynnertonii** —Brenan in Mem. N.Y. Bot. Gard. **9**: 33 (1954).

Leaves densely scabrid-pubescent.

Zambia. C: c. 55 km. NE. of Serenje, fl. & fr. 2.iii.1962, *Robinson* 4980(K). **Zimbabwe**. E: Nyanga
Distr., Nyangani Farm, fl. & fr. 1.iv.1949, *Chase* 1293 (BM; K; SRGH). **Mozambique**. MS: Tsetserra,
2140 m., fl. & fr. 7.ii.1955, *Exell, Mendonça & Wild* 226 (BM; LISC).
In Zimbabwe and Mozambique this variety appears to be restricted to the mountains along their
common border; it is also known from one other collection from Zambia in addition to that cited
above and in both collections the plants bear much larger leaves than those from Zimbabwe, and are
shown as the extremes in the description of the species.

Var. **leiophylla** Brenan in Mem. N.Y. Bot. Gard. **9**: 33 (1954). Type: Zimbabwe, Nyanga Distr.,
   21.i.1948, *Chase* 693 (K, hototype; SRGH, isotype).

Leaves glabrous or very sparsely pubescent.

Zimbabwe. E: Nyanga Distr., Mt. Inyangani, 2100–2500 m., fl. & fr. 3.iii.1956, *Whellan & Davies*
970 (K). C: Makoni Distr., Rusape road, fl. & fr. 13.ii.1961, *Rutherford-Smith* 526 (K; SRGH).
Mozambique. MS: Moribane, 1220 m., Chimanimani Mts., fl. & fr. 2.iii.1907, *Johnson* 233(K).
Not known outside the Flora Zambesiaca area.

2. **Walafrida angolensis** (Rolfe) Rolfe in F.C. **5**, 1: 117 (1901). —Brenan in Mem. N.Y. Bot. Gard. **9**: 33
   (1954). Type from Angola.
       Selago angolensis Rolfe in F.T.A. **5**: 271 (1900). Type as above.
       Walafrida chongwëensis Rolfe in Journ. Linn. Soc., Bot. **37**: 462 (1906). Type: Zimbabwe
   [Victoria Falls], bank of Zambezi R. and islands, 1905, *Gibbs* 117 (BM, lectotype, chosen here; K,
   lectoparatype).
       Selago chongwëensis (Rolfe) Eyles in Trans. Roy. Soc. S. Afr. **5**: 473 (1916). Type as above.

Erect perennial herb, 15–40 cm. tall; stems minutely retrorse hispid-pubescent. Leaves
on main stem below inflorescence branches 18–35(50) × 1–2.25(3) mm., linear, linear-
lanceolate or narrowly oblanceolate, entire, obtuse, glabrous to minutely hispid-scabrid

on margin; axillary fascicles of secondary leaves rarely present but if so then produced just below the inflorescence, very much smaller than main leaves, never throughout main stem. Inflorescence of terminal loose corymbs made up of short racemes c. 5 mm. in diam., lengthening to about 2.5 cm. long or more when fruiting. Pedicels short, 0.5–0.7 mm. long, sparsely short-hispid. Bracts 1.5–2 mm. long, up to 3(4) mm. at maturity, obtuse, somewhat cucullate, glabrous. Calyx (1)1.4–1.8(2) mm. long, (2)3-lobed; lobes 1–1.7 × c. 0.25 mm., linear, obtuse, ciliate, dorsal lobe filiform at times reduced in size to c. 0.5 mm. long, appearing almost obsolete. Corolla white, 2.5–3(4) mm. long, glabrous without; tube 1.5–2 mm. long, c. 0.5 mm. in diam. at base, to 1.5 mm. wide at throat; lobes 1.4–2 × 0.8–1 mm., oblong, rounded, obtuse. Capsule c. 1.2 × 1.1 mm., subglobose, subcordate at base, densely muricate.

Botswana. N: Mohembo, Okavango R., fl. & fr. 8.iii.1976, *Astle* 7493 (SRGH). SW: Ghanzi & Kgalagadi Distr., Kobe Pan, fl. 19.iii.1980, *Skarpe* S-418 (K). Zambia. B: Mongu Distr., Kanda Lake, c. 11.5 km. NE. of Mongu, fl. & fr. 11.xi.1959, *Drummond & Cookson* 6350 (E; K; LISC; SRGH). C: Mumbwa, fl., *Macaulay* 1000 (K). S: Livingstone, shore of Zambezi R., 1000 m., fl. 8.i.1972, *Kornaś* 829 (K). Zimbabwe. C: 24 km. S. of Shurugwe (Selukwe) on Great Dyke, fl. & fr. 16.iii.1964, *Wild* 6378 (K; LISC). S: Mberengwa Distr., near Otto Mine, S. tip of Great Dyke, fl. & fr. 17.iii.1964, *Wild* 6389 (K; LISC; SRGH).
Also known from Angola. Lake- and riversides, and well drained sandy grasslands; up to 1100 m.

3. **Walafrida goetzei** (Rolfe) Brenan in Mem. N.Y. Bot. Gard. **9**: 33 (1954). Type from Tanzania.
   Selago goetzei Rolfe in Engl., Bot. Jahrb. **30**: 402 (1901). Type as above.

Erect woody perennial (20)45–60(100) cm. tall; stems simple or occasionally branched, densely variously pubescent. Leaves on main stem below inflorescence branches, 6–15(18) × 0.5–1.5(4) mm., linear, linear-lanceolate, obtuse, entire or coarsely toothed towards apex, shortly hispid to subglabrous, hispid-ciliate on margin; axillary fascicles of secondary leaves present throughout, subequalling main leaves. Inflorescence of compact, small racemes c. 5 mm. in diam., corymbose at ends of branches, rarely lengthening markedly at maturity. Pedicels very short, 0.2–0.3 mm. long, pubescent. Bracts 1.75–2.5 × 0.75–1 mm., linear-lanceolate, somewhat keeled, acute, ciliate. Calyx 2–3.5 mm. long, (2)3(rarely-5)-lobed; lobes c. 1.1–1.4 × 0.5–0.7 mm., narrowly triangular, acute, ciliate. Corolla c. 6 mm. long, blue, lilac or mauve to pale purple, glabrous without; tube c. 3 mm. long, c. 0.4 mm. wide at base, 0.8–1 mm. wide at throat; lobes to 1–3 × 1.1–1.35 mm., circular-oblong. Capsule 1.15–1.25 × 1.5–1.75 mm., rounded, emarginate, glabrous to slightly muricate.

1. Stem and inflorescence branches densely covered with very long, flexuous, deflexed white hairs   -   -   -   -   -   -   -   -   -   -   -   -   var. *goetzei*
   – Indumentum not as above   -   -   -   -   -   -   -   -   -   -   -   -   2
2. Stem and inflorescence branches densely covered with strict, long, patent white hairs   -   -   -   -   -   -   -   -   -   -   -   -   var. *pubescentior*
   – Stem and inflorescence branches densely covered with very short, usually retrorse, crisped hairs   -   -   -   -   -   -   -   -   -   -   -   var. *brevipila*

Var. goetzei

This variety appears to be restricted to southern Tanzania and is not recorded from the Flora Zambesiaca area.

Var. **pubescentior** Brenan in Mem. N.Y. Bot. Gard. **9**: 34 (1954). Type: Zimbabwe, Chimanimani Mts., Bundi Valley, 1500 m., fl. & fr. 5.vi.1949, *Wild* 2859 (K, holotype; SRGH, isotype).

Stem and main inflorescence branches densely clothed in long, strict, patent white hairs.

Zimbabwe. E: Chimanimani Distr., near Sawerombi boundary, fl. 19.ix.1950, *Crook* M 93 (K; SRGH).
Recorded only from montane grasslands in the Chimanimani Distr. of Zimbabwe; up to 1500 m.

Var. **brevipila** Brenan in Mem. N.Y. Bot. Gard. **9**: 34 (1954). Type: Zimbabwe, Mutare Distr., Stapleford, fl. & fr. 28.vii.1937, *McGregor* 106 (K, holotype).

Stem and main inflorescence branches densely clothed in very short, usually retrorse, crisped hairs.

Zimbabwe. E: Nyanga Distr., near Pungwe View, 2000 m., fl. & fr. 14.ix.1960, *Rutherford-Smith* 88 (K; LISC; SRGH). **Malawi**. S: Mt. Mulanje, fl. 7.ix.1956, *Newman & Whitmore* 691 (BM; SRGH). **Mozambique**. MS: Tsetserra, 2140 m., fl. & fr. 7.ii.1955, *Exell, Mendonça & Wild* 225 (BM; LISC).

Known only from the Eastern Highlands of Zimbabwe. In montane grassland up to 2000 m.

4. **Walafrida paniculata** (Thunb.) Rolfe in F.C. **5**: 127 (1901). —Brenan in Mem. N.Y. Bot. Gard. **9**: 34 (1954). —Merxm. & Roessler in Merxm., Prodr. Fl. SW. Afr. **127**: 6 (1967). TAB. **57**. Type from S. Africa.

　　Selago paniculata Thunb., Prodr. Fl. Cap.: 99 (1800). Type as above.
　　Selago saxatilis E. Mey., Comment Pl. Afr. Austr.: 269 (1837). Type from S. Africa.
　　Selago lacunosa Klotzsch in Peters, Reise Mossamb., Bot.: 255 (1861). —Rolfe in F.T.A. **5**: 272 (1900). Type: Mozambique, banks and islands of the Zambezi R., *Peters* s.n. (B†, holotype; K, isotype).
　　Walafrida lacunosa (Klotzsch) Rolfe in F.C. **5**: 117 (1901). Type as above.
　　Walafrida saxatilis (E. Mey.) Rolfe in F.C. **5**: 126 (1901). Type as above.
　　Walafrida cecilae Rolfe in Bull. Misc. Inf., Kew **1906**: 167 (1906). Type: Zimbabwe, near Bulawayo, xi.1899, *Cecil* 95 (K, holotype).
　　Selago cecilae (Rolfe) Eyles in Trans. Roy. Soc. S. Afr. **5**: 473 (1916). Type as above.

Erect or decumbent woody perennial, 15–30(45) cm. tall, much-branched; stems erect or prostrate, somewhat flexuous, densely retrorse, short-crisped, pilose, or rarely longer shaggy pilose (vidi *van Rensburg* 2327). Leaves (1.5)3–7.5(26) × (0.25)0.5–0.8(3) mm., fasciculate, linear-oblong, obtuse, glabrous or almost so. Inflorescence paniculate of short spreading branches terminating in short, compact, ovoid or oblong spikes. Pedicels where apparent very short, up to 0.15 mm. long, otherwise almost lacking. Bracts (1.3)2– 2.8 mm. long, oblong to ovate-oblong, obtuse, subglabrous. Calyx 0.8–1.3 mm. long, (2)3-lobed; lateral lobes 0.7–1.1 × c. 0.3 mm., oblong, obtuse, ciliate; dorsal lobe very small, 0.15–0.6 mm. long, filiform, or absent. Corolla white, 1.6–2.4 mm. long; tube 1–1.75 mm. long, glabrous without; lobes 0.6–0.8 × 0.5–0.8 mm., elliptic-oblong or circular. Capsule 0.5–0.8 × 0.7–1.2 mm., broadly ovoid, compressed, cordate at base, surface smooth to slightly uneven.

　　Botswana. N: Toromoja, Boteti (Botletle) R., fl. & fr. 26.iv.1975, *Ngoni* 444 (K; SRGH). SW: Ghanzi & Kgalagadi Distr., Kule Pan, fl. 27.i.1977, *Skarpe* S–123 (K; SRGH). SE: Lobatsi Distr., Lobatsi Forest Plantation, 1190 m., fl. & fr. i.1951, *Miller* B1156 (K; PRE). **Zambia**. W: Mwinilunga Distr., Kabompo Gorge, 1200 m., fl. & fr. 23.xi.1962, *Richards* 17474 (K). S: Namwala, c. 1000 m., fl. & fr. 8.i.1957, *Robinson* 2064 (K; SRGH). **Zimbabwe**. N: Bumi River mouth, 457 m., fl. & fr. ix.1955, *Davis* 1489 (K; SRGH). W: Binga Distr., Zambezi R., 16 km. W. of Binga, c. 460 m., fl. & fr. 8.xi.1958, *Phipps* 1401 (K; SRGH). S: Mberengwa Distr., c. 10 km. S. of Bukwa Mt., fl. 4.v.1973, *Pope, Biegel & Simon* 1102 (K; SRGH). **Mozambique**. T: Boroma (Boruma), fl. vi.1891, *Menyhart* 991 (K). GI: Macia, fl. 5.iv.1969, *Balsinhas* 1458 (LISC). M: Marracuene (Vila Luiza), Macaneta Beach, 30 km. N. of Maputo (Lourenço Marques), fl. 4.viii.1967, *Gomes e Sousa & Balsinhas* 4923 (K).

　　Also known from S. Africa and Namibia. Savanna and well grazed grassy areas on well-drained soils, Kalahari Sand and coastal dunes; sea-level to 1400 m.

　　I have chosen here to include *W. saxatilis* (E. Meyer) Rolfe as, following the great diversity within *W. paniculata* as understood by Brenan (1954) which I now accept, I can find no valid reason for excluding it from this species. Admittedly the leaves are apparently more rigid than in the other taxa included here by Brenan, but it is felt that this is due mainly to their size. They are smaller than most in comparison but nevertheless still compare with those shown by *Read* 22 and cited by Brenan. The corolla and fruit sizes also compare with this collection.

　　I have also chosen, for reasons of convenience only, not to include the full synonymy under the name *W. paniculata* by Brenan, but to cite only those known to have been used for plants related to the Flora Zambesiaca area.

5. **Walafrida nachtigalii** (Rolfe) Rolfe in F.C. **5**, 1: 124 (1901). —Merxm. & Roessler in Merxm., Prodr. Fl. SW. Afr. **127**: 6 (1967). Type from south-western Africa.

　　Selago nachtigalii Rolfe in Verhandl. Bot. Ver. Brandenb. **31**: 205 (1890). Type as above (as "*nachtigali*").

Erect or decumbent perennial herbs to 45 cm. tall; stems erect or subprostrate, branching mainly from base, minutely hispid or retrorse crisped-pilose. Leaves 10–15(20) mm. long and up to 1 mm. wide, linear or narrowly linear-lanceolate, acute, glabrous; axillary fascicles of leaves present throughout, half to two thirds the length of main leaves. Inflorescence corymbose of globose or elongate spikes arranged terminally at ends of stem and branches. Spikes up to 30 mm. long, c. 5 mm. in diam.; flowers sessile to subsessile. Bracts 2.5–4 × c. 1.25 mm. at base, lanceolate, somewhat boat-shaped, keeled, glabrous, ciliate on scarious margin at base, markedly longer than flowers. Calyx

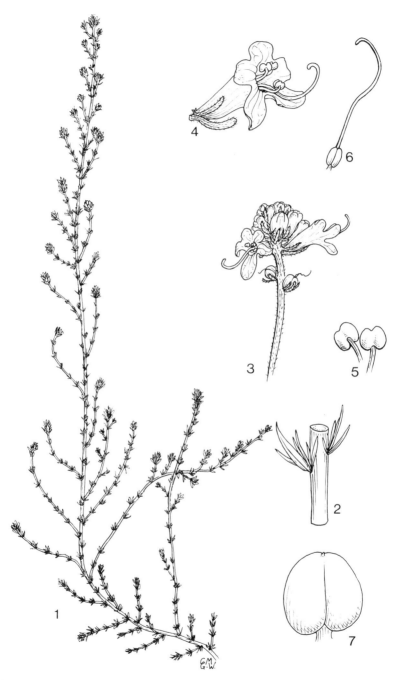

Tab. 57. WALAFRIDA PANICULATA. 1, flowering and fruiting branch (× ⅔); 2, part of stem, to show leaves (× 6); 3, inflorescence (× 6); 4, flower (× 12); 5, stamens (× 24); 6, gynoecium (× 12); 7, capsule with calyx removed (× 30), all from *Hope* 5.

c. 0.8–1.25 mm. long, (2)3-lobed; lateral lobes c. 1 mm. long, narrowly elliptic-oblong, minutely ciliate, dorsal lobe smaller, filamentous to absent. Corolla white, 1.75–2.25 mm. long; lobes elliptic-oblong, about half as long as tube. Capsule 0.8–1.15 × 1.1 mm., somewhat compressed globular, smooth, glabrous, not distinctly muricate.

Botswana. SW: Deception Valley, Kalahari Central Game Res., fl. & fr. 19.iii.1983, *Smith* 4205 (K; SRGH). SE: Takatokwane Pan, fl. & fr. 17.ii.1960, *Wild* 4988 (K; SRGH).
Also known from Namibia and S. Africa. Dry grasslands, pan margins up to about 1000 m.

46. SELAGO L.

Selago L., Syst. Nat., ed. 1 (1735). —Benth. & Hook. f., Gen. Pl. **2**: 1128 (1876).

Small or dwarf, much-branched, heath-like shrubs or undershrubs, or annual herbs. Leaves usually small, narrow, alternate or opposite to nearly so below, or fasciculate, oblong, elliptic or spathulate, entire or occasionally toothed. Flowers sessile or subsessile, or shortly pedicellate, spicate, paniculate or corymbose, bracteate. Calyx 5-lobed. Corolla tubular, 5-lobed; tube short and broad or long and narrow, more or less dilated at throat; limb with subequal lobes or bilabiate, posterior lobes shorter than anterior, central anterior lobe longer than the others. Stamens 4, didynamous, inserted at base of throat, exserted; filaments filiform; anthers monothecal; staminode if present, small. Ovary bilocular; style slender, obtuse, thickened to tridentate at apex. Capsule ovoid or subglobose, often separating into 2 cocci at maturity, included in calyx.

A genus of about 150 species from tropical and southern Africa.

1. Inflorescence corymbose or rounded-capitate - - - - - - - 1. *thomsonii*
- Inflorescence densely or laxly paniculate, of more or less pedunculate heads or short
  spikes - 2
2. Leaves glabrous, linear; panicle usually dense; corolla tube 1.5–3 mm. long; calyx teeth linear-
   oblong, usually densely ciliate - - - - - - - - - - 2. *thyrsoidea*
- Leaves pubescent, narrowly lanceolate or oblong-linear to linear; panicle usually lax; corolla
  tube 3–4 mm. long; calyx teeth triangular-acute or acuminate, sparsely ciliate   3. *welwitschii*

1. **Selago thomsonii** Rolfe in Journ. Linn. Soc., Bot. **21**: 402 (1885); in F.T.A. **5**: 270 (1900). Type from Tanzania.

Perennial woody herb or subshrub 15–50(150) cm. tall: stem erect, ascending or decumbent, branches many-leaved, short white-pubesent. Leaves alternate on inflorescence branches, fasciculate on main stems, 6–12(30) × (0.25)1–2(5) mm., linear-lanceolate, subobtuse, entire, minutely hispid to subglabrous except on margins and prominent major nerves beneath, densely glandular-punctate on both surfaces. Inflorescence corymbose of short, rounded heads; 4–6 mm. in diam., subglobose, terminal, slightly elongated in fruit. Peduncles up to 10 mm. long, slender, pubescent. Flowers sessile. Bracts up to c. 2 mm. long, linear, obtuse or boat-shaped, pubescent. Calyx (1)1.5–2 mm. long, pubescent; lobes 0.5–1 mm. long, about equalling tube, subulate-linear to oblong, obtuse or acute, the 3 lower lobes narrow subequal in width, the 2 upper lobes broader subacute or subobtuse, shortly ciliate. Corolla purple to lilac-blue, rarely white; tube (1)1.5–2(2.5) mm. long, glabrous without; lobes subequalling tube, subcircular-oblong, longest (1.3)1.5–2 × 0.8–1.2 mm. Capsule c. 1–1.5 mm. long, somewhat compressed, ovoid-globose.

Var. **thomsonii** —Brenan in Mem. N.Y. Bot. Gard. **9**: 30 (1954).
    Selago johnstonii Rolfe in Trans. Linn. Soc., Ser. 2, **2**: 344 (1887). Type from Tanzania.
    Selago holstii Rolfe in F.T.A. **5**: 269 (1900). Syntypes: Malawi, Mt. Zomba, 1220–1830 m., *Whyte* s.n. (K); Mt. Chiradzulu, 1220 m., *Whyte* s.n. (K); Mulanje Plateau, *McClounie* s.n. (K); Zomba Plateau, *Whyte* s.n. (K).
    Selago buchananii Rolfe in F.T.A. **5**: 269 (1900). Syntypes: Malawi, *Buchanan* 43, 728 (BM; K); between Lake Shirwa and Lake Chiuta, October 1898, *Cunningham* 19 (K).

Cauline leaves usually 1–1.5 mm. wide, broadly linear, median and upper leaves mostly 4–8 mm. long.

Zambia. E: Nyika, fl. 8.vii. 1962, *Lawton* 918 (K; SRGH). C: c. 128 km. S. of Mpika, fl. 18.v.1973, *Fanshawe* 11852 (K). **Malawi**. N: Nyika Plateau, 2400 m., fl. & fr. 18.viii.1946, *Brass* 17315 (BM; K; SRGH). S: Zomba Distr., Zomba Plateau, 1750 m., fl. 31.v.1946, *Brass* 16116 (K).

Also known from Tanzania and Kenya. Submontane grasslands; up to 2450 m.

Var. **caerulea** (Rolfe) Brenan in Mem. N.Y. Bot. Gard. **9**: 30 (1954). Type: Malawi, summit of Nyika Plateau, 2134 m., *Whyte* 145 (K, holotype).

    Selago caerulea Rolfe in F.T.A. **5**: 267 (1900). Type as above.

    Selago viscosa Rolfe in F.T.A. **5**: 267 (1900). Type from Tanzania.

    Selago tenuicaulis Rolfe in F.T.A. **5**: 268 (1900). Type: Malawi, between Mpata and commencement of Tanganyika Plateau, July 1896, 600–900 m., *Whyte* s.n. (K, holotype).

    Selago melleri Rolfe in F.T.A. **5**: 268 (1900). Type: Malawi, "Manganja Range", Mt. Chiradzulu, September 1861, *Meller* s.n. (K).

    Selago blantyrensis Rolfe in F.T.A. **5**: 268 (1900). Type: Malawi, Blantyre, 1895, *Buchanan* in *Wood* 7009 (K).

    Selago mcclouniei Rolfe in Bull. Misc. Inf., Kew **1908**: 262 (1908). Type: Malawi, Nymkowa, 1950 m., *McClounie* 57 (K, syntype); Panda Peak, 1500 m., *McClounie* 139 (K, syntype).

Cauline leaves mostly (1.5)2–5 mm. wide, narrowly lanceolate to linear-lanceolate.

Malawi. N: Nyika Plateau, Nchenachena (Nchena-chena) Spur, 1700 m., fl. & fr. 10.viii.1946, *Brass* 17149 (K; SRGH). S: Thuchila (Tuchila) Plateau, 1525 m., fl. & fr. viii.1901, *Furnes* 77 (K).
Mozambique. MS: Manica, Serra Zuira, fl. 5.xi.1965, 2200 m., *Torre & Pereira* 12715 (LISC).

Also known from Tanzania. Montane grasslands; up to 2530 m.

Var. **whyteana** (Rolfe) Brenan in Mem. N.Y. Bot. Gard. **9**: 31 (1954). Type: Malawi, Mt. Mulanje, 1980 m., *Whyte* 28 (BM, holotype).

    Selago whyteana Rolfe in Trans. Linn. Soc., **2**, 4: 35 (1894). Type as above.

    Selago milanjensis Rolfe in Trans. Linn. Soc., **2**, 4: 35 (1894). Type: Malawi, Mt. Mulanje, *Whyte* s.n. (BM, holotype).

Cauline leaves 0.25–1 mm. wide, filiform to narrowly linear; median and upper leaves 8 mm. or more long.

Malawi. S: Mt. Mulanje, Likhubula Gorge, c. 1200 m., fl. & fr. 21.vi.1946, *Vernay* in *Brass* 16390 (A; K; SRGH).

Also known from Tanzania. Grasslands; up to 2000 m.

2. **Selago thyrsoidea** Baker in Bull. Misc. Inf., Kew **1898**: 159 (1898). —Rolfe in F.T.A. **5**: 270 (1900). TAB. 58. Type: Malawi, Nyika Plateau, summit, 2135 m., July 1896, *Whyte* 144 (K, holotype).

Perennial woody herb or subshrub, 30–100 cm. or more tall; stem erect or suberect; branches many-leaved, wiry, shortly pubescent. Leaves 5–20(30) × c. 1 mm., linear, subobtuse, entire or sometimes with 1–2 small teeth towards apex, glabrous, fasciculate, or rarely irregularly alternate on inflorescence branches. Inflorescence 3–9(18) cm. long, thyrsoid- paniculate of numerous short pedunculate spikes or heads at ends of branches; spikes or heads c. 5–10 × 4–6 mm., circular to oblong, slightly elongating in fruit. Bracts c. 3 mm. long, linear, obtuse, incurved, pubescent or subglabrous at base, shortly ciliate. Calyx 5-lobed, 1–1.5 mm. long, pubescent, campanulate; lobes c. 0.75 mm. long, subequal, about equalling tube to 3–4 times longer, subacute to obtuse, long ciliate; lower lobes broader than upper. Corolla blue to purplish-blue, tube (1)1.5–2(3) mm. long; lobes equalling or longer than tube, rounded to oblong, largest lobe (1)1.5–2(3) × 1.2–1.5 mm. Capsule not seen.

Var. **thyrsoidea** —Brenan in Mem. N.Y. Bot. Gard. **9**: 31 (1954).

Main stem leaves mostly 1–1.6(2) cm. long; largest corolla lobe c. 3 mm. long; stems without many short lateral leafy branches.

Zambia. E: Nyika, fl. 26.vi.1966, *Fanshawe* 9735 (K; SRGH). **Malawi**. N: Rumphi Distr., Nyika Plateau, road from Chelinda to Kasaramba, just past Chelinda Bridge turning, 2410 m., fl. 14.v.1970, *Brummitt* 10721 (K).

Only from the Nyika Plateau. Submontane grasslands; up to 2450 m.

Var. **austrorhodesica** Brenan in Mem. N.Y. Bot. Gard. **9**: 32 (1954). Type: Zimbabwe, Nyanga Distr., Troutbeck, 2130 m., 21.iii.1948, *Rattray* 1405 (SRGH 20641) (K, holotype; SRGH).

172

Tab. 58. SELAGO THYRSOIDEA. 1, flowering stem apex (×⅔); 2, stem node (× 3); 3, inflorescence (× 4); 4, flower (× 12); 5, gynoecium (× 12); 6, fruiting calyx (× 12), all from *Fanshawe* 9735.

Main stem leaves mostly 1.2–2 cm. long; largest corolla lobe c. 2 mm. long; stems with numerous short lateral leafy branches.

Zimbabwe. E: Nyanga Distr., Nyanga (Inyanga) Downs, 2040 m., fl. 22.iv.1953, *Chase* 4944 (BM; K; LISC; SRGH).
Only known from the Eastern Highlands of Zimbabwe; to about 2400 m.

Var. **nyikensis** (Rolfe) Brenan in Mem. N.Y. Bot. Gard. **9**: 32 (1954). Type: Malawi, Nyika Plateau, near
    Mwanemba, September 1902, *McClounie* 39 (K, lectotype, chosen here).
        Selago nyasae Rolfe in F.T.A. **5**: 270 (1900). Type: Mozambique, mountains E. of Lake Malawi
        (Nyassa), *Johnson* s.n. (K, syntype); German East Africa.
        Selago nyikensis Rolfe in Bull. Misc. Inf., Kew **1908**: 261 (1908). Type as above.

Main stem leaves mostly 0.6–1.2 cm. long; largest corolla lobe 1–2 mm. long; stems more or less simple or with many short lateral leafy branches.

Zambia. N: Isoka Distr., Mafinga Mt., summit, above Chisenga, fl. 22.xi.1952, *Angus* 832 (BM; K).
Malawi. N: Nyika Plateau, c. 25 km. E. of Rest House, 2440 m., fl. 5.vi.1957, *Boughey* 1633 (K; SRGH).
Mozambique. N: mountains E. of Lake Malawi (Nyassa), *Johnson* s.n. (K).
Not known from elsewhere. Submontane grasslands; up to 2440 m.

3. **Selago welwitschii** Rolfe in Journ. Bot. **24**: 175 (1886). Type from Angola.

Scrambling or decumbent perennial woody herb or subshrub to 30 cm. tall; branches lax, erect, suberect or prostrate, densely pubescent, moderately leafy throughout. Leaves 5–15 × 0.8–3 mm., linear, linear-lanceolate or oblong, obtuse, narrowed at base, entire, minutely hispid-pubescent, alternate to nearly opposite. Inflorescence 3.5–10(25) cm. long, elongate-paniculate of many small pedunculate heads; heads 5–10 × 5–7 mm., circular to broadly oblong, appearing secund where flowering branches are prostrate. Peduncles 5–15(20) mm. long, slender, densely pubescent. Bracts 2–3 mm. long, oblong to obovate-oblong, obtuse, minutely hispid, ciliate at base. Calyx 1.5–2 mm. long, campanulate, pubescent, 5-lobed to midway or below; lobes c. 1 mm. long, narrowly triangular to linear-oblong, acuminate to obtuse, ciliate, subequal, equal to or longer than tube. Corolla white, through pink to lavender or pale blue; tube 3–4 mm. long; lobes 1–1.2 × 0.7 mm., circular-oblong. Capsule c. 1.5 × 1.8 mm., broadly ovoid-globose, slightly compressed, valves asymmetric, smooth.

Var. **welwitschii** —Brenan in Mem. N.Y. Bot. Gard. **9**: 32 (1954).
        Selago hoepfneri Rolfe in F.T.A. **5**: 271 (1900) sensu sticto quoad *Hoepfner* 42. —Merxm. &
        Roessler in Merxm., Prodr. Fl. SW. Afr. **127**: 4 (1967). Type from Namibia.
        Selago holubii sensu Rolfe in F.T.A. **5**: 271 (1900) pro parte quoad *McCabe* 28, var. *S. holubii*
        Rolfe.
Stems and leaves sparsely to densely retrorse, appressed pubescent.

Botswana. N: near Lake Ngami, *McCabe* 28 (K). SE: Digkatlong Ranch, 1030 m., fl. & fr. 5.ii.1977, *Hansen* 3027 (K). SW: 11 km. N. of Union End, fl. 15.iii.1969, *Rains & Yalala* 32 (K; LISC; SRGH).
Zimbabwe. W: Matobo Distr., Besna Kobila Farm, fl. 30.i.1973, *Ngoni* 176 (K; SRGH). C: Beatrice, fl. 27.xii.1924, *Eyles* 4433 (K). S: Gutu Distr., Chikwanda Res., fl. xii.1959, *Davies* 2664 (SRGH).
Also known from Angola, Namibia and S. Africa.

Var. **holubii** (Rolfe) Brenan in Mem. N.Y. Bot. Gard. **9**: 32 (1954). Type: Botswana, Damasetshe
    (Tamasetzi), E. Bamangwato Distr., *Holub* 1090 (K, lectotype chosen by Brenan).
Stems and often leaves more or less densely long patent pilose with straight or flexuous hairs.

Caprivi Strip. Katima Mulilo area, c. 23 km. from Katima on road to Linyanti, 26.xii.1958, *Killick & Leistner* 3117 (K; SRGH). **Botswana**. SE: Damasetshe (Tamasetzi), 3.iii.1876, *Holub* 326 (K). **Zimbabwe**. W: Hwange Distr., 8 km. along Shapi Road from Main Camp, Hwange Nat. Park, c. 1000 m., fl. & fr. 26.ii.1967, *Rushworth* 240 (K; LISC; SRGH).
Also known from Angola, Namibia and S. Africa (Transvaal). Grasslands and open woodlands on sandy soils; up to 1000–1320 m.

ALECTRA, 3,**85**
  asperrima, 85,**90**
  aurantiaca, 86
  bainesii, 85,**92**
  barbata, 88
  communis, 90
  cordata, 88
  dolichocalyx, 85,**86**,tab.**33**
  glandulosa, 85,**86**,tab.**32**
  indica, 90
  kilimandjarica, 92
  kirkii, 92
  melampyroides, 88
  orobanchoides, 85,**92**,93
  parasitica, 85,**92**
  parvifolia, 92
  picta, 85,**91**
  pubescens, 85,**90**
  rigida, **85**
  senegalensis, 88
    var. *minima*, 88
    var. *pallescens*, 88
  sessiliflora, 85,**86**
    var. monticola, 88,**90**
    var. senegalensis, **88**
    var. sessiliflora, **88**
  vogelii, 85,**91**
Ambulia bangweolensis, 47
  baumii, 44,45
  ceratophylloides, 45
  dasyantha, 47
  gratioloides, 44
Anarrhinum pechuelii, 14
  veronicoides, 12
ANTHEROTHAMNUS, 2,**20**
  pearsonii, **20**,tab.**9**
  rigida, 20
ANTICHARIS, 1,**7**
  arabica, 9
  linearis, **9**,tab.**2**
Antirrhinum capense, 11
  fruticans, 11
  orontium, 16
APTOSIMUM, 1,**3**
  albomarginatum, 3,**4**
  decumbens, 4,**7**
  depressum var. *elongatum,* 7
  elongatum, 4,**7**
  junceum, 3,**6**
  lineare, 3,**6**
  lugardiae, 3,**4**,tab.**1**
  marlothii, 3,**4**
  procumbens var. *elongatum,* 7
  pubescens, 7
  randii, 6
Aulaya obtusifolia, 95

BACOPA, 2,**48**
  calycina, 49
  crenata, 48,**49**
  floribunda, 48,**51**,tab.**19** fig. **A**
  hamiltoniana, 48,**49**,tab.**19** fig. **B**

  monniera, 49
  monnieri, **48**
  monnieria, 49
  pubescens, 51
Bartsia abyssinica, 156,158
    var. *nyikensis*, 156
    var. *petitiana*, 156
  elgoensis, 156
  mannii, 156
  nyikensis, 156
  petitiana, 156
Benthamistella benthamiana, 106
  nigricans, 105,106
Bonnaya parviflora, 69
Bopusia scabra, 151
BUCHNERA, 2,**96**
  albiflora, 98,**126**
  androsacea, 97,**108**
  arenicola, 96,98,**105**
  asiatica, 134
  attenuata, 98,**125**
  aurantiaca, 29
  bangweolensis, 96,**104**
  bilabiata, 128
  buchneroides, 97,**107**
  candida, 98,**121**
  capitata, 96,**100**
  chimanimaniensis, 98,**121**,tab.**41**
  chisumpae, 97,**113**,tab.**38**
  ciliolata, 97,**107**
  crassifolia, 96,**104**
  cryptocephala, 96,**99**
    var. cryptocephala, **100**,tab.**35** fig. **B**
    var. mwinilungensis, **100**,tab.**35** fig. **A**
  descampsii, 96,**103**
  ebracteolata, 97,**108**,tab.**36**
  euphrasioides, 133
  eylesii, 98,**120**
  foliosa, 96,**99**
  geminiflora, 97,**116**,tab.**39**
  gesnerioides, 131
  granitica, 98,**126**
  henriquesii, 98,**123**,125
  hermonthica, 129
  hispida, 97,98,**119**
  hockii, 97,108,**113**
  humpatensis, 97,**108**
  lastii, 97,**113**,115
    subsp. lastii **115**
    subsp. pubiflora, **115**
  laxiflora, 97,**119**
  leptostachya, 97,**115**
  linearifolia, 128
  lippioides, 97,**110**
  longifolia, 119
  longispicata, 98,**125**
  macrantha, 131
  macrocarpa, 119
  mossambicensis, 115
    var. *usafuensis*, 115,120
  multicaulis, 97,98,**112**
    var. *grandifolia*, 110

namuliensis, 98,**127**
nervosa, 98,**123**,tab.**42**
nigrescens, 105
nigricans, 97,**105**
nitida, 96,**105**
nuttii, 96,**100**
orobanchoides, 131
peduncularis, 96,97,**102**
prorepens, 96,**98**
pulcherrima, 98,**121**
pulchra, 99
pusilliflora, 98,**126**
quadrangularis, 99
quadrifaria, 96,**103**
randii, 96,97,98,**106**,126
rhodesiana, 123
rungwensis, 97,**105**
ruwenzoriensis, 97,**108**
similis, 113,115
speciosa, 98,**120**
splendens, 96,**104**
strictissima, 97,**107**
subglabra, 97,98,**116**,tab.**40**
trilobata, 96,**102**
trinervia, 110
tuberosa, 113
usafuensis, 120
welwitschii, 97,**112**
wildii, 97,98,**110**,tab.**37**
BUTTONIA, 1,3,**141**
hildebrandtii, 141,143
natalensis, **141**,143,tab.**47**

Canscora ramosissima, 59
Capraria lucida, 18
Celsia parviflora, 92
Chaenostoma floribundum, 32
hereroensis, 27
micrantha, 29
CISTANCHE, 1,**158**
tubulosa, **158**,tab. **54**
CRATEROSTIGMA, 1,2,**53**,56
goetzei, 59
gracile, 61
hirsutum, **54**,tab.**21**
lanceolatum, 53,**54**
latibracteatum, 60
monroi, 57
nanum var. *lanceolatum*, 54
ndassekerense, 54
plantagineum, 54,**56**
pumilum, **54**
schweinfurthii, 60
CYCNIUM, 2,**137**
adonense, 137,**140**
subsp. *camporum*, 140,141
buchneri, 140
camporum, 140
carvalhi, 32
carvalhoi, 32
decumbens, 140
filicalyx, 137,**138**
pentheri, 140
rectum, 140
recurvum, **137**
tubulosum, 137,**138**,tab. **46**
subsp. *montanum*, 138

DICLIS, 1,**12**
ovata, **12**,tab.**4** fig. **B**
petiolaris, 12,**14**,tab.**4** fig. **C**
tenella, **12**,tab.**4** fig. **A**
viridis, 14
DIGITALIS, 1,**77**
purpurea, **77**,tab.**29**
DOPATRIUM, 2,**51**
caespitosum, 53
junceum, **51**,tab.**20** fig. **A**
stachytarphetioides, 51,**53**,tab.**20** fig. **B**
Doranthera linearis, 9
Dunalia acaulis, 56

Eylesia buchneroides, 107

FREYLINIA, 2,**20**
tropica, **20**,tab.**8**

Gerardia filiformis, 151
nigrina, 84
obtusifolia, 95
scabra, 151
sessiliflora, 86
tubulosa, 138
Gerardianella scopiformis, 151
GERARDIINA, 3,**154**
angolensis, **156**,tab.**52**
Gerardiopsis fischeri, 9
Glossostylis asperrima, 90
avensis, 90
GRADERIA, 2,**149**
scabra, **151**,tab.**49**
Gratiola juncea, 51
parviflora, 69

HALLERIA, 2,**16**
abyssinica, 16
elliptica, 16,**18**,tab.**6** fig. **A**
lucida, **16**,tab.**6** fig. **B**
HARVEYA, 3,**93**
huillensis, 93,**95**,tab.**34** fig. **B**
obtusifolia, 93,**95**
randii, **93**, tab.**34** fig. **A**
Hebenstreitia, 162,163,
HEBENSTRETIA, 2, **161**
angolensis, 162,**163**,tab.**56**
aurea, 162
bequaertii, 163
comosa, 162,**163**
var. *integrifolia*, 163
dentata, 163
elongata, 163
holubii, **162**,163
integrifolia, **162**,163
oatesii, 162,**163**
subsp. inyangana, **165**
subsp. oatesii, **165**
subsp. rhodesiana, **165**
polystachya, 163
rariflora, 163
scabra, 162
tenuifolia, 162
virgata, 162
watsonii, 162
HEDBERGIA, 3,**156**
abyssinica, **156**,tab.**53**
Herpestis africana, 49

calycina, 49
crenata, 48,49
floribunda, 51
hamiltoniana, 49
monnieria, 49
Hottonia indica, 44
sessiliflora, 45
Ilysanthes conferta, 67
gracilis, 70,72
muddii, 72
nana, 69
parviflora, 69
plantaginella, 6
pulchella, 69,70
subsp. rhodesiana, 70
purpurascens, 69,70
saxatilis, 70
schweinfurthii, 72
stictantha, 65
wilmsii, 72

LIMNOPHILA, 2,3,43
bangweolensis, 43,44,47,tab.18
barteri, 43,44,47
ceratophylloides, 2,43,45
crassifolia, 44,48
dasyantha, 43,44,47
fluviatila, 47
fluviatilis, 43,45
forma fluviatilis, 45
forma terrestris, 45
gratioloides, 44
var. nana, 44
indica, 43,44
LIMOSELLA, 2,73
australis, 73,tab.26 fig. A
capensis, 73,75
macrantha, 73
maior, 73,tab.26 fig. B
subulata, 73
Linaria capensis, 11
fruticans, 11
veronicoides, 12
LINDERNIA, 2,61,63
bifolia, 62,63,64
conferta, 62,67,tab.24
damblonii, 62,67
diffusa, 65
exilis, 62,70,72,tab.25
flava, 62,63
gossweileri, 62,63
gracilis, 72
insularis, 62,63,64
lobelioides, 64
nana, 62,69
nummularifolia, 62,65
oliveriana, 62,64,65
parviflora, 62,69
pulchella, 62,69,70
schweinfurthii, 62,72
senegalensis, 57
stictantha, 62,65
stuhlmannii, 64,65
subreniformis, 62,65,tab.23
whytei, 62,63
wilmsii, 62,72
Lobostemon cryptocephalum, 99
Lyperia aspalathoides, 30
atropurpurea, 30

burkeana, 31
crocea, 30
elegantissima, 27
micrantha, 29,30
multifida, 29
Lysimachia monnieri, 48

Macrosiphon elongatus, 135
fistulosus, 135
MANULEA, 2,34
conferta, 34
crassifolia, 34
floribunda, 32
parviflora, 36
rhodesiana, 34,36,tab.14
MELASMA, 2,82
barbatum, 88
calycinum, 84,tab.31 fig. A
indicum var. monticolum, 90
orobanchoides, 92
pictum, 91
rigidum, 85
scabrum, 84,tab.31 fig. B
zeyheri, 151
MICRARGERIA, 3,151
filiformis, 151,tab.50
scopiformis, 151
MICRARGERIELLA, 1,154
aphylla, 154,tab.51
MIMULUS, 3,36
gracilis, 36,tab.15
MISOPATES, 1,14
orontium, 16,tab.5
Moniera, 48
calycina, 49
floribunda, 51
hamiltoniana, 49
pubescens, 51
Monniera africana, 48
cuneifolia, 48

NEMESIA, 1,9
capensis, 11
divergens, 11
foetens, 11
fruticans, 9,11
var. divergens, 11
lilacina, 9,11
montana, 9,11
zimbabwensis, 9,11,tab.3
Nigrina viscosa, 84
Nortenia thouarsii, 57

OROBANCHE, 1,161
minor, 161,tab.55

Peliostomum leucorrhizum,
var. junceum, 6
var. linearifolium, 4
linearifolium, 4
lugardiae, 4
marlothii, 4
Phelipaea tubulosa, 158
Phyllopodium calvum, 26
POLYCARENA, 2,23
calva, 26
transvaalensis, 26,tab.11
Pyxidaria nummularifolia, 65

Raphidophyllum ramosum, 145
　simplex, 144
RHAMPHICARPA, 2,**135**
　brevipedicellata, 135,**137**
　curviflora, 138
　filicalyx, 138
　fistulosa, **135**,tab.**45**
　montana, 138
　recurva, 137
　serrata, 138
　tubulosa, 138
　　var. *curviflora*, 138

SCOPARIA, 2,**77**
　dulcis, **77**,tab.**28**
SCROPHULARIACEAE, **1**
Selaginastrum rigidum, 20
SELAGO, 2,**170**
　angolensis, 166
　blantyrensis, 171
　buchananii, 170
　caerulea, 171
　cecilae, 168
　chongweensis, 166
　goetzei, 167
　hoepfneri, 173
　holstii, 170
　holubii, 173
　johnstonii, 170
　lacunosa, 168
　mcclouniei, 171
　melleri, 171
　milanjensis, 171
　nachtigalii, 168
　nyasae, 173
　nyikensis, 173
　paniculata, 168
　saxatilis, 168
　swynnertonii, 166
　tenuicaulis, 171
　thomsonii, **170**
　　var. caerulea, **171**
　　var. thomsonii, **170**
　　var. whyteana, **171**
　thyrsoidea, 170,**171**,tab.**58**
　　var. austrorhodesica, **171**
　　var. nyikensis, **173**
　　var. thyrsoidea, **171**
　viscosa, 171
　welwitschii, 170,**173**
　　var. holubii, **173**
　　var. welwitschii, **173**
　whyteana, 171
SIBTHORPIA, 1,**75**
　africana, 75,77
　australis, 75
　europaea, 75,**77**,tab.**27**
　　var. *africana*, 75
Simbuleta pechulii, 14
　veronicoides, 12
SOPUBIA, 3,**143**
　angolensis, 143,**147**
　carsonii, 147
　decumbens, 145
　densiflora, 147
　dregeana, 144
　　var. *tenuifolia*, 145
　eminii, 143,**146**,147

fastigiata, 149
filiformis, 151
karaguensis, 143,**147**
　var. karaguensis, **149**
　var. macrocalyx, **149**
lanata, **147**,tab.**48**
　var. densiflora, 143,**147**
　var. lanata, 143,**147**
laxior, 145
mannii, 143,**144**
　var. mannii, **144**
　var. tenuifolia, 144,**145**
obtusifolia, 95
parviflora, 143,**146**
ramosa, 143,**145**
scabra, 151
scopiformis, 151
similis, 145
simplex, 143,**144**
trifida, 146
　forma *humilis*, 145
　var. *ramosa*, 145
welwitschii var. *micrantha*, 149
Stellularia nigricans, 105,106
STEMODIA, 3,**38**
　serrata, **38**,tab.**16**
Stemodiacra ceratophylloides, 45
　sessiliflora, 45
STEMODIOPSIS, 3,**40**
　buchananii, **40**
　　var. buchananii, **40**
　　var. pubescens, **40**
　eylesii, 40,**42**
　glandulosa, 40,**42**
　humilis, 42
　rivae, 40,**42**,tab.**17**
STRIGA, 1,2,**127**
　angustifolia, 128,**133**
　asiatica, 128,**134**
　aspera, 128,**129**,tab.**43**
　　var. *schweinfurthii*, 129
　barteri, 129
　bilabiata, **128**,129
　canescens, 128
　diversifolia, **135**
　elegans, 128,**134**
　euphrasioides, 133
　forbesii, 128,**131**,tab.**44** fig. **C**
　gesnerioides, 127,**131**
　hermonthica, 128,**129**
　jaegeri, 129
　ledermannii, 129
　linearifolia, **128**,tab.**44** fig. **B**
　lutea, 134
　　var. *bicolor*, 134
　macrantha, 128,**131**,tab.**44** fig. **A**
　orobanchoides, 131
　passargei, 128,**129**
　pubiflora, 128,**133**
　rowlandii, 129
　schimperiana, 119
　strictissima, 128
　thunbergii, 128
　zanzibarensis, 133
SUTERA, 2,**26**
　aspalathoides, 30
　atropurpurea, 27,**30**
　aurantiaca, 27,**29**

blantyrensis, 29,30
brunnea, 27,**30**
   var. *macrophylla*, 30,31
burkeana, 27,**31**
carvalhoi, 27,**32**,tab.**13**
elegantissima, **27**,tab.**12** fig. **A**
fissifolia, 29
floribunda, 27,**32**
fodina, 27,**31**
hereroensis, 26,**27**,tab.**12** fig. **B**
micrantha, 27,**29**,30
palustris, 27,**29**
pulchra, 32
rigida, 20

TEEDIA, 2,**18**
  lucida, **18**,tab.**7**
TORENIA, 2,53,**56**
  brevifolia, 59
  goetzei, 56,**59**
  gracilis, 61
  inaequalifolia, 59
  involucrata, 57,**60**
  latibracteata, **60**
    subsp. parviflora, 57,**60**
  ledermannii, 57,**61**
  monroi, 56,**57**
  parviflora, 57
  plantagineum, 56
  ramosissima, 57
  schweinfurthii, 57,**60**
  spicata, 56,**59**,tab.**22** fig. **B**
  tenuifolia, 57,**61**
  thouarsii, 56,**57**,tab.**22** fig. **A**
Torrenia, 57

Vandellia gracilis, 72
  lobelioides, 64
  nummularifolia, 65
Velvitsia calycina, 84
VERONICA, 2,**80**
  abyssinica, 80,**82**,tab.**30** fig. **B**
  africana, 82
  anagallis-aquatica, **80**
  chamydryoides, 82
  javanica, 80,**82**,tab **30** fig. **A**
  persica, 80
  petitiana, 82
  wogorensis, 82

WALAFRIDA, 2,**165**
  angolensis, **166**
  cecilae, 168
  chongweensis, 166
  goetzei, 166,**167**
    var. brevipila, **167**
    var. goetzei, **167**
    var. pubescentior, **167**
  lacunosa, 168
  nachtigalii, 166,**168**
  paniculata, 166,**168**,tab.**57**
  saxatilis, 168
  swynnertonii, **166**
    var. leiophylla, **166**
    var. swynnertonii, **166**

Zaluzianskia, 23
ZALUZIANSKYA, 2,**23**,tab.**10**
  capensis, 23
  maritima, 23